# Starch and its Components

BY W. BANKS PH.D
SENIOR PRINCIPAL SCIENTIFIC OFFICER
HANNAH RESEARCH INSTITUTE, AYR
AND
C. T. GREENWOOD PH.D, D.SC
DIRECTOR OF RESEARCH
FLOUR MILLING AND BAKING
RESEARCH ASSOCIATION
CHORLEYWOOD, HERTS
FOR THE UNIVERSITY PRESS
EDINBURGH

Edinburgh University Press
22 George Square, Edinburgh
ISBN 0 85224 251 4
Printed in Great Britain by
Aberdeen University Press

## Starch and its Components

BY W. BANKS AND C. T. GREENWOOD

FOR THE UNIVERSITY PRESS, EDINBURGH

## List of Plates

## Preface

Very extensive investigations have been carried out on starch over many years, but much of the literature is contradictory and only the broadest generalizations regarding the chemical and physical character of this biopolymer meet with universal acceptance. Certain milestones mark the progress of the study of starch. The most prominent amongst these are, in our opinion, the now-classical investigations in the 1940s of K. H. Meyer and his co-workers in first establishing the structural non-uniformity of the starch granule, and the subsequent elaboration of a practical method of sub-fractionating starch material evolved by T. J. Schoch.

Only since these developments has the subject been able to progress scientifically, and, furthermore, the past two decades have seen the emergence of new physicochemical and biochemical techniques which have enabled real progress to be made in our understanding of a very complex macromolecule.

Although many problems still remain, it seems appropriate to us to present at this time our view of the *fundamental* aspects of starch chemistry. We think that a coherent picture of the chemical nature of starch can now be developed, and although in the confines of this monograph it has not been possible to deal with any technological aspects, we think that these same concepts can explain much of the technology.

We make no apology for the fact that the emphasis in this monograph is on our own work for we are necessarily most familiar with this aspect of the subject, and we take this opportunity of thanking very sincerely all our numerous collaborators for their dedicated enthusiasm and support.

W.B.

C.T.G.

# 1. Introduction

*Grammatici certant et adhuc sub indice lis est* (HORACE, *Ars Poetica*)

## 1.1. Starch.

Starch constitutes the major food reserve material of all the higher plants. It is laid down inside small cells, the plastids, in the form of particles that are insoluble in cold water, the *starch granules.* The types of plastids, and the corresponding starches, may be classified as described by Badenhuizen (1959) as belonging to one of two types as follows: *Chloroplasts*, which contain chlorophyll and form the site of photosynthesis. They occur predominantly in the leaves, and have a complicated lamellar structure embedded in a proteinaceous medium, the stoma. The starch granules which are formed in the stoma, between the lamellae, are small, and they have a transitory existence. The *amyloplasts*, on the other hand, do not contain any chlorophyll and show very little evidence of internal structure. This type of plastid is found in plant organs such as the endosperm of seeds, roots, and tubers, and the large starch granules formed in the stoma are stored by the plant for long periods.

The starch from the amyloplasts is necessarily of prime interest both academically and commercially, for, being a polymer of α-D-glucose, these polysaccharides are readily assimilated and form an important source of nutritional energy in the human diet. A very high proportion of the world's food energy intake is starch, and cultivation of plant sources rich in this polyglucan are widespread: over 80 per cent of all food crops are composed of cereals and starch root-crops.

The size, shape and properties of the starch granule are of infinite variety, and being controlled by the plant genes are characteristic of the botanical source. Scanning electron micrographs of some typical starch granules are shown later.

## 1.2. Outline of the Historical Background.

Starch has been the subject of very intensive investigation for many years, probably to a greater extent than any other biopolymer. Much of the early work is confusing and unsatisfactory, because the techniques used for study were insufficiently sensitive and the problems of dealing with a macromolecule were not fully understood until the late 1930s.

Without doubt, the most important advance in the chemistry of starch was the realization of the apparent inhomogeneity of the material, followed by its quantitative fractionation into simpler components and their subsequent characterization. Starches can, in general, be separated into at least two chemically distinguishable entities: *amylose*, a mixture of essentially linear polymers, and *amylopectin*, a mixture of very highly branched polymers.

The concept of heterogeneity arises from the now-classical work of Meyer and his collaborators (Meyer, Wertheim and Bernfeld 1940), and the quantitative separation by Schoch (1945), both of which are described in detail later in this monograph.

The basic problem in the subject is to unravel the complexities of the starch granule: what is its chemical and physical architecture? How is the granule synthesized and deposited in the plant tissue? The answers to these questions might lead to an explanation of the granular properties. In view of the well-established heterogeneity of the starch material composing the granule, we have always thought that it is necessary to define the character and behaviour of the two major components, amylose and amylopectin, as an essential prerequisite to establishing the granular structure.

The development within recent decades of more sophisticated physico-chemical and enzymic techniques, has led to their being applied to the study of this biopolymer. But the more our knowledge of starch is increased, the more we realize that few generalizations about its structure and behaviour are possible. This factor makes the study of this polymer academically intriguing, although many practical aspects of this subject are correspondingly made extremely frustrating!

## 1.3. The Problem of Individuality.

The individuality of starch expresses itself at various levels, but perhaps most prominently in the morphology of the granules. As mentioned above, their size and shape is so characteristic that examination of the granules under the optical microscope allows identification of the botanical source of the starch. However, even within a given population of granules, there is much evidence of variation and individuality. For example, our work has shown that if the granules from the mature potato tuber are graded according to size, it is found that the amylose–amylopectin ratio changes in a regular manner, as do the molecular properties of the two components themselves. In fact, it is probably no exaggeration to say that each granule in a population is unique, differing from its neighbours in terms of both fine structure and properties!

## 1.4. Our Approach.

We believe that no single technique, or line of investigation into starch, will be sufficient in itself to solve the outstanding problems, and that all the available methods—chemical, physical, and biochemical—must be simultaneously applied. It may not at first sight appear obvious that those concerned with the hydrodynamic behaviour of amylose should also understand the enzymology of the starch fractions; but without the ability to interpret the observed incomplete $\beta$-amylolysis of amylose, the hydrodynamic study loses its meaning.

Indeed, surveyance of the literature suggests very forcibly that the study of the starch macromolecule has been virtually divorced from the chemistry and physics of *polymers*. For example, as will be described later, the debate that raged among starch enzymologists regarding the mode of action of *beta*-amylase on amylose had its counterpart in the field of synthetic polymers, where the type of chain degradation operative during thermolysis was the subject of some dispute; the confusing influence of that molecular weight distribution known as the 'exponential' or 'most probable' lay at the root of the controversy in both cases. Similarly, studies of the structure of the starch granule have ignored the great mass of detail accumulated in the study of the spherulitic structures of semi-crystalline synthetic polymers.

In the present work, therefore, we have laid emphasis on amylose and amylopectin as *macromolecules*, trying also to view the starch granule as a spherulitic structure, similar to those encountered in synthetic semi-crystalline polymers.

We describe first the isolation of starch, and its subsequent dispersion and fractionation to yield amylose and amylopectin: a procedure which must be accomplished without concomitant degradation in order to obtain meaningful results. Evidence that has led to the formulation of the present model for the fine structure of amylose, as a mixture of linear and lightly-branched molecules, is summarized. The structure of glycogen, chemically very similar to amylopectin, is discussed in conjunction with that of the latter polysaccharide, and attention is drawn to the very different behaviour of the two branched macromolecules in solution. All starch granules contain variable amounts of starch material which is neither amylose nor amylopectin, and this fact is examined in some detail for both normal starches and those of high amylose content.

One of the most characteristic reactions of starch is that with iodine. We compare the simple colorimetric method of measuring this reaction with the technique of semi-micro differential potentiometric iodine titration, and the latter method is discussed in detail. The interactions of iodine with amylose and amylose oligomers, and the enthalpies of these interactions, are detailed. Iodine interaction with the branched polysaccharides, amylopectin and glycogen, is then considered.

The hydrodynamic behaviour of the linear amylose molecule is then discussed. First, an outline is given of the mathematical model of the random coil, and of the ways in which the limiting viscosity number and frictional coefficient are influenced by assumptions regarding the flow of solvent through the coil. Experimental relations, obtained in a number of laboratories, between the limiting viscosity number and the molecular weight of amylose are summarized. Detailed treatment of the hydrodynamic parameters shows that the isolated amylose macromolecule adopts a random structure in aqueous solution. In contrast, in the presence of complexing agents such as iodine, or butan-1-ol, the dissolved molecule is forced into the helical conformation.

A summary of the action pattern of the main groups of starch-degrading enzymes, i.e. exo-acting, endo-acting and de-branching types, is then given. The possible action-patterns of the exo-acting enzymes are considered, and the effect of the molecular weight distribution of the linear substrate on the kinetics of the degradation examined. The endo-acting enzymes (*alpha*-amylases) from different sources are shown to degrade amylose by quite distinct action patterns, and the alternative theories (preferred attack and multiple attack) by which this observation may be explained are discussed. The action of the various de-branching enzymes on amylopectin and glycogen, and the substrate requirements of each of these enzymes, is then given.

The starch granule and its biosynthesis is the last subject with which we deal. Results from techniques such as X-ray diffraction, laser light scattering and scanning electron microscopy are discussed, as are the structural changes brought about by contact with enzymes and mineral acids. The interaction of starch granules with water, and with aqueous salt solutions, as a function of temperature is then dealt with, and the different types of gelatinization behaviour detailed. The action pattern of enzymes involved in this synthetic process are summarized, and the problem of simultaneously synthesizing both a linear and a branched polyglucan discussed. Attention is drawn to the fact that the molecular size and fine structure of the components change during granule development, as do the properties of the granules themselves.

We would like to have presented our views in such a way that no chapter presupposed information given in a later one, but this objective has not proved possible. We appreciate that the presentation is biased to the extent that it represents *our* view of starch, and that many of the statements made and conclusions drawn may not be universally accepted. When we know that this situation is likely, we have tried at least to indicate the alternative arguments without necessarily presenting them in full.

Finally, and regretfully, we are aware that there are many facets of the subject which we have not been able to deal with in this short monograph.

# 2. Fractionation of the Starch Granule, and the Fine Structures of its Components

## 2.1. Introduction.

A study of the fine structures of the amylose and amylopectin components of starch necessarily entails that (a) the starch itself must be isolated from the plant tissues without any inadvertent modification, and (b) the components must be separated from each other by a non-degradative process. All too often, fundamental studies have been carried out on industrial starches which may have been subjected to chemical modification. For example, the commercial production of cereal starches involves steeping the grain in the presence of sulphur dioxide, to denature enzymes and soften the protein–starch matrix (Cox, MacMasters and Hilbert 1944), and there is evidence that this process involves limited hydrolysis and modification of the starch material (Adkins and Greenwood 1969a). We believe very strongly that for fundamental studies it is essential to use a starch which has been very carefully isolated in the laboratory.

*Isolation of Starch from Cereals.* It is, of course, necessary that cereal grains be softened by some type of steeping process: dry milling of the wheat grain, for example, can lead to considerable physical damage of the starch granules. Equally, it is essential that starch-degrading enzymes be inactivated during the steeping process. We have found that steeping in acetate buffer (pH 6.5), which is 0.01M with respect to mercuric chloride, for thirty hours at 6°C (for barley, oats, rye, and wheat), or for forty to fifty hours at 40°C (for maize and its hybrids) has the desired effect of softening the grains and inhibiting amylase action (Adkins and Greenwood 1966a). The softened grains may then be reduced to a porridge-like consistency in a mincer, and the resultant grists slurried in water before being sieved successively and rapidly through 150 $\mu$m and 75 $\mu$m vibrational screens. The steps of grinding, slurrying and sieving are repeated until the material failing to pass through the sieves is free from endosperm.

The starch suspension obtained in this manner is contaminated with proteinaceous material, and this contaminant must be removed without any concomitant degradation of the starch. For this reason we avoid chemical treatment with alkali, or biochemical degradation using impure proteolytic enzymes, and employ rather the physical technique of shaking overnight the aqueous suspension of starch with one-eighth of its volume of toluene; the protein being denatured at the water–toluene interface. Mild centrifugation then separates the toluene layer, bearing denatured protein. Repeated shakings with fresh toluene are usually necessary to produce a starch of low ($<$0.25 per cent) protein content.

Care must be taken to ensure that the loss of starch granules in the rejected toluene layer is not unacceptably high. Such loss may be avoided by adding several volumes of water to the toluene and aerating the mixture, so freeing the starch from its physical entrapment within the denatured protein.

*Isolation of Starches from Tubers.* Tissues having a high water content, such as the potato tuber, do not require the preliminary steep, but it is necessary to carry out the extraction process in the presence of mercuric chloride in order to inhibit enzyme action.

*Isolation of Starches from Pulses.* In general, however, the most difficult starches to extract are those present in dehydrated tissue. For example, it is not necessary to use a steeping process when extracting starch from a young, succulent pea seed, but, once the seed has become dehydrated, some preliminary aqueous swelling is essential for a good recovery of starch.

*General Procedures.* We have employed the same basic procedure of extracting starch in the presence of mercuric chloride, and removing contaminating protein by shaking the aqueous slurry with toluene, in isolating starches from a wide variety of botanical sources. Some variations in procedure may be required in, for example, the duration of the preliminary steeping process, but by suitable adjustment of such parameters, it is possible to obtain a high extraction rate for the starch by a process that does not cause any obvious modification of the granule (Adkins and Greenwood 1966a).

## 2.2. Fractionation.

It is now accepted that starch consists of two major components, amylose and amylopectin. The two basic methods by which separation of these may be achieved are selective leaching, or complete dispersion of the granule into aqueous solution followed by preferential precipitation of the amylose fraction. These two types of fractionation procedure will now be briefly discussed.

### 2.2.1. Selective Leaching.

Selective leaching depends on the fact that amylose, or at least a portion of it, may be removed from the granule by the action of hot water. The technique, evolved by Meyer and his collaborators, consists of holding an aqueous suspension of the starch at, or slightly above, the gelatinization temperature of the granules for a period of about one hour. The residual granular material may then be removed by centrifugation, and amylose precipitated from the supernatant by the addition of a suitable alcohol, usually butan-1-ol (Meyer, Bernfeld and Wolff 1940; Meyer, Bernfeld, Boissonnas, Gürtler and Noelting 1949). Schoch (1945) criticized the method on the grounds that the resultant amylose is heavily contaminated with amylopectin, and that the fraction of amylose remaining within the swollen granule retrogrades *in situ*. Both criticisms possess some substance, but not sufficient to invalidate the technique. There is no doubt that, as the temperature of leaching is raised appreciably above that of gelatinization, amylopectin is also removed from the granule, but it is still possible to obtain a pure amylose fraction by the simple expedient of recrystallizing the initial amylose–butan-1-ol complex from hot aqueous butan-1-ol solution. Retrogradation of the amylose within the residual granular structure does occur, but we have observed that it is a fairly slow process. Thus, if the initial leaching and subsequent dispersion are carried out successively, no evidence of retrogradation is observed. For example, table 2.1 shows the properties of the amylose fractions obtained by a series of successive aqueous leachings followed by dispersion of the granular residue of barley starch.

*Table* 2.1. The properties of amylose samples obtained by the successive leaching and ultimate dispersion of barley starch (Banks, Greenwood and Thomson 1959a).

| procedure | amylose extracted (%) | iodine-binding capacity (%) | $[\eta]$ [1] (ml/g) |
|---|---|---|---|
| leached at 70°C | 22 | 19.6 | 95 |
| leached residue at 80°C | 17 | 19.5 | 140 |
| first leach of residue at 90°C | 25 | 19.5 | 250 |
| second leach of residue at 90°C | 18 | 19.5 | 300 |
| third leach of residue at 90°C | 13 | 19.5 | 370 |
| dispersion of residue | 5 | 19.4 | 380 |

[1] $[\eta]$ is the limiting viscosity number (see section 4.2.2), in this and subsequent tables.

In each case an amylose fraction of high purity is obtained, and, moreover, the molecular size of the material, as measured by the limiting viscosity number, increases during the successive stages. One would certainly not expect retrograded amylose to possess the high values of limiting viscosity number exhibited in table 2.1. These results were obtained by immediate successive leachings; if the swollen granules were stored overnight at refrigeration temperatures (~0°C), subsequent leaching did not remove amylose, indicating that retrogradation had occurred. There is, however, sufficient evidence in the literature to indicate that aqueous leaching is an excellent method of obtaining amylose with a high degree of purity (Meyer and Menzi 1953; Cowie and Greenwood 1957a; Arbuckle and Greenwood 1958a; Banks, Greenwood and Muir 1971a).

Other methods of aqueous leaching have been described. Baum and Gilbert (1956) suspended starch granules in 0.5 M sodium hydroxide under nitrogen and at room temperature for thirty minutes, before subjecting the dispersion to a force-field of 40000*g* for two hours in a preparative ultracentrifuge. Amylose was reported to be obtained from neutralization of the supernatant liquors, whilst the amylopectin component preferentially sedimented. However, a critical evaluation showed it to be an ineffective leaching procedure (Banks, Greenwood and Thomson 1959a).

*Pre-treatment and Aqueous Leaching.* Montgomery and Senti (1958) have described a method by which the starch granules are pre-treated with hot aqueous butan-1-ol, or dioxane, before aqueous leaching. This treatment modifies the starch granule in such a way that the insolubility of the amylopectin fraction is increased relative to that of amylose, so that a single leach at 98°C for eight minutes removes some 75 per cent of the total amylose present; the residual fraction can then be removed by one or more subsequent leaches. An attempt to repeat these observations using potato and wheat starches was not completely successful (Banks, Greenwood and Thomson 1959a). Pre-treatment rendered the cereal starch more amenable to fractionation by leaching, but rendered the potato starch less so. Furthermore, it proved impossible, even on repeated leaching, to remove all of the amylose from the starches, thus yielding a highly impure amylopectin.

The pre-treatment of Montgomery and Senti (1958) produces an effect similar to that obtained by Sair and Fetzer (1944), who showed that the character of potato starch can be drastically altered by heating the granular material at levels of high humidity. Although not affecting the granular appearance or optical birefringence of the treated starch, the pre-treatment caused the X-ray diffraction pattern to change from the normal *B*-type to the *A*-type pattern more usually associated with the cereal starches. Hull and Schoch (1959) demonstrated that an analogous effect could be achieved by heating a slurry of potato starch in 70 per cent aqueous diacetone alcohol under reflux for several hours. They suggested that this medium contained sufficient water to permit the re-orientation of the molecules within the granules, but insufficient to cause gelatinization. As a result of the pre-treatment the gelatinization temperature range of the starch is raised, and upon gelatinization the granules exhibit the restricted swelling typical of cereal starch rather than the large degree of swelling normally typifying potato starch. Hull and Schoch (1959) also reported that their treatment, whilst making the amylopectin more insoluble, increased the insolubility of the amylose fraction.

One other form of pre-treatment should be mentioned. Cowie and Greenwood (1957a) claimed that the *fraction* of amylose leached from potato starch was quantitatively converted into maltose by the action of pure *beta*-amylase, whereas the *total* amylose obtained on complete dispersion gave only 70 per cent of the theoretical maltose on $\beta$-amylolysis. Subsequent work showed that if the former fraction were to be obtained, the granules must be subjected to pre-treatment with methanol before leaching, either by 'Soxhlet' extraction, or by prolonged storage in methanol (Banks, Greenwood and Thomson 1959a). The subtle changes in granular structure brought about by methanol have not been explained—we can only report the phenomenon that amylose leached from untreated granules is incompletely converted into maltose by *beta*-amylase, whereas that obtained after pre-treatment is quantitatively converted.

**2.2.2.** Dispersion of the Granule.

Whilst the technique of aqueous leaching demands the preferential solubilization of one of the starch components, usually the amylose, the second general method of fractionation involves the complete dispersion of the granule: that is, both amylose and amylopectin are brought into solution, and subsequently separated. Although this separation may be achieved by selective retrogradation (Maquenne and Roux 1905), electrophoresis (Samec, Nučic and Pirkmaier 1941) or chromatography (Fischer and Settele 1953; Ulmann and Wendt 1954), the most successful method yet devised is that of selective precipitation (Schoch 1942a). In this last method, the amylose is precipitated from solution as an insoluble complex by the addition of a suitable organic material (substances that act as complexing agents were recorded by Whistler and Hilbert 1945; a more exhaustive list has been provided by French, Pulley and Whelan 1963).

For this technique to be successful, several criteria must be satisfied. The most important of these is that the starch granule is dispersed as fully as possible into solution. In general, no difficulties are encountered with starches from tubers such as the potato and tapioca, for these granules swell and shatter in boiling water to give an apparent dispersion. With normal cereal starches, however, the situation is rather more complex, for even after several hours at 100°C aqueous suspensions of such granules still contain fairly large amounts of undispersed material. These difficulties were overcome originally by carrying out the dispersion at autoclave temperatures, but some degradation of the cereal starch took place under these conditions due to the presence of fatty acids within the granule. For this reason Lansky, Kooi and Schoch (1949) advocated that, prior to dispersion, the starch be thoroughly defatted, and that during dispersion the system be buffered to pH 6.2–6.3.

To avoid the use of the autoclave, various *pre-treatments* have been devised in order to destroy granular structure, and so ensure that the starch could be dispersed into aqueous solution at 100°C. One of the earliest of such pre-treatments made use of liquid ammonia (Hodge, Montgomery and Hilbert 1948): after exposure to this reagent for some fifteen minutes, normal cereal starches lose all degree of order (as shown by their lack of birefringence), and are accordingly easily dispersed in boiling water. The same type of effect is observed on treating potato starch in a similar manner, but the solubility in this case is so greatly increased that the polysaccharide readily dissolves in water at room temperature (Banks, Greenwood and Thomson 1959a). The technique of liquid ammonia pre-treatment has been applied to a number of starches from various botanical sources (Greenwood and Thomson 1960), and has invariably been found to be highly efficient, although some particularly stubborn starches, e.g. that from wrinkled-seeded pea, may require two treatments in order to achieve maximum dispersion.

The second, widely-used method of pre-treatment involves dimethylsulphoxide. This reagent gives a clear starch 'solution', and subsequent addition of ethanol, or butan-1-ol, precipitates a non-granular form of starch which is easily dispersed in boiling water. This technique was pioneered by Killion and Foster (1960), who claimed that potato starch subjected to this pre-treatment could be subsequently dispersed *at room temperature* to yield an amylose significantly greater in molecular weight than that obtained by conventional dispersion of the starch in boiling water. The importance of avoiding degradation of the components, both during the isolation of the starch and its subsequent fractionation, had been stressed earlier (Greenwood

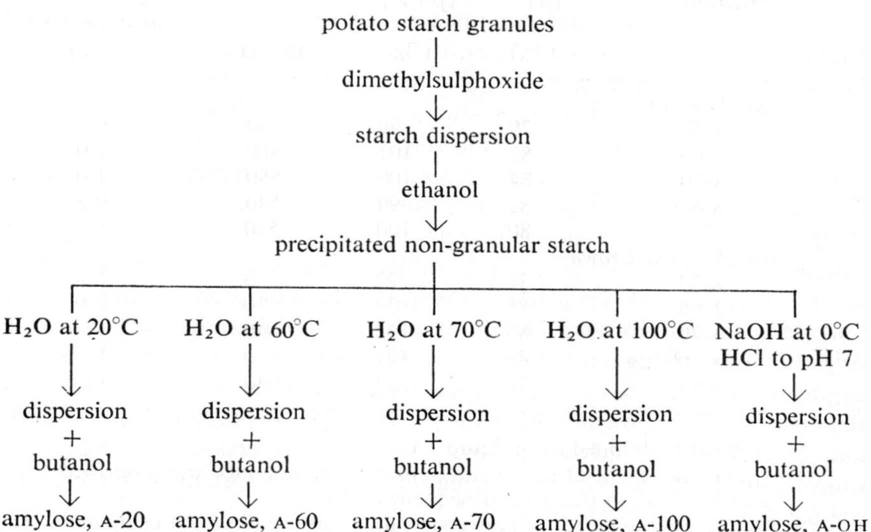


*Figure* 2.1. Fractionation scheme for dimethylsulphoxide-pretreated potato starches. The corresponding amylopectin was obtained from the supernatant liquors after removal of the amylose complex (Geddes, Greenwood, MacGregor, Procter and Thomson 1964).

and Robertson 1954). Evidence was also produced that oxidative modification of amylose took place in both neutral and alkaline aqueous solution at elevated temperatures (Baum, Gilbert and Scott 1956). These observations are easily correlated with the finding of Killion and Foster (1960) that dispersion at low temperatures produces an amylose that has not been degraded. However, repetition of the work using the fractionation scheme shown in figure 2.1 yielded no evidence for this claim, as is demonstrated by the values recorded in table 2.2.

The molecular properties (i.e. the conversion into maltose by *beta*-amylase and the limiting viscosity number) of the amylose fractions are quite independent of the temperature of dispersion. It must, however, be emphasized that in this case all dispersions were carried out under a nitrogen atmosphere, so minimizing the probability of oxidative degradation. These results demonstrate that dispersions and recrystallizations carried out in boiling water will not lead to degradation of the amylose as long as care is taken to rigorously exclude oxygen during operations at high temperatures.

*Table* 2.2. Properties of amyloses obtained from potato starch granules treated with dimethylsulphoxide (Geddes, Greenwood, MacGregor, Procter and Thomson 1964).

| fraction[1] | $[\beta]$[2] (%) | $[\beta+Z]$[3] (%) | $[\eta]$ (ml/g) | amylose in amylopectin[4] (%) |
|---|---|---|---|---|
| *var. Redskin* | | | | |
| A-20 | 79 | 99 | 540 | 0.8 |
| A-60 | 82 | 101 | 560 | 1.0 |
| A-70 | 84 | 100 | 550 | 1.0 |
| A-OH | 82 | 99 | 540 | 0.8 |
| A[5] | 80 | 100 | 550 | 0.8 |
| *var. Pentland Crown* | | | | |
| A-20 | 85 | 99 | 620 | 0.8 |
| A-60 | 85 | 99 | 690 | 1.0 |
| A-70 | 85 | 102 | 640 | 0.3 |
| A-100 | 86 | 101 | 630 | 0.5 |
| A[5] | 88 | 100 | 630 | 0.8 |

[1] Amylose obtained as in figure 2.1.
[2] $[\beta]$ is the extent of conversion into maltose obtained using purified *beta*-amylase, in this and subsequent tables.
[3] $[\beta+Z]$ is the extent of conversion into maltose obtained using *beta*-amylase preparations contaminated with *Z*-enzyme, in this and subsequent tables.
[4] From measurements of iodine-binding capacity.
[5] Amylose from conventional dispersion at 100°C in water; no pre-treatment.

The main use of dimethylsulphoxide pre-treatment is not concerned with the easily dispersed starches such as that from potato, but rather with those that are resistant to dispersion in boiling water, particularly the cereal starches. Additionally, this technique of pre-treatment is especially suited to starches having abnormally high amylose contents, but this particular aspect is dealt with later (see section 2.5.2).

A method of general applicability to cereal starches has been developed (Banks and Greenwood 1967a) in which wet starch (30 g) is dissolved in dimethylsulphoxide (500 ml) with gentle mechanical stirring, for 24 hours; no precipitate should be detected on centrifuging this solution for thirty minutes at 2000*g*. The solution is then poured into butan-1-ol (two volumes), and the resulting non-granular precipitate washed repeatedly with butan-1-ol in order to remove residual dimethylsulphoxide. This butan-1-ol-moist precipitate is added to boiling water (3 l), under oxygen-free conditions, and the boiling continued for one hour. The dispersion is then allowed to cool to 60°C and powdered thymol (1 g/l) added. After standing for three days at room temperature the amylose–thymol complex is removed (the Sharples supercentrifuge is particularly useful for this operation), and subsequently dispersed in boiling oxygen-free water, the boiling being continued for 45 minutes. The solution is then cooled, and butan-1-ol added. After standing overnight, the amylose–butan-1-ol complex may be removed by centrifugation, and stored as such. Alternatively, it may be dehydrated by repeated stirring with butan-1-ol, and dried.

The advantage of using the butan-1-ol complex is that it is easily and completely soluble in cold water, whereas the dried material has a somewhat limited tendency to dissolve even in boiling water; in this latter case it is necessary to use either aqueous caustic alkali or dimethylsulphoxide as a solvent. The use of the wet amylose–butan-1-ol complex excludes any weighing procedures for predetermining the concentration of the solution, but a rapid approximate determination of this parameter may be made using the reaction with iodine—the 'Blue Value' (McCready and Hassid 1943)—and a more accurate value then obtained by hydrolysis of the polysaccharide to glucose by means of acid- or enzyme-catalyzed reaction and subsequent estimation of reducing power (Adkins, Banks, Greenwood and MacGregor 1969), or the specific assay of glucose, using glucose oxidase (Banks and Greenwood 1971a).

The amylopectin fraction is found in the supernatant remaining after removal of the amylose–thymol complex. Excess thymol may be removed by extraction with ether, and the polysaccharide obtained by freeze-drying or by precipitation with alcohol.

The purpose of any technique of fractionating starch is to obtain the amylose fraction free from amylopectin and *vice versa.* Iodine-binding capacities of the amylose and amylopectin fractions obtained from various cereal starches using the above fractionation are shown in table 2.3. The values recorded for the iodine-binding capacities (or iodine affinities) of the amylose fractions fall within the range ($19.5 \pm 0.2$ mg iodine bound/100 mg polysaccharide) expected for an amylose of high purity; the fact that in each case there was 100 per cent conversion into maltose under the concomitant action of *beta*-amylase and *Z*-enzyme (see section 2.3.1), further emphasizes the purity of the fractions. It is rather more difficult to decide the purities of the amylopectin fractions, because it is not possible to state whether this polysaccharide has the intrinsic ability to bind iodine, or whether any finite value for the iodine affinity reflects the presence of contaminating amylose. Even assuming that the latter view is correct, a maximum amylose contamination of approximately 2–3 per cent is indicated for the amylopectin fractions from the results recorded in table 2.3. Thus the method described above for pre-treating and fractionating starch is of general applicability.

*Table* 2.3. The iodine affinities of the amylose and amylopectin fractions of various cereal starches (Banks and Greenwood 1967a).

| source of starch | iodine-binding capacity (%) | |
|---|---|---|
| | amylose | amylopectin |
| barley | 19.6 | 0.6 |
| oats | 19.5 | 0.5 |
| rye | 19.7 | 0.4 |
| wheat I | 19.7 | 0.5 |
| wheat II | 19.3 | 0.6 |

A second advantage accruing from the pre-treatment of starch granules with dimethylsulphoxide is the fact that it provides an excellent method of removing contaminating fatty acid. This extraneous material is soluble in both dimethylsulphoxide and butan-1-ol, so that it remains in solution when the starch is precipitated in a non-granular form. As noted previously, the fatty acids present may cause hydrolytic degradation of the starch components during the process of aqueous dispersion at elevated temperatures. We have found that none of the methods described in the literature for removing these contaminants are nearly so efficient as is pre-treatment with dimethylsulphoxide (Banks, Greenwood and Muir 1971b). The exhaustive removal of these potentially hydrolytic compounds, in conjunction with the use of anaerobic conditions whenever high temperatures are employed, explains, we believe, the fact that the molecular size of the amylose fractions obtained by this technique are the largest reported in the literature (Banks and Greenwood 1967a).

It must be pointed out that an *amylopectin sub-fraction* of high purity can be obtained comparatively simply without any pre-treatment of the cereal starch granules. If such granules are dispersed directly into boiling water, thymol added, and the residual granules removed with the amylose–thymol complex on centrifugation, the supernatant will contain a sub-fraction of pure amylopectin. In this case, however, the 'amylose' fraction may contain up to 50 per cent amylopectin, thus the material in the supernatant represents only some 60 per cent of the total amylopectin. It is our experience that when this initial amylose–thymol complex is very impure, further recrystallizations have no effect—the amylopectin cannot subsequently be removed from the amylose. For this reason, we would emphasize yet again that the initial pre-treatment must rupture all vestiges of granular structure; if it does not, subsequent efforts to purify the *amylose* fraction will be unsuccessful.

The converse fact is also equally true: unless effective disruption of granular structure is achieved initially, the resultant *amylopectin* component will not be obtained in a pure form, and we know of no method of removing traces of contaminating amylose from amylopectin.

The ease of solubility of various starches in dimethylsulphoxide varies with the amylose content, and increases with increasing amylose content—a fact that we shall describe in more detail later (see section 2.5.2). Similarly, as the amylose concentration goes to zero, in the case of the waxy starches, the rate of dissolution of the starch decreases and, moreover, because of the very high viscosities of these solutions, it is necessary to carry out the pre-treatment at a lower concentration of polysaccharide.

Surprisingly, it is almost as difficult to obtain a true molecular dispersion of waxy-maize starch in hot water as it is for the normal dent corn starch—they show very similar behaviour during gelatinization, both in the temperature at which it occurs and in the restricted swelling of the granule. However, in dispersions of moderate concentration (*ca.* 15 per cent) at the temperature of boiling water, the granules of the waxy starch are exceedingly susceptible to the effects of applied shear forces. Thus a solution of waxy-maize starch can be obtained by subjecting a hot, moderately concentrated suspension of swollen granules to shear forces such as those occasioned by vigorous stirring. The solution is then diluted and boiling continued for one hour in the normal manner. Only suspensions in which the swollen granules are in physical contact with one another are amenable to this type of dispersion, for vigorous stirring of dispersions at the usual concentration of 1.5 per cent does not rupture swollen granules. The disadvantage of this particular technique is that the molecular size of the product-amylopectin is considerably reduced, as is demonstrated later in table 2.15, which compares the properties of waxy-maize amylopectin isolated by dimethylsulphoxide pre-treatment and by the application of shear forces. For a study of the fundamental properties of the amylopectin, pre-treatment with dimethylsulphoxide, although more time-consuming, is to be preferred to the method involving shear forces.

Other aids to the fractionation of cereal starches, which have been described, include treatment of the granules with caustic alkali (1.0 M) at 0°C for ten minutes, neutralization, and subsequent dispersion by boiling (Potter, Silveira, McCready and Owens 1953). Even this brief exposure to alkali, however, may degrade the amylose component, for the limiting viscosity numbers of cereal amyloses obtained by this technique are slightly lower than those isolated after pre-treatment of the starch with liquid ammonia. Muetgeert (1961) has described the use of chloral hydrate solution as a dispersive agent. The starch (10 g) is suspended in chloral hydrate solution (200 g, 60 per cent w/w chloral hydrate) for 72 hours in the dark and then centrifuged. The bulk of the amylopectin fraction is found in the centrifugate, whereas the amylose is dissolved in the supernatant. This initial fractionation yields two rather impure fractions, and a rather lengthy purification process has to be employed if the products are to attain a satisfactory state of purity (Greenwood and MacKenzie 1963).

### 2.3. The Fine Structure of Amylose.

The essentially linear nature of amylose was first recognized more than thirty years ago, when Meyer, Wertheim and Bernfeld (1940) reported that the chain length of maize amylose obtained from methylation studies corresponded to the degree of polymerization measured by osmometry. The same conclusion was reached for potato amylose (Meyer, Bernfeld and Hohenemser 1940). Meyer's results were later confirmed independently by Hassid and McCready (1943), although these conclusions were not universally accepted at the time.

#### 2.3.1. Incomplete Degradation by *beta*-Amylase.

Apparently conclusive proof of the linear nature of amylose was obtained when the polysaccharide was found to be quantitatively converted into maltose by the action of *beta*-amylase, for only a chain of unmodified glucose residues linked by α-1:4-bonds could satisfy this observation (see section 5.2). This model gained general acceptance until Peat, Whelan and Pirt (1949) reported that *crystalline* (i.e. pure) sweet potato *beta*-amylase only caused some 70 per cent hydrolysis of amylose. Further work (Peat, Pirt and Whelan 1952a, b) enlarged on this theme, and implicated a second enzyme, the so-called *Z*-enzyme, in the complete degradation of amylose by amorphous *beta*-amylase.

The above conclusions were substantiated in our own laboratories, but additionally it was found that at least a fraction of the total amylose was completely linear (Cowie and Greenwood 1957a; Arbuckle and Greenwood 1958a). These results led to the concept of the presence of two types of amylose in the starch granule; one containing a randomly-situated barrier to the action of *beta*-amylase (and hence having a β-amylolysis limit of 50 per cent conversion into maltose), and the other consisting of a linear chain of unmodified glucose residues linked by α-1:4-bonds. The β-amylolysis limit, $L$, of any amylose would then depend on the relative proportions of the two types, and could be calculated from

$$L = (100 + \%\ \text{linear amylose})/2 \tag{2.1}$$

The presence of anomalous groups or linkages within amylose has still not received universal acceptance, for it has been suggested that the incomplete β-amylolysis may be due to either contamination of the amylose with amylopectin, or to physical association of the macromolecules (retrogradation) making them unavailable to enzymic attack. Because of the importance that we attach to this particular subject, we will deal with both points in some detail.

*Absence of Contaminating Amylopectin in Amylose.* The usual method of demonstrating that a given sample of amylose is pure is based on the fact that the linear polymer binds approximately 20 per cent (w/w) of iodine at 20°C, under conditions in which amylopectin does not absorb iodine to any appreciable amount. It has to be accepted, however, that the sensitivity of this technique, described in section 3.2.2, is insufficient to detect unambiguously the presence of 2–3 per cent of contaminating amylopectin.

A second, more sensitive technique is to carry out the $\beta$-amylolysis of the amylose in the presence of *Z*-enzyme. Peat and his co-workers (1949) showed that, at the relative concentrations of *Z*-enzyme found in contaminated commercial preparations of *beta*-amylase, the simultaneous action of the two enzymes on amylopectin was apparently no different from that of *beta*-amylase acting alone. We have, however, noted that the concomitant action of the two enzymes on amylose causes the quantitative conversion of that polysaccharide to maltose. Taken in conjunction, these observations suggest that if the conjoint action of the two enzymes produces 100 per cent conversion into maltose, contaminating amylopectin must be absent. Conversely, conversions of significantly less than 100 per cent signify the presence of amylopectin. The amount of contaminant can be estimated quantitatively, if both the conversion of the amylose into maltose by the concurrent action of *beta*-amylase and *Z*-enzyme, $\beta_z$, and the $\beta$-amylolysis limit of the amylopectin, $\beta_{Ap}$, are known, using the relation

$$\%\ \text{contaminating amylopectin} = 100[(100\text{-}\beta_z)/(100\text{-}\beta_{Ap})] \qquad (2.2)$$

This method provides the best test devised to date for showing the absence of contaminating amylopectin in amylose preparations.

*Absence of Retrogradation during $\beta$-Amylolysis.* Meyer (1952) suggested that the results of Peat and his colleagues (1949), regarding the incomplete hydrolysis of amylose by purified *beta*-amylase, were due to a physical aggregation of the macromolecules. There is no doubt that, given the appropriate conditions of pH, temperature, concentration, and molecular size, amylose does crystallize from aqueous solution, and furthermore this retrograded material is not susceptible to any further degradation by *beta*-amylase. Although we have never noted any obvious sign of retrogradation in digests incorporating *beta*-amylase, the possibility of some type of incipient retrogradation, in which the aggregates are not sufficiently large to precipitate from solution but nevertheless render the molecules inaccessible to the action of *beta*-amylase, must be considered.

Unfortunately, it is not very easy to prove that such a phenomenon is not occurring. By varying the method of dissolving the amylose, and by altering the concentration of the *beta*-amylase and also the time that the polysaccharide exists in aqueous solution prior to the enzymic reaction being terminated by lack of suitable substrate, it is possible to test whether the aggregation phenomenon is likely to occur during the preparation of the digest, or actually during the course of $\beta$-amylolysis. Such experiments are summarized in table 2.4.

The results in columns (a) and (b) of table 2.4 were achieved using our normal methods of dissolving the amylose–butan-1-ol complex, and the dehydrated complex, respectively: i.e. the butan-1-ol complex is dissolved directly in water at room temperature, and the dried material is dissolved in aqueous caustic alkali (0.15 M) at 0–2°C, and neutralized with dilute hydrochloric acid (0.2 M) prior to the addition of *beta*-amylase. Those in column (c) were obtained by dissolving the moist complex directly in a solution of *beta*-amylase. The results shown in columns (d) and (e) represent the effects of dissolving the moist complex and the dried complex, respectively, in dimethylsulphoxide, and adding these solutions dropwise, with constant stirring, to a solution of *beta*-amylase whose activity was sufficiently high to ensure that the reaction was completed virtually instantaneously.

*Table* 2.4. The $\beta$-amylolysis limit, [$\beta$], as a function of the method of dissolving the amylose (Banks and Greenwood 1967b).

| sample of starch | method of dissolving amylose (see text) | | | | |
|---|---|---|---|---|---|
| | (a) | (b) | (c) | (d) | (e) |
| potato | 83 | 84 | 84 | 83 | 82 |
| oat | 78 | 79 | 77 | 77 | 77 |
| wheat | 72 | 73 | 72 | 71 | 70 |
| barley | 74 | 74 | 71 | 72 | 74 |

It can be seen that in no case did the method of preparation of the digest sensibly affect the extent of conversion into maltose by *beta*-amylase. Although this constancy in the $\beta$-amylolysis limit is not conclusive proof for the absence of aggregation, it does tend to suggest that it is not very likely. We find the results obtained using dimethylsulphoxide of particular significance, for not only does this liquid act as a solvent for starch granules but it also dissolves retrograded amylose. Moreover, when such solutions of retrograded amylose are diluted with water the amylose is again found to be susceptible to hydrolysis by *beta*-amylase. On this basis, we feel that the evidence favours physical aggregation playing little part in the incomplete degradation of amylose by *beta*-amylase.

*Factors Affecting the $\beta$-Amylolysis Limit.* The results shown in table 2.4 refer to newly-prepared amylose samples. It is necessary to stress this fact, because we have found that the $\beta$-amylolysis limit of amylose may well change on storage. Table 2.5 shows the conversion of amylose into maltose by (i) the action of pure *beta*-amylase, and (ii) the simultaneous actions of *beta*-amylase and Z-enzyme, for both the dried amylose–butan-1-ol complex and the aqueous amylose–butan-1-ol complex after storage for different periods of time under various conditions.

After prolonged storage the dried complex is virtually insoluble in water, and although apparently dissolving in caustic alkali, the amylose can no longer be completely digested by the mixture of *beta*-amylase and *Z*-enzyme after neutralization of the alkaline solution. Boiling of the neutralized solution prior to the addition of the enzymes causes a marked increase in subsequent maltose production, but the level obtained for a freshly-prepared sample is not achieved. Only by dissolution of the amylose in dimethylsulphoxide can the original $\beta$-amylolysis and $(\beta+Z)$-amylolysis limits be achieved.

*Table* 2.5. The effect of storage under various conditions on the susceptibility of amylose to degradation by *beta*-amylase, $[\beta]$, and by *beta*-amylase acting in conjunction with *Z*-enzyme, $[\beta+Z]$ (Banks and Greenwood 1967b).

| physical state | storage tem-perature (°C) | storage time in months | method of dispersion (see below) | susceptibility to enzyme attack $[\beta]$ (%) | susceptibility to enzyme attack $[\beta+Z]$ (%) |
|---|---|---|---|---|---|
| dried butan-1-ol complex | 15–20 | 0 | a | 95 | 100 |
| | | 9 | a | insoluble | |
| | | | b | 80 | 89 |
| | | | c | 87 | 97 |
| | | | d | 95 | 100 |
| aqueous butan-1-ol complex | 5 | 0 | a | 90 | 101 |
| | | 6 | a | 81 | 92 |
| | | | b | 84 | 97 |
| | | | c | 89 | 100 |
| | | | d | 88 | 99 |
| aqueous butan-1-ol complex | 15–20 | 0 | a | 100 | 100 |
| | | 40 | a | 100 | 101 |
| aqueous butan-1-ol complex | 15–20 | 0 | a | 88 | 100 |
| | | 6 | a | 88 | 100 |

(a) Heating in water at 98°C for 50 minutes in an oxygen-free atmosphere.
(b) Shaking in 0.5 M KOH for one hour at room temperature, followed by neutralization to pH 6.5.
(c) As in (b), but the neutral dispersion heated to 98°C for 20 minutes in an oxygen-free atmosphere.
(d) Shaken with dimethylsulphoxide for one hour at room temperature.

Prolonged storage of the aqueous butan-1-ol complex at refrigeration temperatures causes a marked decline in the susceptibility of the material, to both *beta*-amylase and the combined actions of *beta*-amylase and *Z*-enzyme, when either water or caustic alkali are used as solvents. (It should be noted that in neither case is there any apparent sign of retrogradation having occurred for the solutions are quite clear, and readily pass through a sintered glass filter. Only the lowered susceptibility to enzymic attack indicates that true molecular solution has not been achieved.) Heating the neutralized alkaline dispersion to boiling for twenty minutes reversed the effects noted on storage, as did the use of dimethylsulphoxide as a solvent.

The decrease in susceptibility to enzymic attack demonstrated by the above results is quite consistent with a model in which some type of incipient retrogradation may occur under specific conditions of storage. In this regard, it is especially noteworthy that dissolution in dimethylsulphoxide effectively destroys these physical aggregates, just as it does for retrograded material. The results in table 2.5 support our claim that $\beta$-amylolysis limits markedly lower than 100 per cent conversion into maltose reflect a chemical property of the molecule rather than being physical in origin.

It should be stressed that storage of the aqueous butan-1-ol complex *at room temperature* causes no change in the observed $\beta$-amylolysis limits, and we recommend this method of storing samples of amylose.

*The General Occurrence of a Barrier to $\beta$-Amylolysis.* It is our contention that the barrier to *beta*-amylase action in amylose is widely distributed in Nature. We have never obtained a total amylose, i.e. one resulting from the dispersion of a starch, that is completely degraded by *beta*-amylase acting alone. For example, table 2.6 shows the iodine-binding capacities and susceptibilities to enzymic degradation of amyloses from a wide variety of starches.

*Table* 2.6. The iodine affinities and susceptibilities to enzymic hydrolysis by *beta*-amylase, [$\beta$], and by *beta*-amylase acting in conjunction with $Z$-enzyme, [$\beta+Z$], of the amylose components of starches from a variety of botanical sources (Greenwood and Thomson 1962a).

| source of starch | iodine-binding capacity (%) | [$\beta$] (%) | [$\beta+Z$] (%) |
|---|---|---|---|
| amylomaize | 19.2 | 77 | 101 |
| apple | 19.0 | 84 | 99 |
| banana | 19.9 | 82 | 100 |
| barley | 19.6 | 74 | 98 |
| broad bean | 19.2 | 82 | 99 |
| iris rhizome | 19.1 | 84 | 100 |
| mango | 19.2 | 77 | 100 |
| oat | 19.5 | 78 | 101 |
| parsnip | 19.4 | 72 | 99 |
| pea (smooth-seeded) | 19.2 | 81 | 100 |
| pea (wrinkled-seeded) | 19.2 | 82 | 101 |
| potato | 19.5 | 76 | 100 |
| rye | 19.7 | 72 | 98 |
| tapioca | 19.5 | 95 | 100 |
| wheat | 19.7 | 72 | 100 |

Three aspects of the results shown in this table are important. First, all the samples show a high iodine-binding capacity; second, the combined action of *beta*-amylase and Z-enzyme gives, within experimental error, quantitative conversion into maltose; third, the extent of conversion into maltose in the presence of purified *beta*-amylase varies from 72–95 per cent, depending on the botanical source of the starch from which the amylose was extracted. Taken together, these results indicate that contaminating amylopectin can play no significant role in the relatively low β-amylolysis limits observed for the samples. For example, the β-amylolysis limit of tapioca amylose is 95 per cent conversion into maltose; assuming that this result were due to the presence of contaminating amylopectin, it is necessary to postulate some 10 per cent of contaminant. The presence of this amount of amylopectin in the sample would give values of 17.5 per cent for the iodine-binding capacity, and 96 per cent conversion into maltose under the combined actions of *beta*-amylase and Z-enzyme, in contrast to the experimental

*Table* 2.7. The properties of various amylose samples, and of their β-amylolysis limit dextrins (Banks and Greenwood 1967b).

| source | iodine-binding capacity (%) | [β] (%) | [β+Z] (%) | [η] (ml/g) |
|---|---|---|---|---|
| barley | 19.6 | 74 | 98 | 355 |
| barley β-limit dextrin | 19.3 | 0 | 99 | 315 |
| potato 1 | 19.5 | 77 | 100 | 430 |
| potato 1 β-limit dextrin | 19.2 | 0 | 100 | 415 |
| potato 2 | 19.6 | 82 | 101 | 455 |
| potato 2 β-limit dextrin | 19.4 | 2 | 98 | 420 |
| wheat | 19.7 | 72 | 100 | 330 |
| wheat β-limit dextrin | 19.4 | 0 | 101 | 300 |

values of 19.5 per cent and 100 per cent respectively. The example chosen (tapioca amylose) for this calculation has the highest β-amylolysis limit in table 2.6; taking the other extreme of the lowest β-amylolysis limit (wheat amylose), a similar calculation shows that if this result were due to the presence of contaminating amylopectin, then some 60 per cent contamination would be required. This value cannot possibly be reconciled with the experimental results shown in table 2.6 for the iodine-binding capacity and extent of conversion into maltose produced by the combined actions of *beta*-amylase and Z-enzyme.

The properties of amyloses isolated after exhaustive treatment with purified *beta*-amylase, i.e. the amylose β-amylolysis limit dextrins, are shown in table 2.7. In all cases, the iodine-binding capacities, the susceptibilities to the concurrent actions of *beta*-amylase and Z-enzyme, and the limiting viscosity numbers of the dextrins are comparable in magnitude to the values obtained with the parent amyloses; only in their susceptibility to purified *beta*-amylase do the $\beta$-amylolysis limit dextrins differ from the parent amyloses, the former polysaccharides not being attacked to any appreciable extent. To ensure that incipient aggregation played no part in the values recorded in table 2.7 for the $\beta$-amylolysis limit dextrins, the dextrins, isolated from the $\beta$-amylolysis digest by complexing with butan-1-ol, were subjected to the various dissolution procedures shown in table 2.4. Again, irrespective of the method by which solution was achieved, the limit dextrins were not hydrolysed with *beta*-amylase.

For the purposes of comparison, the values for the corresponding parameters of various amylopectin $\beta$-amylolysis limit dextrins are shown in table 2.8. These limit dextrins are characterized by having very low iodine-binding capacities, limiting viscosity numbers about half that of the amylose $\beta$-amylolysis limit dextrins, and by being resistant to the concurrent actions of *beta*-amylase and Z-enzyme. A comparison of tables 2.7 and 2.8 demonstrates quite unambiguously that radically different polysaccharides are involved in the two cases.

*Table* 2.8. The properties of the $\beta$-amylolysis limit dextrins of various amylopectins (Banks and Greenwood 1967b).

| source | iodine-binding capacity (%) | $[\beta]$ (%) | $[\beta+Z]$ (%) | $[\eta]$ (ml/g) |
|---|---|---|---|---|
| barley | 0.3 | 2 | 3 | 145 |
| oats | 0.3 | 0 | 0 | 125 |
| potato | 0.2 | 0 | 0 | 130 |
| wheat | 0.2 | 2 | 2 | 140 |

Further evidence that the incomplete $\beta$-amylolysis of amylose reflects a property of the polysaccharide rather than being a mere experimental artefact has been provided by a series of studies relating the properties of starch and its components to the maturity of the botanical source (Greenwood and Thomson 1962b; Geddes, Greenwood and MacKenzie 1965; Banks and Greenwood 1973; Banks, Greenwood and Muir 1973a). In general, as the growth season progresses, the $\beta$-amylolysis limit of the amylose decreases. Table 2.9 shows the relevant results for amylose samples obtained by the complete dispersion of potato starch granules isolated from tubers at various stages of growth.

As this decrease in $\beta$-amylolysis limit is accompanied by an increase in the molecular size of the amylose component during the growth season (a point dealt with in more detail later in section 6.3.1), the barrier to the action of *beta*-amylase may exist preferentially in the molecules of highest molecular weight. Support for this suggestion is provided by the results from leaching experiments at successively higher temperatures, when, as shown in table 2.10, the $\beta$-amylolysis limit of the product-amylose decreases during the sequence of operations. We have already noted that the molecular size of the amylose (as measured by its limiting viscosity number) increases during the sequence (compare table 2.1).

*Table* 2.9. The properties of potato amylose as a function of tuber size (Geddes, Greenwood and MacKenzie 1965).

| tuber size (cm) | iodine-binding capacity (%) | $[\beta]$ (%) | $[\beta+Z]$ (%) |
|---|---|---|---|
| 0–1 | 19.5 | 92 | 101 |
| 2–3 | 19.5 | 90 | 99 |
| 4–5 | 19.5 | 88 | 101 |
| 6–7 | 19.5 | 85 | 101 |
| 8–9 | 19.5 | 84 | 100 |
| 10–11 | 19.5 | 83 | 101 |
| >14 | 19.5 | 81 | 99 |

*Table* 2.10. The properties of the amylose components obtained by successive leachings, followed by dispersion, of starch granules from various sources (Banks and Greenwood 1967b).

| source | fractionation procedure | amylose extracted (%) | $[\beta]$ (%) | $[\eta]$ (ml/g) |
|---|---|---|---|---|
| amylomaize | leached at 80°C | 15 | 98 | 140 |
| | leached at 90°C | 12 | 79 | 170 |
| | residue dispersed | 73 | 74 | 200 |
| pea (smooth-seeded) | leached at 70°C | 2 | 100 | 110 |
| | leached at 80°C | 3 | 88 | 125 |
| | leached at 90°C | 4 | 83 | 170 |
| | residue dispersed | 91 | 80 | 190 |
| potato | leached at 60°C | 35 | 99 | 250 |
| | leached at 65°C | 15 | 82 | 320 |
| | residue dispersed | 50 | 75 | 570 |

To summarize, the results discussed above show that there is present in all the starches examined a polysaccharide fraction which has the iodine-binding characteristics of a linear $\alpha$-1:4-linked polyglucan, but nevertheless is not degraded by the action of pure *beta*-amylase. No evidence supports the postulate that incipient aggregation would account for the incomplete $\beta$-amylolysis of this fraction, and therefore we conclude that the phenomenon is real, and reflects a heterogeneity in either the type of linkage or in the repeat unit. Using an arbitrary definition, based solely on iodine-binding capacity, this anomalous material may be termed 'amylose'. Equally well, however, amylose could be defined as the polysaccharide component of starch that is completely degraded by pure *beta*-amylase. Both definitions are arbitrary and unsatisfactory, but conventionally we shall use the former throughout this section.

The next problem to be considered is to identify the nature of the heterogeneity in amylose. Three likely causes are the presence of ester phosphate groups, oxidative modification of glucose residues, and an anomalous type of linkage, i.e. one other than an $\alpha$-1:4-bond. These possibilities will now be considered in detail.

**2.3.2.** Ester Phosphate Groups.

The presence of ester phosphate groups in the dextrins resulting from the $\alpha$-amylolysis of potato starch was demonstrated by Posternak (1950), who also found that they constituted barriers to the action of *beta*-amylase. Earlier, Schoch (1942b) showed that the major fraction of the phosphate present in tuber starches is associated with the amylopectin component, but Peat, Pirt and Whelan (1952b) reported that amylose contained one phosphate group per 2400 glucose residues. Assuming that these groups are distributed randomly in amylose with a number-average degree of polymerization of 1200 units, the calculated $\beta$-amylolysis limit is 75 per cent conversion into maltose—a value comparable to that obtained experimentally by Peat and his co-workers. Thus the amount of phosphate ester group present in amylose could be sufficient to account for the observed $\beta$-amylolysis limit.

However, Peat *et al.* (1952b) were unable to detect any significant increase in the $\beta$-amylolysis limit of amylose when the polysaccharide was

*Table* 2.11. The $\beta$-amylolysis limits of amylose after treatment of the polysaccharide with acid phosphatase (Banks and Greenwood 1961).

| source of phosphatase | digest conditions(1) | source of substrate | [$\eta$] (ml/g) | [$\beta$] (%) |
|---|---|---|---|---|
| cereal | a | wheat | 260 | 68 |
| | a | wheat-control(2) | 270 | 68 |
| | a | potato | 430 | 83 |
| | a | potato-control(2) | 440 | 83 |
| potato | b | wheat | — | 69 |
| | b | wheat-control(2) | — | 67 |
| | b | potato | — | 84 |
| | b | potato-control(2) | — | 83 |

[1] (a) Action of phosphatase followed by that of *beta*-amylase; (b) concurrent actions of phosphatase and *beta*-amylase.

[2] In the control experiments, phosphatase was omitted.

incubated with purified *beta*-amylase and phosphatase. Subsequent work (Banks and Greenwood 1961) confirmed these findings. The $\beta$-amylolysis limit of amylose did not increase if the polysaccharide was subjected to successive treatments with acid phosphatase, or to the simultaneous action of the two enzymes (see table 2.11).

These results suggest that no significance need be attached to the small amount (*ca.* 0.005 per cent) of phosphate found in amylose, and it is most probably present as a phospholipid contaminant of the polysaccharide.

**2.3.3.** Oxidized Glucose Residues as a Barrier to $\beta$-Amylolysis.

We have stressed earlier (see section 2.2.2) that it is essential to use anaerobic conditions for all those operations which are carried out in boiling water, during the fractionation of starch and recrystallization of the amylose, in order to minimize the possibility that oxidation might modify the polysaccharide structure. However, the relatively small amount of oxidation required in order to reduce the $\beta$-amylolysis limit from 100 per cent to, say, 80 per cent conversion into maltose makes it difficult to guarantee that such modifications do not occur. For example, if the number-average degree of polymerization of the amylose is 2000 residues, a random oxidative reaction affecting only two glucose units per 10000 would be sufficient to reduce the $\beta$-amylolysis limit to 80 per cent conversion into maltose. The probability of detecting a modification of this magnitude by a chemical or physical technique is infinitely small.

Early reports, that heating amylose solutions at 98°C in the presence of sodium hydroxide (0.5M) and an oxygen atmosphere caused (Baum, Gilbert and Scott 1956), or did not cause (Cowie, Fleming, Greenwood and Manners 1957), changes in the observed $\beta$-amylolysis limit, ignored the concominant effects of alkali causing a 'zipping' degradative action and a random hydrolytic scission of bonds (Machell and Richards 1958).

These difficulties may be avoided by subjecting to the oxidative treatment an amylose that is quantitatively degraded by the action of pure *beta*-amylase, for then, irrespective of any hydrolytic action, the introduction of barriers to the progress of the enzyme must be shown by a significant decrease in the $\beta$-amylolysis limit. Such experiments have been carried out (Banks, Greenwood and Thomson 1959b) with the results shown in table 2.12. In all cases (other than the control, in which anaerobic conditions were maintained), treatment with oxygen caused the $\beta$-amylolysis limit to decrease from the original value of *ca.* 100 per cent conversion into maltose. Also, after 'oxidation', all the samples were quantitatively degraded by the concurrent action of *beta*-amylase and Z-enzyme.

These experiments show that oxidative barriers may be introduced into amylose, but we do not consider that this effect is significant in the context of the fractionation and recrystallization of amylose. For example, even after two hours at 100°C, *oxygen* caused only a decrease in the $\beta$-amylolysis limit to 93 per cent—a value far removed from those (70–80 per cent) recorded in table 2.6—and furthermore, under the *anaerobic* conditions of fractionation, any such modification would be expected to be very much less. Indeed, we have shown that neither the $\beta$-amylolysis limit, nor the limiting viscosity number, of an amylose sample change after three successive one-hour recrystallizations from boiling aqueous butan-1-ol.

*Table* 2.12. The effect of oxygen treatment at 95°C on samples of linear amylose (Banks, Greenwood and Thomson 1959b).

| source | conditions pH | time (min) | initial $[\eta]$ (ml/g) | final $[\eta]$ (ml/g) | $[\beta]$ (%) |
|---|---|---|---|---|---|
| potato 1 | 9.2 | 20 | 290 | 260 | 86 |
| potato 2 | 9.2 | 20 | 260 | 220 | 86 |
| potato 3 | 9.2 | 20 | 230 | 195 | 87 |
| Iris germanica 1 | 9.2 | 20 | 190 | 120 | 86 |
| | 9.2[1] | 20 | 190 | 180 | 98 |
| | 7.0 | 20 | 190 | 130 | 91 |
| | water | 120 | 190 | 160 | 95 |
| Iris germanica 2 | water | 120 | 150 | 150 | 93 |

[1] Control experiment, in which the heating was carried out in an atmosphere of nitrogen.

In section 2.2.1 attention was drawn to the finding that, after pre-treatment with dimethylsulphoxide, amylose may be isolated from potato starch at temperatures as low as 20°C. The $\beta$-amylolysis limit of such an amylose was identical to that for samples isolated at higher temperatures (see table 2.2). Again, this finding is not to be expected if all the barriers to the action of *beta*-amylase were indeed oxidative artefacts introduced as a result of the use of elevated temperatures during isolation of the polysaccharide.

The fact that barriers deliberately introduced into amylose by oxidation can be effectively removed by the concurrent action of *beta*-amylase and Z-enzyme, in the same way as can the natural barriers, makes the concept of Z-enzyme acting *specifically* rather unlikely: because of its importance, this enzyme is considered in detail in the next section.

### **2.3.4.** The Nature of Z-Enzyme.

Peat, Whelan and Thomas (1952a) isolated Z-enzyme as a freeze-dried powder—free from amylase activity—with *beta*-glucosidase activity. Moreover, they found that when Z-enzyme was replaced by emulsin (a crude mixture of enzymes, including *beta*-glucosidase), amylose could be completely hydrolysed by the action of *beta*-amylase. From these results, they concluded that amylose possessed single-glucose-unit side-chains, joined to the main chain by a $\beta$-linkage. However, Hopkins and Bird (1953) reported that they could not detect glucose in the initial stages of the attack of Z-enzyme and *beta*-amylase on amylose $\beta$-amylolysis limit dextrin, from which they concluded that the model of amylose suggested by Peat and co-workers (1952a) was incorrect, and that *Z*-enzyme was merely an *alpha*-amylase present in very small amounts. Peat and Whelan (1953) then re-identified *Z*-enzyme as a *beta*-glucosidase, but modified their model of amylose by suggesting that the side chains contained a number of glucose residues.

Later, Neufeld and Hassid (1955) showed that the *Z*-enzyme activity of emulsin could be destroyed without reducing the *beta*-glucosidase activity, and Baba and Kojima (1958) showed that emulsin contained an *alpha*-amylase, which they suggested could be the so-called *Z*-enzyme.

The random, non-specific *α-amylolytic* nature of *Z*-enzyme was eventually proved by Banks, Greenwood and Jones (1960). In these experiments *Z*-enzyme was not isolated, but its properties were established by preferentially inhibiting the *beta*-amylase in mixtures of the two enzymes, using mercuric chloride solution ($1.5 \times 10^{-6}$M). The changes in limiting viscosity number and $\beta$-amylolysis limit of amylose treated in this way with *Z*-enzyme are summarized in table 2.13.

*Table* 2.13. The action of *Z*-enzyme on amylose (Banks, Greenwood and Jones 1960).

| source | pH of incubation | initial $[\eta]$ (ml/g) | final $[\eta]$ (ml/g) | initial $[\beta]$ (%) | final $[\beta]$ (%) |
|---|---|---|---|---|---|
| potato 1 | 3.6 | 440 | 440 | 83 | 83 |
| | 4.6 | 430 | 290 | 84 | 89 |
| | 5.5 | 420 | 120 | 84 | 94 |
| potato 2 | 4.6 | 250 | 150 | 98 | 99 |
| wheat | 3.6 | 190 | 190 | 71 | 73 |
| | 4.6 | 195 | 160 | 74 | 77 |
| | 5.5 | 195 | 80 | 73 | 87 |

As originally defined by Peat and his co-workers (1952a), *Z*-enzyme was inactive at pH 3.6. Table 2.13 shows that in unpurified preparations of *beta*-amylase there is an enzyme, capable of reducing the molecular size of amylose and of increasing the susceptibility of the polysaccharide to $\beta$-amylolysis, that is active at pH 4.6 and pH 5.5 but not at pH 3.6, i.e. an enzyme which is behaving as *Z*-enzyme.

However, this *Z*-enzyme caused a marked decrease in the limiting viscosity number of amylose, even when the substrate contained no barriers to the action of *beta*-amylase (potato 2 amylose). Moreover, exposure of amylose to the action of *Z*-enzyme alone did not lead to all these barriers being removed. It may appear anomalous that the concurrent actions of the two enzymes bring about the quantitative degradation of the polysaccharide, whereas *Z*-enzyme followed by *beta*-amylase cannot do so. In fact, the observation is quite consistent with *Z*-enzyme exerting an α-amylolytic action. In mixtures of the two, the α-amylolytic activity is present merely as a minor contaminant, so that the *beta*-amylase will exhaust its potential substrate before the Z-enzyme acts. Each bond hydrolysed by the latter enzyme must then offer a new substrate for the *beta*-amylase, i.e. each hydrolytic event involving Z-enzyme must of necessity lead to an increase in the β-amylolysis limit. When acting on its own, however, the Z-enzyme may hydrolyse an amylose molecule that contains no barrier to the action of *beta*-amylase, and this event does not cause any change in the susceptibility of the polysaccharide on subsequent β-amylolysis. Therefore, a given number of scissions by Z-enzyme acting in the presence of *beta*-amylase is much more effective in increasing the β-amylolysis limit than is the same number when the substrate is subjected to the successive actions of the two enzymes.

The α-amylolytic nature of Z-enzyme was further demonstrated by its action on amylopectin and on amylopectin β-amylolysis limit dextrin. Light-scattering measurements (see section 4.2.2) carried out on these two substrates showed that at pH 3.6 there was no change in the molecular size of either with time, whereas degradation was quite marked at pH 4.6 and pH 5.5. All the observations were consistent with the view that Z-enzyme was merely a trace of *alpha*-amylase except the fact that it had no apparent effect on glycogen, as measured by light scattering. However, later work by Cunningham, Manners, Wright and Fleming (1960) not only confirmed that Z-enzyme was an *alpha*-amylase, but showed that it did cause limited hydrolysis of glycogen.

More recently, purified, highly active forms of Z-enzyme from barley (Greenwood and MacGregor 1965) and soya bean (Greenwood, MacGregor and Milne 1965a, b) have been prepared, and have been shown to have the typical properties of plant *alpha*-amylases.

From the point of view of our enzymic assay for the purity of amylose (see section 2.3.1), it is a useful coincidence that the concentration of Z-enzyme (or *alpha*-amylase) contaminating all but the most highly purified preparations of *beta*-amylase is so low that, under normal conditions, it does not appreciably increase the β-amylolysis of amylopectin. Of course, by increasing the concentration of *beta*-amylase, and therefore that of Z-enzyme, it is possible to obtain a measurable increase in the β-amylolysis limit of amylopectin. For the enzymic assay to be used, it is necessary to adjust the concentration of Z-enzyme so that no change in the β-amylolysis limit of amylopectin occurs.

The fact that Z-enzyme does not specifically remove the barrier to β-amylolysis in amylose, but merely by-passes it, makes the identification of the barrier much more difficult. In fact, the α-amylolytic nature of Z-enzyme implies that the barrier may vary from one amylose to another, and also that there may be more than one type of barrier within a given sample. However, our work suggests that the structural modifications that constitute these barriers are native to the amylose molecule, and are not merely physical or chemical artefacts resulting from aggregation or oxidative modification, respectively. One possible structural modification is the occurrence of the occasional α-1:6-branch linkage. It is perhaps not unreasonable to expect that when amylopectin and amylose are produced biosynthetically in close proximity, a few branch points may be introduced into the amylose.

**2.3.5.** Branching within the Molecule.

Early evidence that some amyloses may be branched was provided by Potter and Hassid (1951), from comparisons of chain lengths (from periodate oxidation) with number-average degrees of polymerization (from osmotic pressure measurements on amylose acetate). In view of the difficulty of measuring fairly large chain lengths accurately by periodate oxidation at that time, these results cannot be regarded as definitive.

Physical or enzymic methods provide much more sensitive techniques for the determination of very limited branching than do chemical methods. In fact, a hydrodynamic study, in which a total amylose was sub-fractionated by precipitation from dimethylsulphoxide, provided the first evidence that the solution behaviour of fractions containing barriers to β-amylolysis was quite different from that of linear fractions (Greenwood 1960; Banks and Greenwood 1967b).

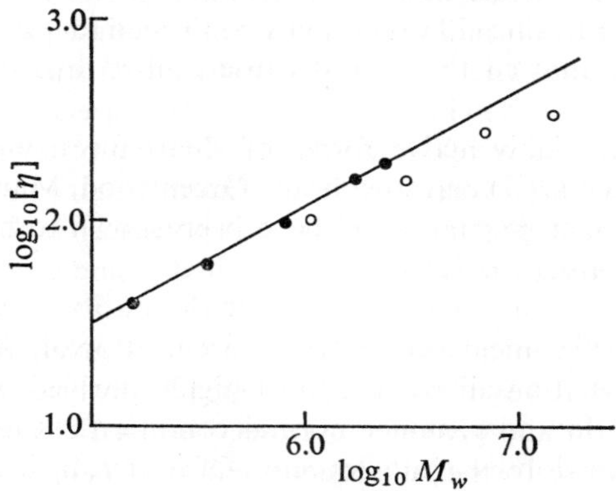


*Figure* 2.2. Graph of $\log_{10}[\eta]$ *versus* $\log_{10}\bar{M}_w$ for amylose sub-fractions: linear amylose (dots), and 'modified' fractions of amylose (open circles). (Greenwood 1960.)

Figure 2.2 shows the relation between limiting viscosity number and molecular weight for the amylose sub-fractions. A linear relation has been drawn through the points representing the linear fractions, and those fractions containing the barrier to β-amylolysis lie below this line; i.e. for a given value of molecular weight, the hydrodynamic radius is less for a fraction having a barrier to the action of *beta*-amylase than for one completely degraded by the enzyme. This effect would be expected if some of the molecules were branched.

These hydrodynamic studies also provide some idea of the nature of the branching. For example, short-chain branching would not be expected to alter the hydrodynamic behaviour of the macromolecule greatly from that observed for its linear counterpart. The results shown in figure 2.2 indicate long-chain branching, in which the branches are several hundreds, or even thousands, of glucose residues in length. Moreover, we have found that the hydrodynamic properties of the $\beta$-amylolysis limit dextrins of the anomalous amylose fractions also are not the same as those of the linear macromolecules, suggesting that the material resistant to $\beta$-amylolysis is multiply branched.

Other hydrodynamic evidence for heterogeneity was obtained from ultracentrifugation studies of aqueous potassium chloride solutions of those fractions incompletely converted by *beta*-amylase. The results indicated that two quite distinct molecular weight ranges were present (Geddes, Greenwood, MacGregor, Procter and Thomson 1964). The sub-fractions of amylose were prepared by fractional precipitation from dimethylsulphoxide solution, i.e. the basis of separation is the solubility of the species. In general, when linear and lightly-branched molecules of the same chemical composition are present, the lightly-branched ones are more soluble than their linear counterparts for the same molecular weight. Thus, at a given solvent–non-solvent composition, the molecular weight of the branched molecules precipitated from solution will be greater than that of the linear molecules. In the case of these particular fractions of amylose, the fact that analytical ultracentrifugation showed two quite distinct ranges of molecular weight to be present is in agreement with what might be expected if amylose were composed of a mixture of linear and lightly-branched macromolecules. Moreover, it was found that only one molecular weight range remained after treating the polysaccharide with *beta*-amylase, and linear sub-fractions invariably exhibited a single range of molecular weight species.

The hydrodynamic evidence in favour of branching was given further support when it was found that yeast isoamylase, which hydrolyses $\alpha$-1:6-branch linkages in amylopectin and glycogen (see section 5.4.3), increased the susceptibility of the amylose to the subsequent action of *beta*-amylase, and that the two enzymes acting in conjunction caused a virtually quantitative conversion of the polysaccharide into maltose (Kjølberg and Manners 1963). In subsequent work, using pullulanase as the de-branching enzyme (see section 5.4.2), these results were fully substantiated (Banks and Greenwood 1966). It was found that pullulanase acting on amylose caused a decrease in the limiting viscosity number of the polysaccharide and an increase in its subsequent susceptibility to the action of *beta*-amylase, although quantitative conversion to maltose was not always attained. (Control experiments demonstrated that the enzyme preparation had no effect on the molecular size of linear amylose, eliminating any possibility of the observed results being due to the presence of a trace of *alpha*-amylase.) The magnitude of the decrease in limiting viscosity number was consistent with that expected of the de-branching of a structure exhibiting long-chain branching. Analogous results to those of Kjølberg and Manners (1963) were also found on studying the concurrent actions of *beta*-amylase and pullulanase—the amylose samples from wheat and potato starches were hydrolysed quantitatively into maltose (Banks and Greenwood 1966). Subsequent studies have confirmed that the combined actions of these two enzymes invariably cause complete degradation of amyloses from a wide variety of starches.

### **2.3.6.** Conclusions.

We have presented here, in some detail, the experimental basis on which our view of the amylose molecule is based. It is our contention that this polysaccharide—as obtained from a total dispersion of the starch—is composed of two quite distinct populations of molecules, one of which is completely hydrolysed by the action of *beta*-amylase, whilst the other is not. Although barriers to $\beta$-amylolysis may be introduced artificially by oxidation, such artefacts are easily avoided by maintaining anaerobic conditions during fractionation and recrystallization operations carried out at high temperature. The barriers are therefore native to amylose. Hydrodynamic studies implicate branching as the source of incomplete $\beta$-amylolysis, and enzymic studies confirm that $\alpha$-1:6-branch points are present in certain fractions of amylose. The evidence suggests that a range of multiply-branched structures are present in the non-linear amyloses.

## **2.4. The Fine Structures of Amylopection and Glycogen.**

Early methylation studies indicated that the amylopectin structure possessed some 4–5 per cent of $\alpha$-1:6-branch points, while the corresponding value for the glycogen structure was about 10 per cent.

Whilst enzymic hydrolysis of amylopectin yielded isomaltose (Montgomery, Weakley and Hilbert 1949), the final proof of the presence of the $\alpha$-1:6-branch linkage was provided by the isolation of panose from a partial acid hydrolysis of waxy-maize amylopectin (Thompson and Wolfrom 1951). Occasional reports have appeared of the presence of sugar residues other than glucose and of linkages other than $\alpha$-1:4 and $\alpha$-1:6 in these branched polymers, but none of these, in our opinion, should be seriously considered.

*Model Structures.* Three models for amylopectin and glycogen advanced on the basis of early structural studies are shown in figure 2.3. The *laminated structure* (a), proposed by Haworth, Hirst and Isherwood (1937), was the simplest consistent with methylation studies, but was not intended as a complete representation of the macromolecule (Haworth 1947; Hirst and Manners 1954). In the *herring-bone structure* (b), suggested by Staudinger and Husemann (1937), a single main chain carries all the branch linkages. The *randomly-branched* (or tree-like) *structure* (c) was proposed by Meyer and Bernfeld (1940).

Myrbäck and Sillén (1949) emphasized that these various models contain different arrangements of the same chains; a concept developed by Peat, Whelan and Thomas (1952a) with the introduction of the terminology *A*-,

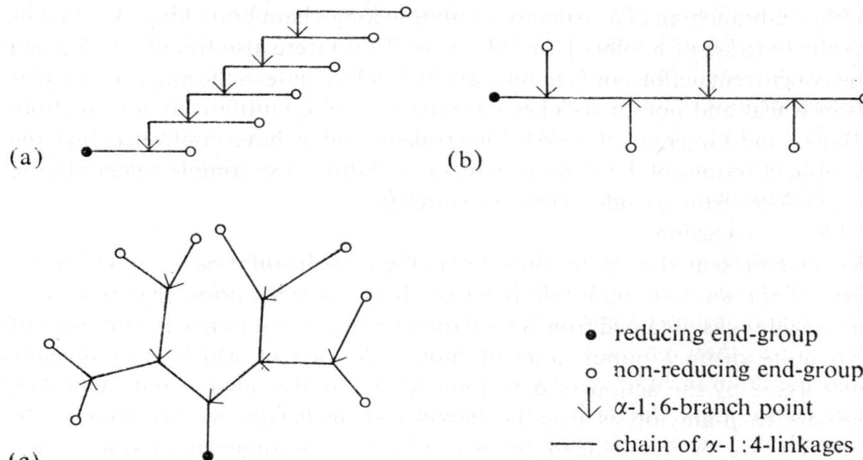


*Figure* 2.3. Proposed models for amylopectin: (a) laminated structure (Haworth, Hirst and Isherwood 1937); (b) herring-bone structure (Staudinger and Husemann 1937); (c) randomly-branched structure (Meyer and Bernfeld 1940).

*B*- and *C*-chains. The *A*-chain is linked to the rest of the molecule only through its reducing end-group; the *B*-chain, in addition to being linked as an *A*-chain, is also substituted through the $C_6$-hydroxyl in one or more of its constituent glucose units; the *C*-chain carries the reducing end-group.

**2.4.1.** Enzymic Exploration of Fine Structure.

Although each of the model structures proposed for glycogen and amylopectin contains a single *C*-chain, they differ in the ratio of *A*- to *B*-chains. If the total number of unit-chains per molecule is $n$, then the Haworth structure contains one *A*-chain, ($n$-2) *B*-chains, and one *C*-chain; the Staudinger model has $(n-1)$ *A*-chains, no *B*-chains, and one *C*-chain; the Meyer structure possesses $(n/2)$ *A*-chains, $(n-1)/2$ *B*-chains, and one *C*-chain, i.e. neglecting the single *C*-chain this model has equal numbers of *A*- and *B*-chains. Thus the ratio of *A*- to *B*-chains, summarized below for the different models, constitutes a diagnostic test for structural type:

| *model* | $A:B$ |
|---|---|
| Haworth | $\rightarrow 0$ |
| Staudinger | $\infty$ |
| Meyer | 1 |

The experimental determination of this ratio was first carried out for amylopectin by Peat, Whelan and Thomas (1952b) who used *beta*-amylase to shorten the external chains, with the specific aim that all the *A*-chains should be reduced to either two or three units in length depending on whether the chain had originally contained an even or odd number of glucose residues. This $\beta$-amylolysis limit dextrin was then subjected to the action of *R*-enzyme (see section 5.4.1); the yield of maltose and maltotriose (obtained from the remnants of the *A*-chains), in conjunction with the presence of linear maltodextrins containing six or more glucose units, was such as to favour the model proposed by Meyer. [The original paper (Peat, Whelan and Thomas 1952b) contained an arithmetical error, which was subsequently amended (Peat, Whelan and Thomas 1956); the conclusions remained unaltered.]

Simultaneously, the structure of glycogen was deduced (Larner, Illingworth, Cori and Cori 1952) to be of the randomly-branched type. In this investigation, the polysaccharide was subjected to the successive actions of phosphorylase and amylo-1:6-glucosidase (see section 5.4.4). These workers assumed that phosphorylase reduced the *A*-chains to a single glucose residue, which was then removed by the de-branching enzyme. If this were the case, and the de-branching activity was confined to the single residues, then only one chain in the Haworth model would become susceptible to further phosphorolysis, whereas the entire molecule would be available if the Staudinger–Husemann concept were valid. In the Meyer model, on the other hand, removal of the residual glucose units would generate a structure very similar to the starting material; the extent of phosphorolysis of this material would then be deduced to be similar to that of the parent glycogen. In fact, the 'stripping' of the glycogen macromolecule by the successive actions of phosphorylase and amylo-1:6-glucosidase gave successive phosphorylase limits that could be interpreted only on the basis of the Meyer model.

Recently, Lee and Whelan (1971) have pointed out that in this early work of Larner *et al.* (1952) it was not realized that the single enzyme possessed both de-branching and transferase activity (see section 5.5). Present knowledge of this transferase activity suggests that some of the *A*-chain stubs, rather than being removed, may have increased in length; whilst equally some of the *B*-chains could have been converted into *A*-chains and then removed. These unforeseen side-reactions may complicate the interpretation of the results, but whether the complication is sufficiently serious to warrant the claim of Lee and Whelan (1971), that the experimental observations of Larner *et al.* (1952) could be equally well interpreted on the basis of the models of both Haworth and Staudinger, is another matter. We believe that rather more would have to be known regarding the kinetics of the transfer activity before such a far-reaching conclusion could be substantiated.

Bathgate and Manners (1966) measured the release of maltose and maltotriose from various $\beta$-amylolysis limit dextrins, using pullulanase (see section 5.4.2). They found virtually equimolar amounts of the two sugars, and used these quantities to calculate the proportion of *A*-chains; their results were again compatible with the Meyer model. However, Lee and Whelan (1971) have also criticized this work, on the basis that no proof was given that all the *A*-chains stubs were removed by the pullulanase. It must, of course, be recognized that in exploring polysaccharide structure by means of enzymes, the simplifying assumption is made that the action of the enzyme on the polymeric substrate is identical to that on small, model compounds, and that steric effects do not therefore measurably influence the pattern of the enzyme. This assumption may not always be warranted, as has been clearly demonstrated by French and his co-workers in their study of the action of *beta*-amylase on model substrates (Summer and French 1956; Kainuma and French 1970). These authors found that if the main chain has an odd number of residues at the non-reducing side of the branch point, *beta*-amylase will reduce it to a single unit, irrespective of whether the side chain is a glucose or maltose unit. However, when the main chain contains an even number of glucose residues to the non-reducing side of the branch point, the structure of the $\beta$-amylolysis dextrin is critically dependent on whether there is one or two units in the side chain—a single glucose unit in the side chain does not block the action of the enzyme until the branch point is reached, whereas the presence of a maltose side-chain confers resistance on this linkage.

On the basis of this type of evidence it follows that, if it is accepted that the bulk of the side chain is the dominant factor in determining the resistance of certain linkages to enzymic hydrolysis, the result of enzyme action on a branched macromolecule may be radically different from that observed in the available model compounds. This criticism could be applied, however, to all enzymic studies of structure.

A revision of the Meyer model of glycogen and amylopectin has been proposed by Gunja-Smith, Marshall, Mercier, Smith and Whelan (1970). This work was based on the inability of *Cytophaga* isoamylase (see section 5.4.3) to remove branch linkages joining maltosyl units to the main chain. By exposing amylopectin and glycogen first to phosphorylase and subsequently to *beta*-amylase, all the *A*-chains should be reduced to two-unit stubs (a phosphorylase–$\beta$-amylolysis limit dextrin). The idealized Meyer structure would yield, after de-branching with the *Cytophaga* enzyme, a series of chains each carrying a maltosyl side-chain on the third glucose residue from the non-reducing end. As a result, the de-branched structure would not be susceptible to the action of *beta*-amylase. In fact, Gunja-Smith *et al.* reported that the phosphorylase–$\beta$-amylolysis limit dextrins of shellfish glycogen and waxy-maize amylopectin had $\beta$-amylolysis limits of 44 per cent and 29 per cent conversion into maltose, respectively, after the action of the *Cytophaga* enzyme. The authors' revised model (shown in figure 2.4) is that approximately half the *B*-chains carry, on average, two *A*-chains, whilst the remaining *B*-chains each carry two *B*-chains. Thus the structure overall possesses the equality in numbers of *A*- and *B*-chains which experiment indicates to be the case, and which is demanded by the Meyer model. This equality is, however, achieved in a completely different manner in this revised structure, in that half the *B*-chains are postulated to have their non-reducing chain-ends inside the macromolecule rather than at the surface. The concept of such 'buried' chains is due to French (1957). The Meyer model is a representation of idealized dichotomous branching in which each chain in the system has exactly the same probability for growth and branching as all the others (see section 2.4.2). It is not inconceivable that, during the enzymic synthesis of glycogen and amylopectin, steric effects preclude several chain ends from growing. Such chain ends would, of course, be buried within the macromolecular domain, and may even be protected from hydrolysis by enzymes such as phosphorylase and *beta*-amylase. The presence of substantial numbers of such buried chains would invalidate the reasoning on which this revised model is based, as was indeed recognized by Gunja-Smith and her colleagues. However, they prefer to explain their results on the basis of a new model structure rather than by employing the concept of buried chains, principally because of the extents to which the de-branched phosphorylase–$\beta$-amylolysis limit dextrins are attacked by *beta*-amylase.

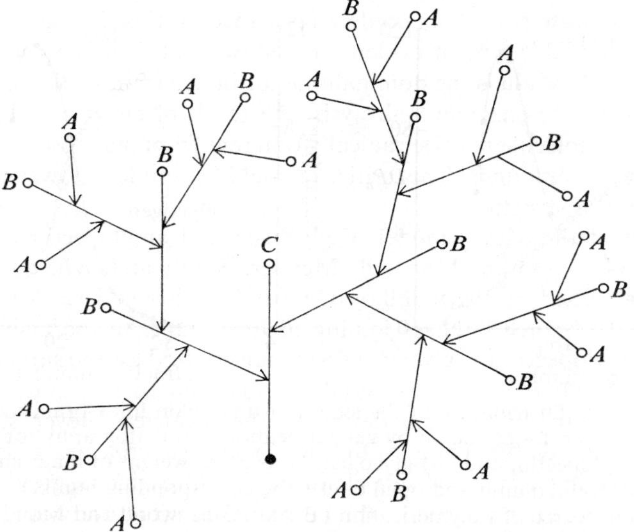


*Figure* 2.4. Revised randomly-branched structure (Gunja-Smith, Marshall, Mercier, Smith and Whelan, 1970). Nomenclature as in figure 2.3 with the addition of the indication of *A*-, *B*- and *C*-chains.

It should be noted, however, that Gunja-Smith *et al.* found and reported small but finite $\beta$-amylolysis limits for the phosphorylase–$\beta$-amylolysis limit dextrins prior to de-branching. This result of course is a contradiction in terms: if the starting material is truly the limit dextrin, it cannot be susceptible to further $\beta$-amylolysis. Conversely, if it is not the limit dextrin, the subsequent de-branching and exposure to *beta*-amylase is not meaningful. Gunja-Smith *et al.* also make the assumption, inherent in all enzymic techniques for exploring polysaccharide structure referred to above, that the reaction with the branched macromolecules proceeds in the same way as it does with model compounds. In this specific instance, they must assume that all *A*-chains in the phosphorylase–$\beta$-amylolysis limit dextrin are reduced to two glucose residues in length. Their chromatographic analysis of the de-branched products of the limit dextrins shows surprisingly large amounts of maltotriose, particularly in the case of that derived from glycogen. No evidence was given that this maltotriose did not arise because the *A*-chain stubs had not reached their limiting length of four units when the polysaccharide was exposed to phosphorylase. Since maltotriosyl side-chains can be removed by the action of *Cytophaga* isoamylase, the presence of such chains would upset the interpretation of the results on which the revised model is based. We believe that, in the light of the above criticisms, the revised Meyer model should be regarded with some caution. Cori (1973) has also pointed out that little is known of the transferase activity of the de-branching enzyme, and that therefore some doubt must exist about its use.

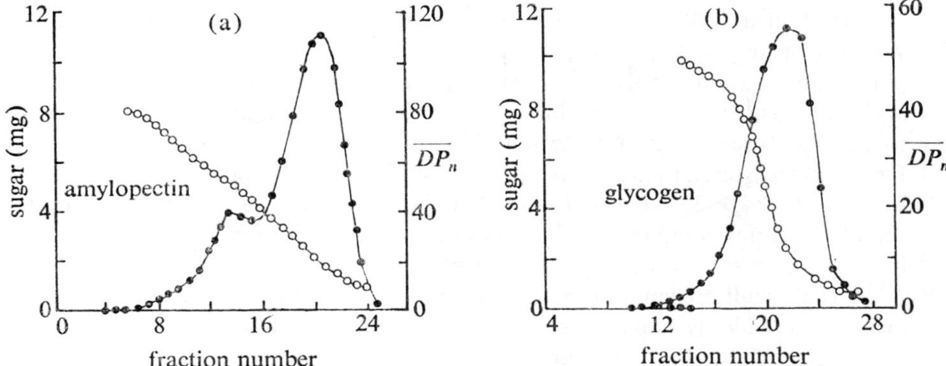


*Figure* 2.5. Distribution of the length of unit chain in de-branched polysaccharides as shown by gel permeation chromatography for (a) amylopectin, and (b) glycogen. Dots show weight of material eluted from column and open circles the corresponding number-average degree of polymerization (Banks, Greenwood and Muir, unpublished work).

We would not expect the structures of glycogen and amylopectin to be identical. It has, for example, been known for some time that the hydrodynamic behaviour of the two polymers is completely different, and we shall deal with this aspect in rather more detail in section 2.4.4. More recently, the use of de-branching enzymes in conjunction with gel permeation chromatography has enabled the unit-chain profiles of the two macromolecules to be obtained (Lee, Mercier and Whelan 1968; Akai, Yokobayashi, Misaki and Harada 1971; Gunja-Smith, Marshall, Mercier, Smith and Whelan 1970; Harada, Misaki, Akai, Yokobayashi and Sugimoto 1972). There is a broad measure of agreement between the various workers when isoamylase is used as the de-branching enzyme, and typical results obtained in our own laboratory are shown in figure 2.5. [Lee, Mercier and Whelan (1968) employed pullulanase as the de-branching enzyme; their results are more difficult to interpret because of the inability of the enzyme to completely degrade amylopectin (Abdullah, Catley, Lee, Robyt, Wallenfels and Whelan 1966).]

Whilst the unit-chain profile of glycogen is characterized by a single major peak occurring at a degree of polymerization of about eight, that of amylopectin exhibits two overlapping distributions, in which the most populous species (by weight) occur at degrees of polymerization of 19 and 60. The reason for the large difference between the structures of glycogen and amylopectin as reflected by figure 2.5 is obscure, but the fact remains that the differences are so great as to lead us to believe that the modes of synthesis of the two polysaccharides differ fundamentally. We suggest, therefore, that in the light of existing evidence it is probably misleading to look for a single model to represent the structures of both amylopectin and glycogen.

Marshall and Whelan (1970) have reported that the quantitative conversions of amylopectin and glycogen to glucose obtained on using amyloglucosidase may in certain instances be due to the presence of contaminating *alpha*-amylase; using a crystalline preparation of amyloglucosidase, shown to be free from *alpha*-amylase activity by means of a procedure (Smith, Drummond, Marshall and Whelan 1970) similar in principle to that devised by Ziese (1934, 1935), it was found that the conversion of certain glycogen samples was not complete. In fact, a value as low as 82.2 per cent conversion into glucose was recorded for cat liver glycogen.

It is difficult to gauge the significance of this observation, because no mention is made by the authors of the loss of enzyme activity in the digests, nor do they show that a true amyloglucosidase limit dextrin (i.e. a structure resistant to further hydrolysis by the enzyme) has been formed. Certainly, it should not be very difficult to detect an anomalous linkage or unit in the resistant material, since it must be present at a concentration of more than one per cent. (Even if the block to enzyme action occurs at the non-reducing end of the *C*-chain of the idealized Meyer structure, the total length of the resistant chain is unlikely to exceed 100 glucose residues.)

### **2.4.2.** Probability Models.

We noted previously that the models (see figure 2.3) proposed for these branched polysaccharides contain different arrangements of the same basic chains. Although at first sight these models appear very different, there is a close relation between them which is most easily understood if we consider the growth of the molecules on a probability basis (see figure 2.6). Thus if the probability of a chain growing once it is initiated is unity until a branch point is introduced when it decreases to zero, the result will be the comb structure of Haworth. On the other hand, if the probability of growth remains unity after a branch point is introduced, and the probability of the side chain growing is zero, the Staudinger structure results. The idealized Meyer structure results when the probability of any chain growing is unity, before and after branch points are introduced. To obtain the Meyer structure, but with buried chain-termini, it is necessary not only to maintain equality of probability of the main and side chains growing, but to have that probability less than unity. By suitably altering the probabilities of growth of the main and side chains, it is possible to generate any structure intermediate between the extremes of the Haworth and Staudinger structures.

This situation appears to us to be the one most likely to be occurring in Nature. Both macromolecules are likely to be composed of a variety of similar, but not identical, branched structures arising from variations in probabilities of growth—perhaps through steric factors.

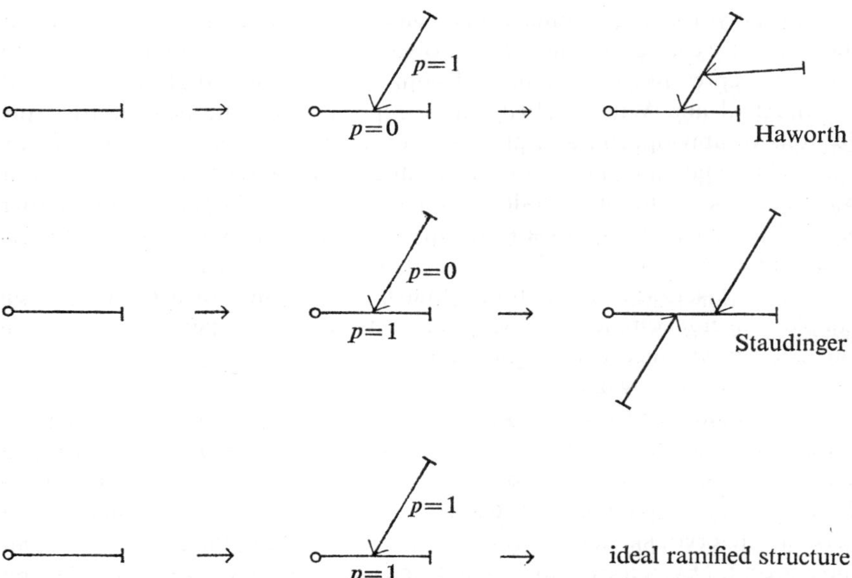


*Figure* 2.6. Schematic representation of formation of branched polymers on a probability basis (see text for details).

Again, remarkably little attention appears to have been given to the kinetic schemes necessary to produce these branched polymers. Rather, the concept of idealized structures has been employed to predict molecular properties. For example, the idealized Meyer structure implies dichotomous branching with a fixed number of glucose residues between branches, and the molecular weight $M$ of this structure is given by

$$M=(2^n-1)m \tag{2.3}$$

where $n$ is the number of tiers in the structure, and $m$ is the molecular weight of the chain between branch points. It is quite easy to show, for this type of structure and for the extent of branching typical of glycogen, that the surface

4

area available to each chain end must decrease with decreasing molecular weight. Madsen and Cori (1958) employed this type of calculation, and predicted an upper limit for the molecular weight of glycogen of $20 \times 10^6$: at this level there was barely enough space at the surface of the macromolecule to accommodate all the chain ends. This difficulty is easily circumvented—if the surface becomes too crowded, the population pressure of chain ends may be relieved by a fraction of them not growing any further, thus giving rise to buried non-reducing chain-ends, a fact recognized by Cori (1973).

### **2.4.3.** Heterogeneity of Branching.

Experimental evidence that the distance between branch points is variable rather than fixed was first provided by Roberts and Whelan (1960), who isolated, by the action of human salivary *alpha*-amylase on amylopectin and glycogen, various α-amylolysis limit dextrins which were resistant to further α-amylolysis. Further analysis, using the de-branching *R*-enzyme, showed that these dextrins had internal chain-lengths (i.e. the number of glucose units between branch points) of only one or two residues. This value is much smaller than the average internal chain-lengths of typical amylopectins and glycogens (approximately seven and four units, respectively), indicating that there is considerable heterogeneity in branching density within both polysaccharides.

The demonstration of regions within amylopectin where the branching density was *less* than average, was provided by Ohashi (1959, 1961), whose iodine-binding studies demonstrated the existence of fairly long linear regions of up to 60 glucose units.

Further studies by Schramm and his co-workers, using glycogen from rabbit liver and shellfish, showed the presence of a range of multiply-branched structures, called *macrodextrins*, that were resistant to hydrolysis by porcine pancreatic *alpha*-amylase (Heller and Schramm 1964; Cabantchik and Schramm 1965). The distribution of such regions within the glycogen macromolecule was the subject of a detailed study by Brammer, Rougvie and French (1972). The action of porcine pancreatic *alpha*-amylase again showed that the regions of high branching-density were scattered throughout the molecule, rather than being concentrated at the centre. On average, 77 regions of macrodextrin per $10^6$ daltons of the glycogen α-amylolysis limit dextrin were found and these regions accounted for some 25 per cent by weight of the dextrin. In order for the porcine enzyme to hydrolyse between branch points, the minimum requirement is that there should be at least two α-1:4-linked glucose residues between the α-1:6-branch points (Kainuma and French 1969). If this minimum requirement is not attained, and the branch points are separated by less than two glucose residues, the structure cannot be hydrolysed by the *alpha*-amylase (Abdullah and French 1970). A local concentration of such high-density branching gives rise to regions that are converted into multiply-branched oligosaccharides on α-amylolysis; as the size of the resistant region increases, the term macrodextrin is introduced to describe the phenomenon. Thus there is present in glycogen a continuous spectrum of $\alpha$-amylolysis limit dextrins of increasing molecular weight and complexity of branching, reflecting the marked heterogeneity in branching density throughout the macromolecule.

Babor, Kaláč and Tihlárik (1968), in an investigation of the fine structure of amylopectin, also prepared a series of macrodextrins having internal chain-lengths of three or four glucose units, i.e. approximately half the average value for the initial amylopectin. They concluded that heterogeneity of branching is a characteristic feature of the structure of amylopectin.

### **2.4.4.** Hydrodynamic Behaviour.

We have earlier indicated that the same model is not likely to be appropriate for both amylopectin and glycogen. In many respects the behaviour of the two polysaccharides is very similar, e.g. the differences in their susceptibility to enzymic attack are of degree rather than kind. Certain de-branching enzymes (*R*-enzyme and pullulanase) admittedly have more difficulty in hydrolysing glycogen than they do amylopectin; however, the fact that pullulanase when acting in conjunction with *beta*-amylase completely hydrolyses both polysaccharides suggests that steric effects within the close-packed glycogen macromolecule may be the source of the difficulty.

One line of evidence for a fundamental difference in the structures of the two polymers is derived from the analysis of the unit-chain profiles of the polysaccharides (see figure 2.5); another is from hydrodynamic studies.

Early solution studies showed that the limiting viscosity number of amylopectin was rather more than an order of magnitude greater than that of glycogen, for the same molecular weight. An even more striking difference is to be found on ultracentrifugation of the two polysaccharides. Figure 2.7 shows the sedimentation data for amylopectin and a sub-fraction of glycogen (the glycogen sub-fraction, prepared by differential ultracentrifugation, was chosen because it had almost the same value as did the amylopectin for the sedimentation coefficient at infinite dilution). The concentration dependence of the sedimentation coefficients is completely different in the two cases. Some workers (Bridgman 1942; Madsen and Cori 1958) have reported that the sedimentation behaviour of glycogen is ideal (that is, $S$ is independent of concentration), whilst others (Larner, Ray and Crandall 1956; Bryce, Greenwood, Jones and Manners 1958) have found a slight departure from ideality, in agreement with the results shown in figure 2.7. In the case of amylopectin, $S$ shows extreme dependence on concentration, the value changing by a factor of three in going from a concentration of 0.3 g/dl to infinite dilution, whereas the value for glycogen changes by only 10 per cent in the same concentration interval. In the concentration range 0.3–1.0 g/dl the sedimentation coefficient of amylopectin is virtually independent of concentration, whereas, in the same concentration range, the results for glycogen may be represented by a continuation of curve (a) of figure 2.7. Also the exceedingly sharp ultracentrifugal patterns observed for amylopectin indicate that the molecules are packed so closely together that they exhibit gel-like properties rather than those of independent molecules.

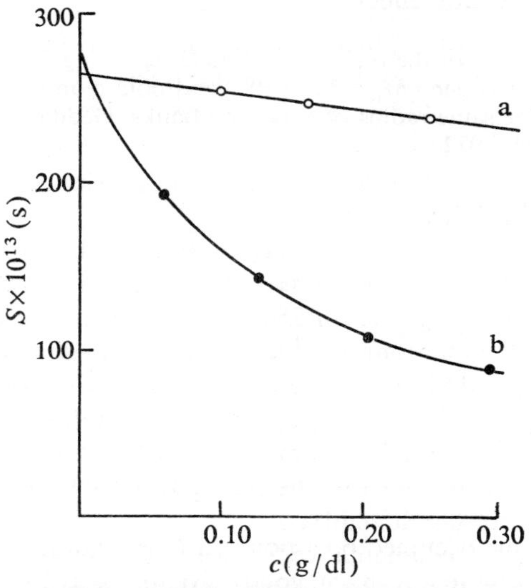


*Figure* 2.7. Graph of the sedimentation coefficient, *S*, as a function of concentration, *c*, for (a) glycogen, and (b) amylopectin (Banks, Geddes, Greenwood and Jones 1972).

One explanation for the difference in hydrodynamic behaviour is that amylopectin and glycogen differ in overall molecular shape. The relative shapes of the two polysaccharides may be obtained by comparing their particle scattering factors ($P_\theta$; the relevant equation is given in section 4.2.2). This comparison has been made, using amylopectin and glycogen sub-fractions prepared by differential ultracentrifugation and chosen so as to have approximately the same value (*ca.* 1200Å) for their root-mean-square radii of gyration. Table 2.14 shows the values of $P_\theta^{-1}$ for both samples as a function of angle $\theta$. Effectively, the values of $P_\theta^{-1}$ are the same for both polysaccharides over the entire angular range. The particle scattering factor is related to the shape of the molecule, hence the data of table 2.14 indicate that the polysaccharides have the same general shape in solution. This result shows that the difference in hydrodynamic behaviour does not arise from gross differences in molecular shape. But whilst both polysaccharides have the same shape and radius of gyration, and are therefore presumably similar in volume, their molecular weights differ by a factor of six; consequently their densities differ by a similar factor. Thus one major difference, that is immediately obvious, is that the arrangement of chains within the glycogen macromolecule allows it to adopt a much more compact form than can be achieved by amylopectin. The compact nature of the glycogen molecule is further emphasized by the low value recorded for its limiting viscosity number ($8 \pm 1$ ml/g), and by the fact that it is virtually independent of solvent. In the case of amylopectin, however, the more open molecular structure would be expected to make it susceptible to solvation effects.

*Table* 2.14. The variation of the reciprocal particle scattering factor $P_{\theta}^{-1}$ with angle, for glycogen ($\bar{M}_w = 325 \times 10^6$) and potato amylopectin ($\bar{M}_w = 50 \times 10^6$) of the same radius of gyration (Banks, Geddes, Greenwood and Jones 1972).

| angle | amylopectin $P_{\theta}^{-1}$ | glycogen $P_{\theta}^{-1}$ |
|---|---|---|
| 30 | 1.33 | 1.34 |
| 45 | 1.71 | 1.73 |
| 60 | 2.20 | 2.14 |
| 80 | 3.01 | 2.96 |
| 90 | 3.45 | 3.43 |
| 120 | 4.66 | 4.69 |
| 135 | 5.13 | 5.22 |

The hydrodynamic volume occupied by a polymer (see section 4.2.2) is a function of its interaction with the solvent: in good solvents the hydrodynamic volume, and therefore the limiting viscosity number, is greater than in poor solvents. The same phenomenon will be displayed by lightly-branched polymers, but as the density of branching increases the amount of space within the molecular domain available for solvent–polymer interactions decreases, and hence the amount of solvent-induced swelling is also restricted. In the limiting case of a very high degree of branching the individual monomer units could be so crowded together that inter-monomer flexibility is destroyed, with the result that the volume occupied by the polymer would be independent of the nature of the solvent. Glycogen may be regarded as just such a polymer.

According to studies carried out on amylose (see section 4.3.8), the solvent power of various aqueous systems is a function of both pH and the type of electrolyte present. Figure 2.8 shows that the limiting viscosity number of waxy-maize amylopectin varies as a function of pH in the presence and absence of potassium chloride in a very similar manner to that for amylose, confirming the observations of Erlander, Purvinas and Griffin (1968). (Care has to be exercised in the choice of amylopectin for this experiment as some, notably those derived from tuber starches, contain esterified phosphate groups, so enabling them to behave as polyelectrolytes; see section 2.4.6. Such behaviour complicates the interpretation of the relation between limiting viscosity number and pH.) The reasons for the two types of curve are dealt with in detail later (see section 4.3.8), but figure 2.8 does illustrate the fact that the chains within the amylopectin macromolecule are arranged in such a way that considerable solute–solvent interaction occurs, again in contradistinction to glycogen.

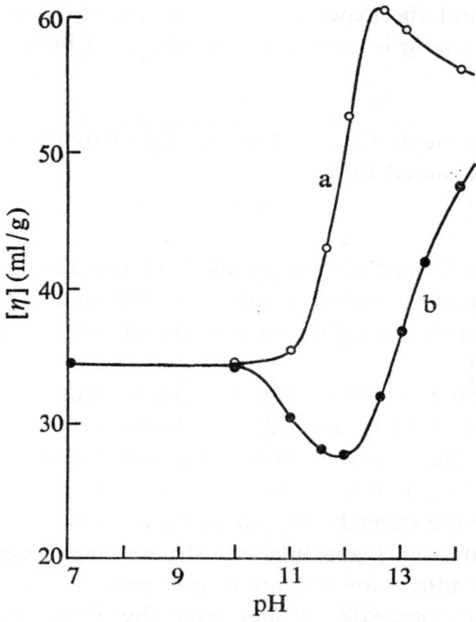


*Figure* 2.8. The limiting viscosity number $[\eta]$ of waxy-maize amylopectin as a function of pH: (a) no supporting electrolyte, and (b) 0.33 M KCl as supporting electrolyte (Banks, Geddes, Greenwood and Jones 1972).

It is important to realize, however, that branching does severely restrict the solvent-induced swelling of which a polymer is capable. For the amylopectin used in this work, the values of the limiting viscosity number in 0.2 M alkali and in neutral aqueous solution are 59 ml/g and 34 ml/g respectively. It can be calculated, using equations 4.83 and 4.80, that the corresponding values for an amylose of the same weight-average molecular weight (approximately $10^7$) are 1970 ml/g and 363 ml/g respectively. Thus, for this particular molecular weight, solvent effects increase the hydrodynamic volume of the linear polymer by a factor of 5.5, whereas they cause an increase of only 70 per cent in the case of amylopectin. This comparison shows the great reduction in hydrodynamic volume as a consequence of branching, and the decreased opportunity for solvent-induced swelling in branched structures.

According to the theory of Zimm and Kilb (1959), the limiting viscosity numbers of a branched polymer $[\eta]_{br}$ and of a linear one $[\eta]_{lin}$ of the same chemical constitution and identical molecular weight are related by

$$g^{1/2} = [\eta]_{br}/[\eta]_{lin} \tag{2.4}$$

Equation 2.4 is valid only when polymer–solvent contacts are reduced to a minimum, a condition satisfied for both amylose and waxy-maize amylopectin by the use of neutral aqueous solvents. Substituting the calculated value of $[\eta]_{lin}$ (363 ml/g) and the experimental value for $[\eta]_{br}$ (34 ml/g) gives $g=9\times10^{-3}$. The parameter $g$ is related to the number of branches $f$ by

$$g=(3f-2)/f^2 \tag{2.5}$$

Using the derived value of $g$ yields $f=330$. The average chainlength $\overline{CL}$ of the polymer may then be calculated from

$$\overline{CL}=M/M_0f \tag{2.6}$$

where $M$ and $M_0$ are the molecular weights of the polymer and monomer unit respectively. This calculation yields a value for the chain length of approximately 190 glucose units—a value too high by virtually an order of magnitude.

Repeating the above calculation for glycogen, using a value for $[\eta]_{br}$ of 8 ml/g gives $g\approx5\times10^{-4}$, $f\approx0.6\times10^4$, and $\overline{CL}\approx10$ glucose units.

For glycogen, therefore, the theory of Zimm and Kilb (1959) correctly predicts the chain length, whereas in the case of amylopectin the predicted chain length is greater than the experimental value by a factor of ten. The theory is open to criticism on many grounds, but we are encouraged in its use by the correctness of the prediction applied to glycogen.

If it is accepted that these somewhat inexact calculations are significant, then we must conclude that the known branching characteristics of amylopectin cannot be reconciled with its hydrodynamic behaviour. We believe that the hydrodynamic evidence is strongly in favour of glycogen and amylopectin having markedly different arrangements of the unit chains. It is our contention that amylopectin is synthesized at a surface and is a *two-dimensional* structure, whereas glycogen is a *three-dimensional* structure (Greenwood 1956; Geddes and Greenwood 1969). The arrangement of chains within amylopectin is therefore constrained so as to produce a plate-like structure, a constraint lacking in the case of glycogen. In solution, this type of amylopectin molecule would be expected to curl up to produce an approximately spherical shape. No solution property of the polysaccharide is in any way incompatible with this suggested model; the ability of amylopection to form crystalline regions within the starch granule (Whelan 1958), or on crystallizing in a gelatine matrix (Doi and Doi 1969), would certainly favour a two-dimensional structure.

It might be argued that the variations in solution behaviour between amylopectin and glycogen are due solely to their very different chain lengths rather than the way in which the chains are arranged. However, the solution behaviour of an amylopectin having a chain length of 13–14 units (very similar to some glycogens) is quite different from that of glycogen (Banks and Greenwood 1959a). Furthermore, the differences in *shape* between glycogen and amylopectin are further emphasized when attempts are made to extend the outer chains of both polysaccharides, using glucose-1-phosphate and potato phosphorylase (Banks, Greenwood and Khan 1970a, b). Measurement of chain length and iodine stain showed that virtually all the chain ends acted as sites for further growth in the case of amylopectin, whereas only a very small number of the chain ends were extended in the case of glycogen. Consequently, although the average external chain-length of the glycogen increased only from 8 to 38 glucose units, the iodine-binding capacity increased from zero to 13.3 per cent, at 20.4°C.

Our conclusion regarding the shape of the glycogen molecule with extended external chains was independently confirmed by Bittiger, Husemann and Pfannemüller (1971). They studied, by means of the electron microscope, the complex formed by the reaction of iodine–iodide with glycogen, the outer chains of which had been extended by means of potato phosphorylase and glucose-1-phosphate. The 'star-shaped' molecules observed are quite consistent with the model we postulated. Drochmans (1973) shows electron microscope pictures of the reaction product of glycogen with glucose-1-phosphate and phosphorylase; the pictures may be interpreted as the physical interaction of the long linear chains joining together the glycogen particles.

The different ways in which the outer chains of glycogen and amylopectin lengthen when exposed to enzymic synthesis certainly suggests that the surfaces of the two macromolecules are quite dissimilar. Whether this dissimilarity can be explained purely on a density basis, or whether it reflects a subtle difference in the position of chain-ends relative to branch points, is not yet obvious.

**2.4.5.** Molecular Weight.

*Amylopectin.* Early light-scattering studies showed that the weight-average molecular weight of amylopectin was of the order of $10^7$ or $10^8$, making it one of the largest of the natural polymers (Witnauer, Senti and Stern 1955). These high values were subsequently confirmed (Stacy and Foster 1956; Erlander and French 1958a), and it was shown that the ramified nature of this polysaccharide resulted in it having a very broad distribution of molecular weights (Erlander and French 1958b).

The molecular weights of a number of amylopectin samples, obtained from starches isolated from a fairly wide spectrum of botanical sources, are shown in table 2.15. It will be noted that most of the molecular weights are in excess of $10^8$. The lowest molecular weight ($10^7$) is recorded for a sample of amylopectin isolated from waxy-maize starch using the technique involving the application of shear forces (see section 2.2.2). Solubilizing this same sample of starch directly by means of dimethylsulphoxide resulted in a forty-fold increase in the molecular weight of the amylopectin, confirming that the polysaccharide was susceptible to degradation by shear forces. Subjecting this high molecular weight material to shear forces, using a concentration of 15 per cent polymer in aqueous solution, resulted in a dramatic decrease in molecular weight; repetition of the experiment using a more dilute solution (0.5 per cent) did not cause any degradation of the polymer. Potato amylopectin showed exactly the same type of behaviour, for vigorous stirring of concentrated solutions caused degradation but had no effect on the dilute solutions which are typically employed in fractionating starch.

*Table* 2.15. Molecular parameters, measured in 0.2 M NaCl, for various amylopectin samples (Banks, Geddes, Greenwood and Jones 1972).

| amylopectin | $[\eta]$ (ml/g) | $\bar{M}_w \times 10^{-6}$ | $\langle R_G^2 \rangle^{1/2}$ (Å) | $S_0 \times 10^{13}$ (s) |
|---|---|---|---|---|
| barley | 95 | 400 | 4000 | 260 |
| potato 1 | 95 | 65 | 1600 | 255 |
| potato 2 | 105 | 440 | 4800 | 290 |
| smooth-seeded pea | 100 | 500 | 4000 | 285 |
| tapioca | 110 | 450 | 4500 | 270 |
| waxy maize 1 | 35 | 10 | 1050 | 55 |
| waxy maize 2 | 100 | 400 | 4200 | 270 |
| wheat | 90 | 400 | 4000 | 280 |
| wrinkled-seeded pea | 105 | 500 | 4500 | 290 |

(For parameter definitions, see section 4.2.2.)

Within any one botanical species there appears to be some variation in the molecular size of amylopectin, and this variation is real in the sense that repeated fractionation of the starch invariably yields an amylopectin of constant molecular weight (within experimental error). Thus samples of amylopectin isolated from potato 1 and potato 2 (different cultivars) consistently yielded molecular weights of $65 \times 10^6$ and $440 \times 10^6$, respectively. We have some evidence that the molecular weight of the amylopectin from a given cultivar varies from year to year, but have not yet determined the cause of the variation.

From table 2.15, it is obvious that both the limiting viscosity number and the sedimentation coefficient become insensitive to molecular weight as this latter parameter increases.

It has been suggested that molecular weights of the order of those shown in table 2.15 indicate some type of physical aggregation, but light-scattering measurements in a number of disaggregating solvents, such as 8 M urea and 15 per cent magnesium chloride, result in the same high molecular weight values. Experiments in which the dissolution of the amylopectin was achieved in different ways (shaking with cold water, heating in boiling water, cold dilute caustic alkali, and dimethylsulphoxide) always yielded identical values for the molecular parameters. Similarly, the molecular weights of all the $\beta$-amylolysis limit dextrins prepared from the amylopectin samples shown in table 2.15 were found to have molecular weights that were approximately 45 per cent of those of the starting materials—this is the expected value, as some 55 per cent by weight of each polysaccharide is removed by the action of the enzyme. To show that the isolation of the limit dextrin did not itself lead to physical aggregation, a number of experiments were carried out in which the *beta*-amylase was added to the light-scattering cells immediately on completion of the measurement of $\overline{M}_w$ of the parent polysaccharide, and the measurement repeated. Invariably, the value of the parameter $(1/R_\theta)_{\theta=0}$ decreased by a factor of four to five during this experiment, in agreement with theory (both the molecular weight and the concentration of the polymer decrease by a factor of rather more than 50 per cent).

A potential complication in light-scattering measurements is that a small amount of *microgel* may dominate the measurements, and cause misleadingly high values for the molecular weight. From this viewpoint, it would be satisfying to duplicate the light-scattering values of $\overline{M}_w$ by some alternative technique, but none exists. However, an approximate measure of the molecular weight may be obtained from a combination of sedimentation and viscosity data, using an equation derived by Flory and Mandelkern (1952). According to this relation

$$M = S_0^{3/2}[\eta]^{1/2}[\eta_0 N_A/\beta(1-\overline{V}\rho)]^{3/2} \tag{2.7}$$

where $\eta_0$ is the solvent viscosity, $N_A$ is Avogadro's number, $\beta$ is a constant whose value is dependent on the shape of the macromolecule, and $\overline{V}$ is the partial specific volume of the polymer in solution of density $\rho$. Assuming that the molecule may be represented hydrodynamically as a sphere, then $\beta = 9.75 \times 10^6$ for $[\eta]$ in ml/g. Using average values from table 2.15 for the samples of high molecular weight, $[\eta] = 100$ ml/g, $S_0 = 270 \times 10^{-13}$ s and $\overline{V}\rho = 0.62$ (Banks, Geddes, Greenwood and Jones 1972); substitution into (2.7) gives $M = 80 \times 10^6$. Whilst we recognize that the Flory–Mandelkern relation is strictly applicable to monodisperse systems, the fact that it generates a molecular weight of the same order of magnitude as direct measurement using light scattering gives some confidence to the values obtained from the application of the latter technique. It should be noted that in this particular case, the dominant factor of (2.7) is $S_0$, the sedimentation coefficient at infinite dilution. As the method of calculating this parameter depends only on the most populous species present, very low concentrations of extremely high molecular weight material (microgel) would not affect the measured value of $S_0$.

Whelan (1971) has pointed out that the chain length of the material obtained on the de-branching of amylopectin (60) cannot be reconciled with molecular weights of the order of $10^8$. In fact, other workers have found the maximum to be higher at about 80 or 90 glucose units (Harada, Misaki, Akai, Yokobayashi and Sugimoto 1972). Nevertheless, it has to be admitted that if the molecule does have a molecular weight of the order we suggest, the maximum length of unit-chain should, in the case of the *ideal* Meyer structure, be greater than 100 residues. We show in table 2.16 the number of chains $n_i$

having a given chain length $P_i$ for a multiply-branched perfect Meyer structure with eight residues between branch points (this is the typical value for the internal chain-length of amylopectin).

*Table* 2.16. The number $n_i$ of chains of given chain-length $P_i$ in an idealized Meyer structure in which there are eight units between branch points, and in which the molecule contains 17 tiers, corresponding to a molecular weight of $1.70 \times 10^8$ (Banks and Greenwood, unpublished work).

| $n_i$ | $P_i$ | $n_i$ | $P_i$ | $n_i$ | $P_i$ |
|---|---|---|---|---|---|
| 1 | 136 | 32 | 88 | 2048 | 40 |
| 1 | 128 | 64 | 80 | 4096 | 32 |
| 2 | 120 | 128 | 72 | 8192 | 24 |
| 4 | 112 | 256 | 64 | 16384 | 16 |
| 8 | 104 | 512 | 56 | 32768 | 8 |
| 16 | 96 | 1024 | 48 | | |

This structure, containing seventeen tiers, was chosen to give a molecular weight of $1.70 \times 10^8$. It can be seen that the proportion of chains having a degree of polymerization of 56 glucose residues or greater is rather small (1.6 per cent by number, and 6.25 per cent by weight). Furthermore, the number-average degree of polymerization of this high-molecular-weight tail of the distribution is only 64. Since most determinations of the lengths of these de-branched products depend on the reaction of the reducing end-unit, the experimentally-accessible parameter is the number-average chain length. (The use of the wavelength of maximum absorption of the iodine complex gives an ill-defined average that is intermediate between number- and weight-averages.) Gel permeation is effectively a measure of the volume of the permeating particle rather than its molecular weight. As the chains lengthen the degree of coiling increases, and hence the volume changes more slowly than the molecular weight. Accordingly, it is more difficult to carry out a separation of the longer chains by gel-permeation chromatography, and this effect, in conjunction with the form of the distribution of chains in this high molecular weight region, would lead to the erroneous conclusion that the maximum length of unit chain present in amylopectin contains only 70–80 glucose residues, whereas in fact the maximum length is well in excess of 100 units.

*Glycogen.* The analogous study of the molecular weight of glycogen has been made more difficult, because it is only comparatively recently that methods have been evolved for isolating glycogen from animal tissue without causing severe degradation of the polysaccharide. The classical method, due to Pflüger (1904), involved heating the tissue with 30 per cent aqueous sodium or potassium hydroxide at 100°C for about three hours. Such extreme conditions cause degradation. For example, after extraction with cold trichloracetic acid, Stetten, Katzen and Stetten (1956) found the molecular weight of glycogen to be an order of magnitude higher than hitherto supposed; a result shortly thereafter confirmed (Bryce, Greenwood and Jones 1958).

A number of techniques are now in use to avoid degradation of the polysaccharide during its isolation—these include extraction with cold water (Bueding and Orrell 1961), dimethylsulphoxide (Whistler and BeMiller 1962), phenol–water (Laskov and Margoliash 1963), glycine buffer (Bueding and Orrell 1964), and aqueous mercuric chloride (Mordoh, Krisman and Leloir 1966). These techniques yield values for the weight-average molecular weight of glycogen that are comparable to those of amylopectin, i.e. of the order of $10^8$.

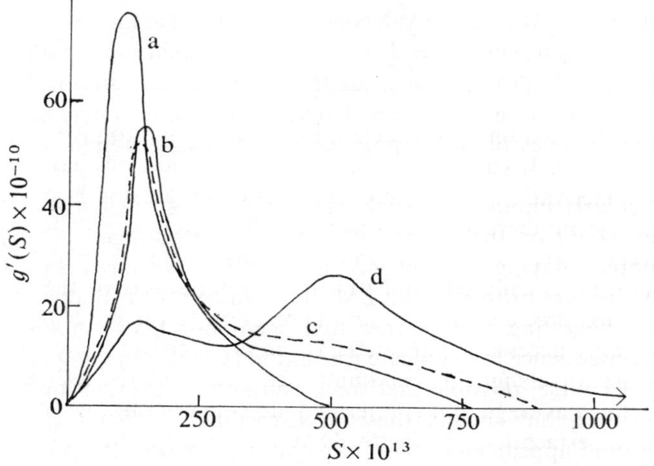


*Figure* 2.9. Apparent distribution of sedimentation coefficients for: (a) OH-glycogen; (b) TCA-glycogen; (c) DMSO-glycogen; (d) Ph-glycogen. See text for details (Geddes 1965).

The virtually ideal solution behaviour of glycogen has enabled the molecular weight distribution of the polymer to be investigated in detail. Because of the limited dependence of the sedimentation coefficient on concentration, it is a comparatively simple matter to construct a distribution of sedimentation coefficients from an analysis of ultracentrifugal data. This distribution is then converted into a molecular weight distribution, by measuring the relation between the sedimentation coefficient and the molecular weight for a number of fractions. However, for most purposes the distribution of sedimentation coefficients is satisfactory; figure 2.9 shows this type of distribution for glycogen samples isolated from a single bovine liver, using hot caustic alkali (OH-glycogen), cold trichloracetic acid (TCA-glycogen), dimethylsulphoxide (DMSO-glycogen), and phenol–water (Ph-glycogen).

All of the extraction methods give distributions having peaks in the region $(100\text{–}150)\times 10^{-13}$s; the proportion of the total material represented by this region, however, varies enormously with the method of extraction. In consequence, the amount of material of very high sedimentation coefficient also varies. Thus whilst the OH-glycogen possesses no material of $S_0$ greater than $500\times 10^{-13}$s, the upper limit is reached in the case of the TCA- and DMSO-glycogen at $800\times 10^{-13}$s and $900\times 10^{-13}$s respectively; the Ph-glycogen actually has a second peak at $500\times 10^{-13}$s and a maximum value in excess of $1000\times 10^{-13}$s. Figure 2.9 shows that the molecular size of glycogen is critically dependent on the method by which it is extracted.

Subjecting the various samples to the Pflüger treatment of 30 per cent aqueous potassium hydroxide at 100°C for three hours was found to degrade all the samples of high sedimentation coefficient to a distribution identical to that of OH-glycogen. At a high alkali concentration and at an elevated temperature this degradation will proceed by means of an anaerobic pathway, and the distribution will stabilize when all the reducing end-groups have been converted to saccharinic acid residues. Thus the OH-glycogen may be regarded as a polydisperse series of stable polysaccharide acids (Stetten and Katzen 1961).

It is particularly important to ask why the form of the distribution is so markedly dependent on the method of extraction. Prolonged incubation of glycogen samples having very broad distributions with reagents such as 8 M urea, 8 M lithium bromide, and 2 M thiocyanate failed to alter the initial distributions, suggesting that neither hydrogen- nor hydrophobic-bonding were of any consequence in aggregate formation (Orrell, Bueding and Reissig 1964). It is, of course, possible that trace amounts of protein link glycogen sub-units together, but such a thesis is exceedingly difficult to prove or disprove. It would appear more reasonable to us to accept the postulate that only glucosidic bonding holds the macromolecule together, and that the susceptibility to degradation results from the mechanical stresses placed on the few chains that pass from one highly-branched region to another.

Glycogen structure has also been extensively studied by means of the electron microscope (Drochmans 1963, 1973; Orrell, Bueding and Reissig 1964). Using the nomenclature of Drochmans (1963), liver glycogen shows particles having a fairly wide size distribution, with a number-average diameter of approximately 130 nm and a weight-average diameter of 176 nm (Drochmans 1973)—these are *alpha-particles.* On treatment with boiling solutions of concentrated caustic alkali, these large particles are reduced in diameter to approximately 34 nm—the so-called *beta-particles.* These beta-particles have a sedimentation coefficient of $100\times 10^{-13}$s, and therefore may be formally identified with the distribution of molecular weights obtained by the classical Pflüger extraction process. Orrell, Bueding and Reissig (1964) also reported that the degradation resulting from extreme conditions used in the Pflüger extraction technique could be caused by somewhat milder conditions, and they suggested that irreversible degradation resulted if the glycogen were exposed to pH values greater than 13 or less than 3.5. Indeed, electron-microscope studies have shown that exposure of the alpha-particles to acids (pH 3.0) caused extensive degradation, with the formation of a virtually homogeneous population of beta-particles within approximately three hours (Drochmans 1973). Geddes (1965) also reported extensive changes in the form of the distribution of sedimentation coefficients as a result of exposing Ph-glycogen to buffer solutions (pH 2.4 and pH 13.0) for several hours at 35°C. It should be noted that alpha-particles are not characteristic of all forms of glycogens: extraction of skeletal muscle tissue, for example, invariably yields only beta-particles with an average diameter of 37 nm.

The electron microscope studies have provided elegant confirmation of the results obtained from ultracentrifugal analysis, and the combination of the two techniques promises an even deeper understanding of the make-up of the glycogen particle in the future.

**2.4.6.** Polyelectrolyte Behaviour of Amylopectin.

Some starches, particularly those from roots and tubers, contain a measurable phosphate content, most of which is found as an ester in the amylopectin fraction. The ester phosphate confers on such amylopectins the properties of a polyelectrolyte. Furthermore, the limited degree of branching confers considerable flexibility on the molecule, and enables it to respond to its solvent environment by changing the volume it occupies. This effect can be demonstrated by passing the isolated amylopectin through a cation-exchange resin in the hydrogen form to remove contaminating sodium and potassium ions, and studying the hydrodynamic properties of the resultant 'amylopectin acid' as a function of the concentration of supporting electrolyte.

*Table* 2.17. The root-mean-square radius of gyration $\langle R_G^2 \rangle^{1/2}$ of potato amylopectin in aqueous solution as a function of the concentration of supporting electrolyte (sodium chloride) present (Jones 1959).

| solvent | $\langle R_G^2 \rangle^{1/2}$ (Å) |
|---|---|
| water | 5500 |
| $10^{-3}$M NaCl | 5300 |
| $10^{-1}$M NaCl | 4800 |

Table 2.17 shows the light-scattering values of the radii of gyration of amylopectin acid in water, and in various solutions of sodium chloride. The molecule exhibits its largest dimensions in water, and as salt is added it collapses on itself. The amylopectin acid will be fully ionized in solution, and, despite the fact that there is only one phosphate group per three or four hundred residues, the mutual repulsion of the charged groups forces the

molecule to expand. Addition of neutral supporting electrolyte effectively reduces the mutual repulsion by means of a screening effect, and the molecular domain shrinks. Calculations, on the basis that amylopectin in solution can be adequately represented by a sphere, show that the volume in water is 80 per cent larger than in 0.1 M sodium chloride.

These polyelectrolyte effects make it imperative to use fairly high concentrations of supporting electrolyte when comparing the hydrodynamic properties of a series of amylopectins, to eliminate the effect of the presence of possible esterified phosphate groups.

It should be noted that removal of these phosphate groups by enzymic hydrolysis is not possible without first degrading the polysaccharide. Although more than 90 per cent of the total phosphate is hydrolysed from the α-amylolysis limit dextrins by the action of acid monophospho-esterase, only some 10 per cent is removed from the intact polysaccharide; the β-amylolysis limit dextrin occupies an intermediate place, with approximately 35 per cent of the total phosphate residues susceptible to enzymic hydrolysis (Fukui 1958; Greenwood and MacGregor 1969). The steric requirements of the phosphatase cannot be satisfied, therefore, in the interior of the branched polysaccharide.

### 2.4.7. Conclusions.

Despite an immense amount of research effort, it must be apparent that the fine structures of amylopectin and glycogen are not yet established.

We think that it is illogical to consider that these two polysaccharides have the same basic type of branched structure: indeed, all the physico-chemical evidence would strongly indicate otherwise. Our contention is that amylopectin is synthesized in an essentially two-dimensional form, whilst glycogen is more truly three-dimensional.

All the three model structures proposed for these glucans have the disadvantage of indicating a regularity in degree of branching, which is not in agreement with recent enzymic evidence. But, again, all enzymic methods of structural analysis have basic shortcomings, which are not always appreciated.

Complications arise because of the very large molecular weight ($10^8$) of these biopolymers, and it is our contention that the two materials each are composed of a range of closely related, but non-identical, branched structures.

In our view, progress in this field of starch chemistry will only be made by the simultaneous application of *all* the available techniques—enzymic, physical and chemical.

## 2.5. Intermediate Material.

### 2.5.1. Introduction.

Many reports exist of the presence of a material having properties different from those of both amylose and amylopectin. In some cases, the suggestion of intermediate material comes from the difficulty in equating properties of the separate fractions, such as iodine affinity and β-amylolysis limit, with the corresponding properties of the granular starch. On this basis, Lansky, Kooi and Schoch (1949) postulated that maize starch contained some 5–7 per cent of a material not defined by the terminology 'amylose' or 'amylopectin': similarly, Peat, Pirt and Whelan (1952a) suggested that potato starch granules may contain material with a higher iodine-binding capacity, but a lower $\beta$-amylolysis limit, than the amylopectin fraction. Perlin (1958) isolated an anomalous amylopectin fraction (termed 'amylopectin C') from wheat starch, and Erlander and French (1958a) reported the presence of a second component, of lower molecular weight than amylopectin, in waxy-maize starch.

The type and amount of intermediate material varies considerably, but, for the mature starch, it appears to be primarily dependent on the amylose content of the starch.

### 2.5.2. Intermediate Material in Normal Starches.

Intermediate material can be readily isolated from normal mature starches, i.e. those with an amylose content of 20–30 per cent, by examination of the supernatant liquor from the recrystallization of amylose. In the case of potato starch Cowie and Greenwood (1957b) found that the initial amylose–thymol complex was only some 70 per cent pure, and that on recrystallization the fraction remaining in the supernatant, the 'thymol–amylopectin', had different properties from the bulk amylopectin. The characteristic properties of thymol–amylopectin isolated from potato and rubber seed starch are shown in table 2.18.

*Table* 2.18. Properties of amylopectin-type polysaccharides obtained from potato and rubber seed starches (Banks and Greenwood 1959a).

| starch | polysaccharide | amount (% of starch) | contaminating amylose (%) | $\overline{CL}$[1] | $[\beta]$ (%) | $[\eta]$ (ml/g) |
|---|---|---|---|---|---|---|
| potato | amylopectin | ~70 | 0.5 | 23 | 57 | 190 |
| | thymol-amylopectin | 5–10 | 1.0 | 13 | 53 | 170 |
| rubber seed | amylopectin | ~70 | 0.5 | 23 | 64 | 170 |
| | thymol-amylopectin | 10 | 2.0 | 16 | 61 | 160 |

[1] $\overline{CL}$ is the average length of unit chain, in this and subsequent tables.

The thymol–amylopectins have very much shorter chain lengths but similar $\beta$-amylolysis limits to those of the bulk materials. Indeed, the thymol–amylopectin fractions exhibit the degree of branching reminiscent of glycogen, but have the limiting viscosity number of normal amylopectin, i.e. an order of magnitude greater than the value for glycogen. Although these materials exhibited somewhat complex sedimentation patterns, the concentration dependence observed was typical of amylopectin rather than glycogen. It is not clear why these particular fractions co-precipitate with the thymol complex; in fact, once the thymol–amylopectin has been isolated, it cannot subsequently be complexed with thymol.

We have also carried out a detailed study of the thymol–amylopectins obtained from the dispersion of cereal starches made soluble by treatment with dimethylsulphoxide (Banks and Greenwood 1967a), as shown in table 2.19.

*Table* 2.19. Properties of amylopectin and intermediate material obtained from some cereal starches (Banks and Greenwood 1967a).

| starch | polysaccharide[1] | amount (%) | IBC[2] (%) | $\overline{CL}$ | $[\beta]$ (%) | $[\beta+Z]$ (%) | $[\eta]$ (ml/g) |
|---|---|---|---|---|---|---|---|
| barley | amylopectin | ~65 | 0.6 | 20 | 57 | 56 | 160 |
| | thymol-amylopectin | 4 | 2.6 | — | — | — | — |
| | anomalous amylopectin | 4 | 2.6 | 27 | 62 | 63 | 165 |
| | anomalous amylose | 0 | — | — | — | — | — |
| oat | amylopectin | ~65 | 0.5 | 18 | 54 | 55 | 145 |
| | thymol-amylopectin | 9 | 4.3 | — | — | — | — |
| | anomalous amylopectin | 4.5 | 0.8 | 25 | 60 | 60 | 145 |
| | anomalous amylose | 4.5 | 8.5 | — | 61 | 93 | 170 |
| rye | amylopectin | ~65 | 0.4 | 20 | 58 | 57 | 150 |
| | thymol-amylopectin | 4 | 2.6 | — | — | — | — |
| | anomalous amylopectin | 4 | 2.4 | 27 | 57 | 60 | 170 |
| | anomalous amylose | 0 | — | — | — | — | — |
| wheat I | amylopectin | ~65 | 0.5 | 21 | 56 | 56 | 150 |
| | thymol-amylopectin | 9 | 5.7 | — | — | — | — |
| | anomalous amylopectin | 5 | 1.2 | 27 | 62 | 64 | 135 |
| | anomalous amylose | 4 | 11.8 | — | 56 | 94 | 180 |
| wheat II | amylopectin | ~65 | 0.6 | 20 | 54 | 55 | 150 |
| | thymol-amylopectin | 6 | 5.0 | — | — | — | — |
| | anomalous amylopectin | 4 | 0.2 | 25 | 55 | 55 | 155 |
| | anomalous amylose | 2 | 12.5 | — | 66 | 97 | 185 |

[1]See text for full description.
[2]IBC is the iodine-binding capacity, in this and subsequent tables.

The high values for the iodine affinities of the thymol–amylopectins compared to the bulk amylopectins suggested that the butan-1-ol had not removed all the amylose fraction. However, repeated recrystallization from saturated butan-1-ol solution, or use of the critical concentration of this reagent (Muetgeert 1961), failed to effect any separation. Unlike the potato thymol–amylopectin, it was found that the addition of more thymol to a solution of the polysaccharide at 60°C did achieve fractionation in the case of three starches; oat, wheat I and wheat II. The complexed material was termed *anomalous amylose*, and the polysaccharide remaining in solution *anomalous amylopectin*. The properties of these intermediate materials are also given in table 2.19.

In contrast to material isolated from potato and rubber seed starches, the

5

anomalous amylopectins from the cereals have chain lengths some 30–40 per cent *greater* than do the bulk amylopectins, and also slightly larger $\beta$-amylolysis limits. However, the concurrent action of *beta*-amylase and Z-enzyme produces no more hydrolysis than does *beta*-amylase acting alone. This, of course, is the criterion we use for defining amylopectin in the enzymic assay for the purity of amylose (see section 2.3.1). Thus we conclude that the anomalous amylopectin fraction is similar in nature to the bulk amylopectin, but rather more lightly branched.

The properties of the anomalous amylose fractions shown in table 2.19 do not enable the exact structure to be established. The iodine affinities of the three samples are high, equivalent to an amylose content of 40–60 per cent, but for at least one sample (wheat I) the $\beta$-amylolysis limit is the same as that of the amylopectin fraction, and for the other two samples the $\beta$-amylolysis limits are only slightly higher. The concurrent action of Z-enzyme and *beta*-amylase, however, causes a large increase in the extent of conversion into maltose relative to *beta*-amylase acting alone, a characteristic property of amylose. The extent of conversion resulting from the concurrent actions of the two enzymes is in all three samples less than the 100 per cent predicted for a normal amylose structure, suggesting that the fractions contain amylopectin in addition to the amylose-like material. It was noted that the wavelength of maximum absorption ($\lambda_{max}$) of the iodine complex of the residual limit dextrins (obtained by the concurrent action of Z-enzyme and *beta*-amylase) was for all three samples higher than is usually observed with the limit dextrins of amylopectin (for the limit dextrin of the anomalous material $\lambda_{max}$=5550Å; the corresponding value for cereal amylopectin limit dextrin is 5400Å). Moreover, the anomalous materials failed to reach a true limiting value for maltose production on being exposed to the combined activities of *beta*-amylase and Z-enzyme, and the conversions asymptotically approached 100 per cent on prolonged incubation. From these observations, we suggest that the material designated 'anomalous amylose' is composed of a range of structures with various degrees of branching intermediate between amylose and amylopectin.

Some support for this conclusion is obtained from a study of the iodine-binding properties (see section 3.2.2) of the various polysaccharide materials; the relevant curves for the various fractions of wheat starch II are shown in figure 2.10.

Of particular interest are the forms of the curves for the anomalous amylopectin fraction (curve e) and the anomalous amylose fraction (curve b). The former is quite typical of amylopectin (cf. curve d), but there is little similarity between the curves for anomalous amylose and for amylose (curve a). In particular, the iodine binding by the anomalous material does not reach a limiting value in the range of free iodine employed in this experiment (up to $7\times10^{-6}$M), and the concentration of free iodine at which iodine binding is most rapid is moved to higher values for the anomalous amylose, relative to the amylose fraction obtained by complexing with butan-1-ol. These characteristics indicate the presence of a short-chain material—the anomalous amylose cannot, therefore, consist of an admixture of normal amylose with a second polysaccharide. It should be noted that curve (b) typifies the iodine-binding characteristics of certain fractions obtained from the fractionation of starches of abnormally high amylose content (see below).

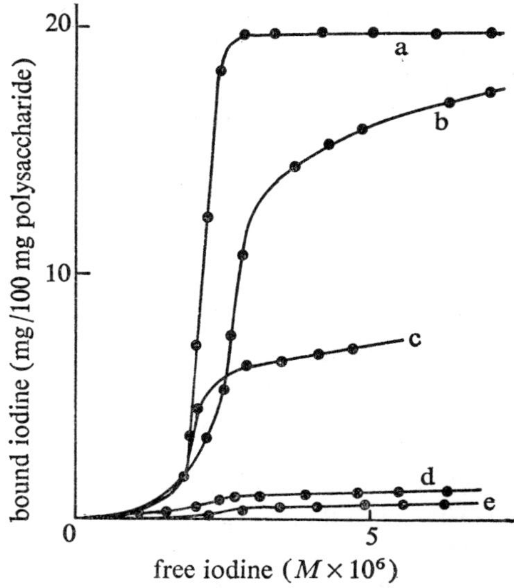


*Figure* 2.10. Potentiometric iodine-titration curves for fractions and intermediate material of wheat starch II (see text for details): (a) amylose; (b) anomalous amylose; (c) thymol-amylopectin; (d) amylopectin; (e) anomalous amylopectin (Banks and Greenwood 1967a).

**2.5.3.** Intermediate Material in Starches of High Amylose Content.

Three common cultivars—pea, maize and barley—have been found to possess genotypes characterized by an abnormally high amylose content of the starch granules. Additionally, each starch has been found to contain a large amount of intermediate material.

*Early Studies.* The first starch to be recognized as containing an exceptionally high (65–75 per cent) content of amylose was that from the wrinkled-seeded pea (Nielsen and Gleason 1945; Hilbert and MacMasters 1946). It was also reported that the granules were swollen, but not gelatinized, in boiling water. Fractionation is therefore difficult, and it is necessary to destroy granular structure prior to dispersion of the starch, using alkali (Potter, Silveira, McCready and Owens 1953), liquid ammonia (Hilbert and MacMasters 1946; Hodge, Karjala and Hilbert 1951; Greenwood and

Thomson, 1962b), or dimethylsulphoxide (Muir 1973). The work of Potter *et al.* (1953) indicated that the material in the dispersion that could not be complexed with amyl alcohol, and classified as amylopectin, had an abnormally large chain-length of approximately 36 glucose units, whereas the smooth-seeded pea (the genotype having the starch of normal amylose content) yielded, by the same procedure, a polysaccharide fraction with the usual amylopectin chain-length of 27 glucose residues. In addition, the amylopectin fraction from the wrinkled-seeded pea showed a higher capacity to bind iodine, but possessed a lower molecular weight than did its counterpart from the smooth-seeded pea. These observations, in conjunction with the inflexionless nature of the iodine-binding curve, led to the conclusion that this material was a new type of branched glucan with an average unit-chain length of 36 glucose units.

The work on maize hybrids led eventually to the development of a genotype characterized by a starch of high amylose content (Deatherage, MacMasters, Vineyard and Bear 1954). Intensive breeding programmes have now allowed the development of a series of maize mutants, the *amylomaizes*, the starches of which reputedly contain some 50–80 per cent of amylose. Like the starch from the wrinkled-seeded pea, amylomaize starch is characterized by very limited swelling in boiling water, and therefore it is difficult to achieve a complete dispersion of the starch. For this reason, therefore, the granules were pretreated with alkali or liquid ammonia (Wolff, Hofreiter, Watson, Deatherage and MacMasters 1955). The amylose fraction obtained from such dispersions was in all respects similar to that from maize starch; the amylomaize amylopectin, on the other hand, had a greater chain length, possessed an inflexionless potentiometric iodine titration curve, and was of a smaller molecular size than maize amylopectin. The $\beta$-amylolysis limits of the amylomaize amylopectin and the maize amylopectin were on this occasion reported to be very similar. Wolff *et al.* (1955) therefore concluded that amylomaize amylopectin was similar in structure to that from the wrinkled-seeded pea.

*Later Work*. In 1962 Greenwood and Thomson fractionated a number of pea starches, from both smooth- and wrinkled-seeded varieties, and confirmed the overall findings of Potter *et al.* (1953). The relevant characteristics of both amylopectin fractions are shown in table 2.20, and it was again noted that the interaction of iodine with the wrinkled-seeded pea amylopectin could not be explained on the basis of contaminating amylose. However, when both types of amylopectin were examined in the ultracentrifuge the polysaccharide from the wrinkled-seeded variety was found to consist of two quite distinct molecular weight ranges, whereas that from smooth-seeded pea showed but a single molecular weight range. The sedimentation patterns themselves indicated, furthermore, that the wrinkled-seeded pea amylopectin was contaminated with material of low molecular weight. Differential ultracentrifugation of aqueous solutions of the polysaccharide resulted in separation into a sediment and a supernatant fraction, the properties of which are also shown in table 2.20, together with the corresponding values for the smooth-seeded pea amylopectin.

*Table* 2.20. Properties of the amylopectin fractions and sub-fractions (see text) obtained from wrinkled-seeded peas (WSP) and smooth-seeded peas (SSP) (Greenwood and Thomson 1962b).

| starch | amylopectin fraction | amount (%) | IBC | $\overline{CL}^{(1)}$ (i) | (ii) | [$\beta$] (%) | [$\eta$] (ml/g) |
|---|---|---|---|---|---|---|---|
| WSP | total | 100 | 2.7 | 42 | 36 | 65 | 120 |
| | sediment | 80 | 2.4 | 31 | 27 | 57 | 150 |
| | supernatant | 20 | — | — | — | 100 | 30 |
| SSP | total | 100 | 1.0 | 27 | 26 | 58 | 150 |
| | sediment | 98 | 0.6 | 28 | 27 | 57 | 160 |
| | supernatant | 2 | — | — | — | — | — |

[1] $\overline{CL}$ (i) as measured; (ii) calculated assuming IBC is due to contaminating amylose.

These results showed that the wrinkled-seeded pea amylopectin was a mixture of two polysaccharides—amylopectin of normal chain length, $\beta$-amylolysis limit, and iodine-binding capacity, together with amylose of such short chain length that it could not be complexed with amyl alcohol, thymol, nor butan-1-ol, but was still capable of binding a limited amount of iodine (Greenwood and Thomson 1962b).

A similar examination of an amylomaize amylopectin (Greenwood and Thomson 1960) also demonstrated that it could, by differential ultracentrifugation, be split into two components, one a typical amylopectin and the other an amylose of short chain length.

In the case of amylomaize starch, the above results have been disputed by Montgomery, Sexson, Dimler and Senti (1964), but, as pointed out by Adkins and Greenwood (1966b), the sedimentation patterns shown by Montgomery *et al.* were misinterpreted by these authors, and in fact demonstrate very clearly a fairly high contamination of their samples with low molecular weight material.

Studies of amylomaize starch are complicated because investigators seldom use *exactly* the same techniques for the dispersion and fractionation of the starch, but even so the reported yields of the butan-1-ol complexable and non-complexable fractions from these starches certainly emphasized the fact that recovery of the complexed material is much lower than the reputed amylose contents of the starches. Table 2.21 shows that not only is the yield of amylose in all cases appreciably lower than expected (and the yield of amylopectin correspondingly higher), but the two major fractions account only for some 90 per cent of the polysaccharide content. The butan-1-ol-complexable fractions exhibited, in all cases, the high iodine-binding capacities typical of amylose, and they may therefore be regarded as substantially free from amylopectin; the non-complexable fractions had iodine-binding capacities of 5–9 per cent, the value generally increasing with the reputed amylose content of the starch.

*Table* 2.21. Experimental yields of butan-1-ol complexable and non-complexable fractions from amylomaize starches.

| starch[1] | % complexable with butan-1-ol | % non-complexable | reference |
|---|---|---|---|
| HA50 | 36 | 58 | Montgomery, Sexson and Senti (1961) |
| HA63 | 45 | 45 | |
| HA66 | 48 | 43 | |
| HA67 | 48 | 42 | Montgomery, Sexson, Dimler and Senti (1964) |
| HA68 | 49 | 42 | |
| HA71 | 50 | 43 | |
| HA71 | 50 | 39 | Montgomery, Sexson and Senti (1961) |

[1] HA = 'high amylose content'; the number designates the reputed amylose content of the starch.

The values shown in table 2.21 fully support our contention that the amylomaize starches contain large amounts of anomalous material. In comparison the non-complexable fraction from wrinkled-seeded pea constitutes only 23 per cent of the initial weight of the starch, with an iodine-binding capacity of but 2.7 per cent. This fact is demonstrated by comparing the iodine-binding curves of the two starches (see figure 2.11). Whilst the wrinkled-seeded pea starch contains a component which gives an inflexionless iodine-binding curve, it is not present in sufficient concentration to materially affect the interaction of the starch with iodine. On the other hand, the iodine-binding relation for amylomaize starch itself is inflexionless, suggesting that considerable amounts of the intermediate material are present.

It is necessary to devise fractionation schemes which are capable of showing whether or not the intermediate material is chemically homogeneous. We have already noted that ultracentrifugation showed that the non-complexable material from amylomaize could be separated into two quite distinct fractions. Repetition of this work confirmed these observations in general (Greenwood and MacKenzie 1966), and also showed that a comparable separation could be achieved by forming an iodine complex; a technique which was later extended in the same laboratories.

The problem of effectively disrupting the granular structure of the amylomaize was solved by the use of dimethylsulphoxide (Adkins and Greenwood 1969b). We have previously noted that dimethylsulphoxide is useful for dispersing a variety of starches, but we found that the solubility in this reagent *increases* with increasing apparent amylose content, thus making it a particularly suitable solvent for amylomaize starches. This increased solubility makes it possible to obtain 5 per cent solutions of amylomaize merely by gentle stirring of the hydrous granules for some three hours in dimethylsulphoxide; under similar conditions, it is difficult to obtain solutions of greater than 3 per cent normal maize starch. The amylomaize polysaccharide may then be precipitated from solution by the addition of ethanol, to yield a non-granular product.

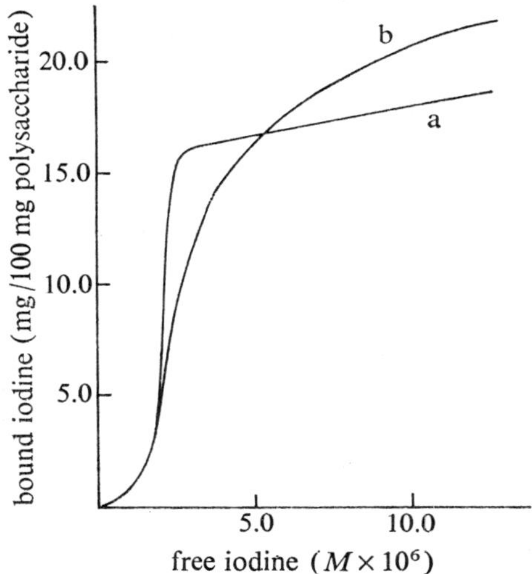


*Figure* 2.11. Potentiometric iodine-titration curves showing the characteristic behaviour of starches of high amylose-content from (a) the wrinkled-seeded pea containing 70% of amylose, and (b) amylomaize with a reputed content of 70% amylose.

Adkins and Greenwood (1969b) devised a fractionation scheme in which starches of various maize genotypes were dispersed into dimethylsulphoxide in this manner, and the butan-1-ol-complexable material was obtained directly by addition of the precipitant to the dimethylsulphoxide–starch dispersion. The yield of complex (fraction I) and of the material in the supernatant liquor (fraction II) are shown in table 2.22. Two recrystallizations of the butan-1-ol complex under standard conditions led to a product of constant properties (fraction III), and to the isolation of material from the recrystallization supernatant liquors (fraction IV). The material in the original supernatant liquor (fraction I) was finally separated into a fraction which did not complex with iodine (fraction V), and one which did complex with this reagent (fraction VI).

The main significance of this fractionation scheme is that it provided the first *quantitative* assessment of the yields of all the different fractions in a range of amylomaize starches.

The properties of fractions III and V for waxy and regular maize corresponded to amylose and amylopectin fractions respectively, whilst fractions IV and VI were equivalent to the intermediate materials encountered in the fractionation of cereal starches (see section 2.5.2). In this respect, it is interesting to note that the yield of this material for regular maize starch is 6 per cent of the total, a value comparable with those obtained for the other cereal starches (see table 2.19). Perhaps the most surprising aspect of these results is the very small amount of fraction V (amylopectin) found in amylomaize starches, and the large amounts of anomalous materials, corresponding to 30–40 per cent of the total polysaccharide. However, these amounts of amylopectin are comparable to those obtained by Adkins and Greenwood (1966c) from the direct ultracentrifugation of dispersions of amylomaize starches.

*Table* 2.22. Yields of the components obtained on fractionation of starches of various maize genotypes, using the technique of Adkins and Greenwood (1969b).

| fraction | percentage of original starch[1] | | | | | | |
|---|---|---|---|---|---|---|---|
| | WM | RM | HA57 | HA62 | HA67 | HA70 | HA75 |
| I. first BuOH-precipitate | 0 | 27 | 55 | 59 | 64 | 68 | 71 |
| II. supernatant from I | 99 | 73 | 45 | 41 | 36 | 32 | 29 |
| III. final BuOH-precipitate | 0 | 25 | 51 | 55 | 59 | 63 | 65 |
| IV. supernatant from III | 0 | 2 | 4 | 4 | 5 | 5 | 6 |
| V. non-$I_2$-complex from II | 96 | 68 | 9 | 6 | 6 | 4 | 4 |
| VI. $I_2$-complex from II | 2 | 4 | 36 | 35 | 30 | 27 | 25 |

[1] WM=waxy maize; RM=regular maize; HA=high amylose maize, with the number signifying the reputed amylose content.

In all cases, the properties of fraction III corresponded to those of amylose, but it is of interest that the limiting viscosity numbers of the amylomaize amyloses were approximately half those of the amylose isolated from regular maize. The properties of the intermediate material from the recrystallization of the butan-1-ol complex from the amylomaize starches (fraction IV) are similar, in their susceptibility to *beta*-amylase and *Z*-enzyme, to those from regular maize. Thus this particular intermediate material is common to both maize and amylomaize starches, but occurs to a greater extent in the latter type. However, even in amylomaize starch of the highest reputed amylose-content (HA75), it does not exceed 6 per cent of the total polysaccharide.

The properties of the fractions and sub-fractions that cannot be complexed by butan-1-ol are shown in table 2.23. Fractions designated II in the above fractionation scheme, would normally be described as 'amylopectin'. This terminology is indeed correct for the material obtained from waxy and regular maize starches, but, in the case of the amylomaize starches, the anomalously high values for the iodine affinities, $\beta$-amylolysis limits, and chain lengths again present us with the problem of whether we are dealing with a homogeneous polysaccharide, or with an admixture of two quite distinct polysaccharide types. The fractionation of this material, using iodine to complex the larger chains, yields two materials, V and VI, the properties of which are also shown in table 2.23.

*Table* 2.23. Properties of fractions and sub-fractions, which cannot be complexed with butan-1-ol, of starches of various maize genotypes (Adkins and Greenwood 1969b).

| starch[1] | fraction[1] | % of starch | IBC[2] (%) | $\overline{CL}$ | [$\beta$] (%) | [$\beta$+$Z$] (%) | [$\eta$] (ml/g) |
|---|---|---|---|---|---|---|---|
| WM | II | 99 | — | 18 | 56 | 57 | 170 |
| | V | 96 | 0.2 | 16 | 54 | 54 | 175 |
| | VI | 2 | 5.2 | 46 | 95 | 96 | 18 |
| RM | II | 73 | — | 29 | 58 | 61 | 120 |
| | V | 68 | 0.3 | 24 | 55 | 56 | 159 |
| | VI | 4 | 4.1 | 58 | 94 | 96 | 22 |
| HA57 | II | 45 | — | 34 | 53 | 68 | 84 |
| | V | 9 | 0.3 | 26 | 57 | 57 | 120 |
| | VI | 36 | 10.8 | 88 | 92 | 93 | — |
| HA62 | II | 41 | — | — | — | — | — |
| | V | 6 | 0.4 | 24 | 56 | — | 160 |
| | VI | 35 | — | 52 | 95 | 95 | 24 |
| HA67 | II | 36 | — | 42 | 65 | 77 | 70 |
| | V | 6 | 0.3 | — | 54 | 55 | 152 |
| | VI | 30 | 13.8 | 52 | 89 | 93 | — |
| HA70 | II | 32 | — | 45 | 64 | 77 | 70 |
| | V | 4 | 0.4 | 27 | — | 57 | — |
| | VI | 27 | 14.7 | 110 | 92 | 93 | 48 |
| HA75 | II | 29 | — | 44 | 65 | 79 | 68 |
| | V | 4 | 0.3 | 22 | 54 | 54 | 154 |
| | VI | 25 | 14.7 | 61 | 90 | 90 | 46 |

[1] As in table 2.22. [2] Measured at 2°C.

The properties shown for fraction V indicate that this sub-fraction is *amylopectin*; the iodine-binding capacity of the material, even when measured at low temperatures that accentuate the capacity to bind iodine (see section 3.2.8), is consistently low, and the combined action of *beta*-amylase and *Z*-enzyme produces no more maltose than does *beta*-amylase acting alone. However, whilst the amylopectin fraction accounts for some 96 per cent of the polysaccharide material of waxy-maize starch, and 68 per cent of that of regular maize starch, the various amylomaize starches contain 4–9 per cent, the amount decreasing with increasing amylose content. Thus the dominant feature of these maize starches of high amylose content is that they contain much smaller amounts of amylopectin than would be expected from their reputed amylose contents. The discrepancy arises because, in addition to the amylose and amylopectin fractions, they contain very large amounts (25–35 per cent) of intermediate material that does not complex with butan-1-ol but does with iodine. It is the presence of this fraction, we believe, that gives rise to the anomalous properties of amylomaize starch.

It is difficult to find a meaningful classification for fraction VI, i.e. the iodine-complexing material, which possesses the ability to bind considerable amounts of iodine but which has iodine-binding capacities far removed from the asymptotic value (22.5 per cent) observed for pure amylose at low temperature. The high $\beta$-amylolysis limits demonstrate that the material is rather more closely related to amylose than to amylopectin, whilst the chain lengths show that if this fraction is indeed amylosic in nature, then the degree of polymerization of the material is quite small ($<100$ anhydroglucose units). Molecules of this size are indeed known to bind considerably less iodine than do molecules of degree of polymerization of 1000 units or more (see section 3.2.7).

The results of Adkins and Greenwood (1969b) support the postulate that the so-called 'long-chain amylopectin' of amylomaize is a mixture of amylopectin of normal properties and short-chain amylose. The fact that this conclusion has been reached using two quite distinct methods of subfractionating the 'long-chain amylopectin'—ultracentrifugation and complexing with iodine—certainly makes unlikely the possibility that this material is homogeneous with respect to structure.

Another technique has also been developed to show the presence of amylose-like material in the fractions II, i.e. the 'long-chain amylopectins' from starches of high amylose content (Banks and Greenwood 1968a). This technique depends on the fact that when *beta*-amylase acts on a *mixture* of linear chains and normal amylopectin, the enzyme will release glucose from linear chains with an odd number of glucose residues (half of the total number of linear chains, on a statistical basis); this enzyme cannot release glucose from a branched substrate. Glucose oxidase enables glucose to be quantitatively determined in the presence of maltose and branched polysaccharide, and hence $\beta$-amylolysis followed by estimation of the liberated glucose provides a method for the detection of amylosic material in the presence of amylopectin. When this technique is applied to various samples of 'long-chain amylopectin' from amylomaize starch they always yield glucose, a product absent when either glycogen or potato amylopectin are used as substrates.

The evidence therefore indicates that short-chain amylosic material is an important component of amylomaize starch, but other types of intermediate material may also be present. This concept arises from calculations based on the relation

$$1/x = a/y + (1-a)/z \qquad (2.8)$$

where $x$ is the chain length of the 'long-chain amylopectin' fraction, $a$ is the fraction of linear material having a degree of polymerization of $y$, and $z$ is the chain length of the normal amylopectin fraction (Banks and Greenwood 1968a). Unpublished calculations, using the data in table 2.23, show that a *simple* admixture of linear and amylopectin structure is not occurring exclusively in the amylomaize starches.

*Current Studies*. A successful investigation of the problem of intermediate material in such starches required a new, rapid method of obtaining *large* amounts of the polysaccharide. We had observed that although amylomaize starch does not show a large increase in viscosity in boiling water, considerable amounts of material were leached from it under these conditions (Banks, Greenwood and Muir 1971a; see figure 2.12). The fairly rapid initial increase of soluble material enabled a satisfactory fractionation scheme to be developed. The leached fraction is then separated into material which will complex with butan-1-ol, and the corresponding non-complexable material.

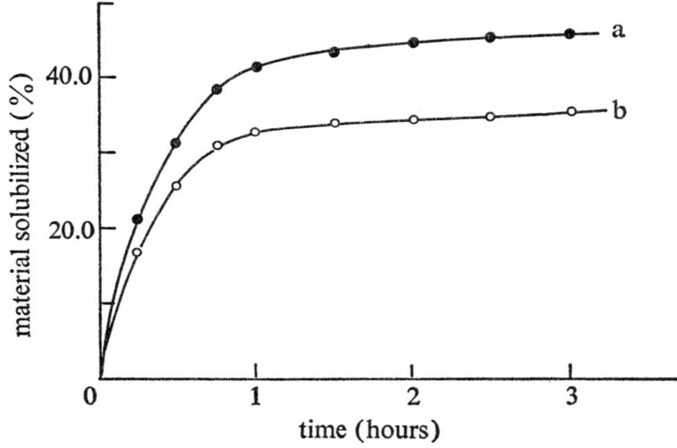


*Figure* 2.12. Leaching of amylomaize starches in boiling water: material solubilized (%) as a function of time for (a) HA50, and (b) HA70 (Banks, Greenwood and Muir 1971a).

The iodine-binding curves for the resultant intermediate material are shown in figure 2.13. Curve a is in most respects typical of that expected of an amylose, both in the concentration of free iodine at which iodine binding occurs, and in the high value (19.0 per cent) of the iodine affinity, which is perhaps marginally low. However, more importantly, the iodine-binding capacity does not reach a limiting value within the range of free iodine concentration used in this work. The noticeable increase in iodine binding at concentrations of free iodine greater than $5 \times 10^{-6}$M is due, we believe, to the presence of a contaminating intermediate fraction. Recrystallization of the amylose–butan-1-ol complex left 7 per cent of the polysaccharide in solution, but the material forming the complex then had an iodine affinity of

19.4 per cent, and, furthermore, the iodine-binding capacity reached a limiting value when the concentration of free iodine reached $5 \times 10^{-6}$M. Also, it has been observed that the amount of this intermediate material found as a contaminant in the amylose fraction is grossly dependent on the details of the fractionation protocol. For example, using thymol as the initial precipitant, or allowing a period longer than two hours for complexing with butan-1-ol, increased the amount of contamination.

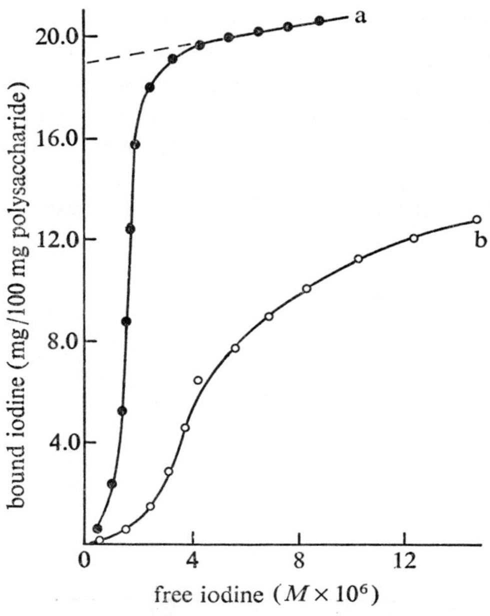


*Figure* 2.13. Iodine-titration curves (at 20°C) for products leached from amylomaize starch (HA 50): (a) material complexed with butan-1-ol; (b) material not complexed with butan-1-ol (Banks, Greenwood and Muir 1971a).

More important, however, is the form of curve b, the iodine-binding curve of the non-complexable fraction, which amounts to some 15 per cent by weight of amylomaize HA 50. This graph is inflexionless, which is generally regarded as being a characteristic of 'long-chain amylopectin'. To obtain an amylopectin by a leaching process that does not cause fairly pronounced disruption of the granule would, by comparison with most other cereal starches, be most unusual. Additionally, ultracentrifugation for 2.5 hours at 50000 *g* of an aqueous solution (0.1 per cent) of this non-complexable fraction did not lead to any loss in concentration; under comparable conditions, waxy-maize amylopectin would be completely removed from solution. Thus this intermediate material is of comparatively low molecular weight, several orders of magnitude removed from that (*ca.* $10^8$) normally associated with amylopectin. We therefore classify this material merely as a non-complexable fraction.

The concurrent action of *beta*-amylase and pullulanase causes complete degradation into maltose; *beta*-amylase acting alone causes only 68 per cent conversion into maltose, and in conjunction with *Z*-enzyme 78 per cent conversion. From these results, we conclude that this non-complexable fraction has a structure similar to that of amylopectin, i.e. it is composed of glucose units linked by $\alpha$-1:4-bonds, but with a small number of trifunctional residues in which the hydroxyl group on $C_6$ is also involved in an $\alpha$-glucosidic linkage. However, the presence of glucose (0.21 per cent) after $\beta$-amylolysis indicates that linear chains are present. Thus the anomalous material is itself heterogeneous with respect to its structure. Of further interest is the fact that the $\beta$-amylolysis limit dextrin was found to possess a relatively high capacity to bind iodine *at low temperature* (see section 3.2.8), again yielding an inflexionless iodine-binding curve. Thus, the *internal* chain-length of this branched material also appears to be greater than that associated with normal amylopectin; a conclusion supported by calculations based on the various observed $\beta$-amylolysis limits and average length of unit chain (Banks, Greenwood and Muir 1974a).

These results are in accord with recent work of Mercier (1973). She fractionated the de-branched $\beta$-amylolysis limit dextrins of normal, waxy, and high-amylose maize, using gel-permeation chromatography, and found that the material from the high-amylose maize had a much greater proportion of long chains (degree of polymerization $<30$) than did that from either the normal or waxy maize. This observation can be interpreted only on the basis of there being present in the high-amylose maize a branched material having an internal chain-length greater than that associated with normal amylopectin. However, it should be noted that Mercier was unable to perform the experiment using the de-branched starch, because of excessive retrogradation during the chromatographic separation of the chains from the high-amylose starch. The fact that this effect was absent when the de-branched $\beta$-amylolysis limit dextrin was used suggests that the anomalous properties of amylomaize starch do not arise solely from the long internal chains.

*Summary*. Amylomaize starch contains amylose (defined with respect to its ability to bind iodine), a small amount of amylopectin (defined with respect to its branching characteristics), and a considerable quantity of intermediate material, some of which is linear and some lightly branched. Many of the discrepancies in the literature regarding the properties of this intermediate material are due to the fact that it may, depending on the fractionation conditions, be divided between the 'amylose' and 'amylopectin' fractions. However, we believe that this intermediate material must be considered as a separate entity, quite distinct from either amylose or amylopectin, and that it is this material that gives amylomaize starch its anomalous characteristics—high instability in aqueous solution and inflexionless iodine-binding curves. The intermediate polysaccharide fraction is itself heterogeneous with respect to structure, there being present a range of linear chains (50 < degree of polymerization < 200) and lightly-branched molecules of relatively low molecular weight ($<10^6$), the internal, and probably also the external, chain-length being greater than that of normal amylopectin. There is some evidence that such fractions occur in all cereal starches, but only in significant amounts when the genetic background is manipulated to produce a starch of high amylose content. An intermediate fraction in high-amylose barley starch appears to possess similar anomalous properties to those exhibited by material from amylomaize, but is present to a much lower extent; a similar observation applies in the case of pea starch of high amylose content.

Although not yet fully understood, the fascination of this intermediate material lies in the fact that it may supply the answer to that most puzzling of questions, namely the manner in which the components of starch are synthesized.

# 3. The Reaction of Starch and its Components with Iodine

## 3.1. Introduction.

Only three years after the discovery of iodine, the reaction of this element with starch in aqueous solution to yield a brilliant blue colour was noted. This phenomenon excited much attention, but an adequate explanation for its occurrence was not provided until rather more than a century later when Hanes (1937) suggested that the complex consisted of a helical polysaccharide molecule, with iodine occupying the helical cavity. Freudenberg, Schaaf, Dumpert and Ploetz (1939) supported this model, because it explained their proposed cyclic structure for the Schardinger dextrins.

Subsequent studies by Rundle and his co-workers fully confirmed the correctness of the Hanes model. Flow dichroism measurements (Rundle and Baldwin 1943) demonstrated quite unequivocally that the iodine atoms within the complex were stacked in a linear array, as demanded by the helical model. A study of the optical properties of crystals of the amylose–iodine complex (Rundle and French 1943a), showed them to be consistent with a helical structure, whilst X-ray diffraction studies (Rundle and French 1943b; Rundle and Edwards 1943) enabled the unit cell dimensions of the complex to be obtained. It was concluded that the helix consisted of six glucose residues per turn, with a pitch of 8Å. Later, Rundle (1947) carried out a Fourier projection of the data from his study of the amylose–iodine complex, and found it to be compatible with the proposed helical model.

The work of Rundle and his colleagues deserves much more attention than it was ever accorded. It demonstrated beyond reasonable doubt that Hanes' hypothesis of a helical conformation was correct, and, moreover, it was clearly shown that this ordered structure was maintained in solution. Only in the following decade were similar conclusions reached in the case of the polypeptides and nucleic acids (Pauling, Corey and Branson 1951; Watson and Crick 1953), but this later work has quite overshadowed what was one of the earliest and most brilliant studies of polymer conformation.

As well as being a study in itself, the amylose–iodine reaction provides a diagnostic tool for starch chemists, for conditions can be so arranged that amylose binds some 20 per cent of its own weight of iodine whilst amylopectin binds none, thus enabling the ratio of these two polysaccharides present in any sample to be determined. Experimentally, either the optical density of the solution is measured, or the amount of iodine complexed by the amylose is determined by an amperometric, or potentiometric, technique. Although measurement of optical density provides the most rapid method for determining the proportion of amylose, our own preference has been for the semi-micro differential potentiometric technique. This latter method enables the full iodine-binding curve to be elucidated, i.e. the relation between the amounts of iodine bound and that of free iodine in solution, although a continuous photometric titration system with an automatic recording facility has been claimed to be the most sensitive method of obtaining the iodine-binding curve (Richter and Szejtli 1966; Szejtli 1966; Szejtli, Richter and Augustat 1967a, b; Szejtli, Augustat and Richter 1967; Holló and Szejtli 1968).

## 3.2. Semi-Micro Differential Potentiometric Iodine Titration, and the Factors affecting it.

### 3.2.1. The Apparatus.

The technique of potentiometric titration as a means of studying the amylose–iodine reaction was originally developed by Bates, French and Rundle (1943). However, the accuracy of the direct titration method can be greatly improved by using the elegant differential potentiometric technique of Gilbert and Marriott (1948).

Anderson and Greenwood (1955) adapted this technique for routine analyses, but a laboratory-built electrometer had to be employed to deal with the high impedance of the circuit, and this factor mitigated against the

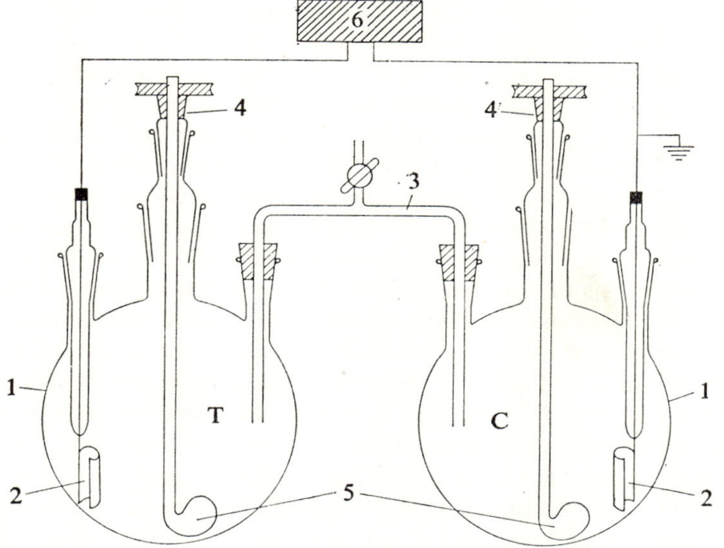


*Figure* 3.1. Diagrammatic representation of the apparatus for semi-micro differential potentiometric iodine titration: (1) half-cell—1 litre round-bottom Pyrex flasks, fitted with one B34 and two B19 sockets (B10 sockets for the addition of iodine are fitted at the front of the flasks and therefore are not shown on the diagram); (2) bright platinum electrode; (3) liquid bridge; (4) stirrer gland for pulley system; (5) stirrers; (6) digital voltmeter. Flask T contains starch solution under test, flask C contains control buffer solution. (For full details of the experimental set-up see Banks, Greenwood and Muir 1971b.)

widespread use of the apparatus. However, more recently, the apparatus has been redesigned (Banks, Greenwood and Muir 1971b) to make use of a commercial digital voltmeter; this resulted in increased stability and sensitivity, as well as ensuring ready availability. A diagrammatic representation of the apparatus is shown in figure 3.1.

The principle of the method is that the test half-cell T contains the polysaccharide sample in iodide solution, buffered to a slightly acid pH; the control half-cell C contains the identical iodide–buffer solution. Thus we have an iodine concentration cell without transference, represented by

$$I_2[(a_{I_2})_c] \,|\, I^-(a_{I^-}) \,|\, I_2[(a_{I_2})_t] \qquad (3.1)$$

where $a$ is the activity. For one electrode, the potential $E$ is given by

$$E = E_0 - \frac{RT}{F} \ln [(a_{I_2})^{1/2}/(a_{I^-})] \qquad (3.2)$$

where $R$ is the gas constant, $T$ the temperature, and $F$ the Faraday constant. Therefore, the overall potential is

$$E = -\frac{RT}{F} \ln [(a_{I_2})_c/(a_{I_2})_t] \qquad (3.3)$$

When no potential exists between the electrodes

$$(a_{I_2})_c \equiv (a_{I_2})_t \qquad (3.4)$$

or

$$[I_2]_c \equiv [I_2]_t \qquad (3.5)$$

at the concentrations of free iodine employed. An aliquot of iodine is added to the test half-cell, and sufficient iodine to give null potential metered into the control half-cell. At this point, the amount of iodine added to the control half-cell corresponds to the amount of free molecular iodine in the test half-cell; the difference between this latter amount and that originally added gives the quantity of iodine bound by the polysaccharide. Repetition of this titration procedure then enables the curve of 'bound iodine' as a function of 'free iodine' to be obtained. (Unless stated to the contrary, we shall use the term 'free iodine' to describe that portion of the iodine not bound by the polysaccharide; it is the sum of the molecular iodine and the tri-iodide ion.)

### **3.2.2.** Definition of Iodine-Binding Capacity.

Typical titration curves for amylose (curve a), starch (curve b) and amylopectin (curve c) are shown in figure 3.2. We define the *iodine-binding capacity* as the weight (mg) of iodine bound by 100 mg polysaccharide *at zero concentration of free iodine*. The necessary extrapolations are shown in figure 3.2 by broken lines; the iodine-binding capacities are designated by the $X$-values.

Other workers define iodine-binding capacity as the point at which the extrapolated maximum and minimum slopes meet, denoted by the $Y$-values in figure 3.2. However, only when the limiting slope is high, as in the case of the amylomaize starches (see section 3.2.8), is there an appreciable difference between the two procedures.

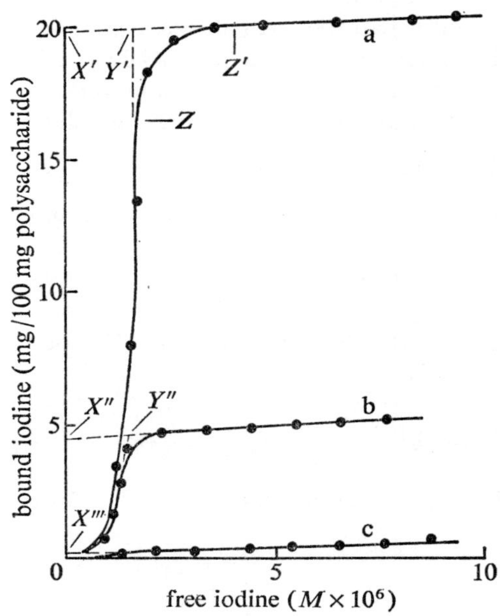


*Figure* 3.2. Iodine-binding curves at 20.4°C for potato starch and its components: (a) amylose, (b) starch, and (c) amylopectin. The dotted lines show alternative methods of obtaining the iodine-binding capacity; in the region Z–Z′, equation 3.8 is valid.

It is generally accepted that the slow increase in iodine-binding capacity which occurs at high concentrations of free iodine reflects an adsorption process on the surface of the amylose–iodine micelles. If this process starts as soon as iodine is bound within the helix, then extrapolation of the results to zero concentration of free iodine is justified. Alternatively, if it is argued that all the helices must be filled before the secondary adsorption process commences, the *Y*-values would represent a more accurate measure of the iodine bound within the helices.

It must be emphasized that both of these methods of obtaining the iodine-binding capacity of a polysaccharide are equally arbitrary—no theoretical basis has in fact been provided for either process. Eliassaf and Lewin (1963) attempted to provide a theoretical interpretation of the potentiometric iodine titration of starch, by applying to the starch–iodine system a model similar to that used to describe the binding of salt to proteins. Their model assumed that one molecule of iodine and *n* anhydroglucose units were involved in a single bond, which was mutually independent of its neighbours. One may then write

$$k(\text{conc. free sites})(\text{conc. free iodine}) = (\text{conc. bound sites}) = (\text{conc. bound iodine}) \qquad (3.6)$$

where $k$ is the equilibrium constant. For a solution containing $m$ moles of anhydroglucose units per litre

$$k(m/n - c_B)c_F = c_B \qquad (3.7)$$

where $c_B$ and $c_F$ are the bound and free iodine respectively. Rearrangement of (3.7) gives

$$m/c_B = (n/k)(1/c_F) + n \qquad (3.8)$$

Thus a graph of $m/c_B$ as a function of $1/c_F$ should be linear, with an intercept of $n$ and a slope of $n/k$. Eliassaf and Lewin (1963) reported that the value of $n$ was the same for both amylose and amylopectin ($n = 6.8$), but that $k$ was five times larger for amylopectin than it was for amylose, thus indicating that the difference between amylose and amylopectin in their reaction with iodine is not in their binding capacity but in the bond strength.

We have attempted to analyse our results by means of (3.8), but only with limited success. It is our experience that the extrapolation can be made only from a very limited range of the iodine-binding curve, i.e. the portion between the points designated $Z$ and $Z'$ in figure 3.2. A value for $n$ of six or seven glucose residues can be obtained in this region, but the portion of the curve fitting (3.8) is too small for it to be claimed that a satisfactory theoretical basis has been developed for interpreting the results of iodine titration experiments. Eliassaf and Lewin (1963) did not, in fact, specify the portion of the titration curve they used in their analysis.

In interpreting iodine titration curves, therefore, it must be remembered that the existing definitions of iodine-binding capacity are quite arbitrary.

**3.2.3.** Preparation of the Polysaccharide Sample.

For an iodine titration to have quantitative significance, all the amylose present must be in a state capable of binding iodine. It is known that as amylose aggregates (i.e. retrogrades), the iodine-binding capacity decreases considerably (Loewus and Briggs 1957), and that the presence of other agents having the ability to complex amylose causes a similar decrease. In this latter respect, Schoch (1942b) has shown that many starches contain sufficient amounts of lipid material to cause serious interference in the iodine titration. The preparation of the sample for titration must therefore be carried out in such a way as to destroy aggregates of amylose, and remove any potential complexing agent.

Dimethylsulphoxide has proved to be the most generally satisfactory method of achieving dissolution of starch granules. Using hydrous starch granules and gentle stirring, it is possible to dissolve starches of high amylose content in this solvent within a period of thirty minutes, as signified by the absence of any gel at the end of that time. As the proportion of amylose decreases, dissolution in dimethylsulphoxide becomes more difficult and longer periods of stirring are required (Adkins, Greenwood and Hourston 1970). The time required to obtain gel-free solutions also depends on the type of starch, for tuber and root starches dissolve much more slowly than do cereal starches. For example, potato starch granules require exposure to dimethylsulphoxide (with gentle stirring) for approximately seven hours before a limiting value is obtained for the iodine-binding capacity; that value is then maintained during subsequent lengthy periods of storage of the dimethylsulphoxide–starch solution (Banks, Greenwood and Muir 1971b).

The use of dimethylsulphoxide as a solvent for starches generally leads to slightly higher values for the iodine-binding capacity than is obtained using caustic alkali.

*Table* 3.1. The iodine-binding capacity of potato starch related to the method of dispersion of the granules (Banks, Greenwood and Muir 1971b).

| reagent for dispersion | iodine-binding capacity (mg iodine bound/100 mg starch) |
|---|---|
| dimethylsulphoxide | 4.55 |
| 0.2 M KOH | 4.27 |
| 0.5 M KOH | 4.41 |

Table 3.1 shows the iodine-binding capacity of a potato starch dissolved by overnight stirring in (a) dimethylsulphoxide, (b) 0.2 M potassium hydroxide, and (c) 0.5 M potassium hydroxide (samples dissolved in alkali are brought to pH 5.8 by the addition of phosphoric acid before titration). Dispersion of the granules in dimethylsulphoxide gives the sample having the highest iodine-binding capacity. Moreover, that value was reproducible to $\pm 1$ per cent, whereas the variation was $\pm 3$ per cent in the case of dissolution in alkali. The higher concentration of alkali is more efficient for the dispersion of the potato starch than is the lower one, in agreement with the results of Adkins and Greenwood (1966d).

In comparing the effect of dissolution procedure on the iodine-binding capacity of starch, it is necessary to avoid the complication of contamination by lipid. Schoch (1942b) advocated that fat be removed from the starch by repeated extraction with 85 per cent aqueous methanol, or 80 per cent aqueous dioxane. Although relatively effective, this extraction procedure is tedious and difficult to use when only small amounts of sample are available. However, both dimethylsulphoxide and ethanol are solvents for fat, and a dissolution procedure based on this fact has been devised (Banks, Greenwood and Walker 1970). The starch is dissoived ln dimethylsulphoxide, precipitated with ethanol, and the precipitate washed repeatedly with ethanol and dried. The product, a fine powder, is *non-granular starch*, and is easily soluble in dimethylsulphoxide or dilute solutions of alkali. The effectiveness of this procedure to remove contaminating fat is shown in table 3.2, where the iodine-binding capacities of four starches are given for (I) dissolution of the starch in dimethylsulphoxide, followed by titration, and (II) dissolution of the non-granular starch in dimethylsulphoxide followed by titration.

*Table* 3.2. The effect of fat removal on the iodine-binding capacity of starch (Banks, Greenwood and Muir 1971b).

| cultivar | iodine-binding capacity: (mg iodine bound/100 mg starch) procedure[1] | |
|---|---|---|
| | I | II |
| barley (var. Pentland field) | 5.35 | 8.20 |
| barley (var. Ymer) | 3.73 | 5.66 |
| wheat | 4.22 | 5.40 |
| rye | 4.70 | 5.51 |

[1] Procedure (I) iodine-binding capacity of the undefatted starch; (II) iodine-binding capacity of starch precipitated from dimethylsulphoxide solution by the addition of ethanol.

The amount of dimethylsulphoxide used in sample preparation does not affect the characteristics of the iodine titration curve, when identical amounts of dimethylsulphoxide are present in both half-cells; we have noted that some slight interaction between iodine and dimethylsulphoxide can result in error if there is an imbalance in the amounts of dimethylsulphoxide present in the two half-cells (Banks, Greenwood and Muir 1971b).

**3.2.4.** Supporting Electrolyte.

It has long been recognized that the iodide ion is of considerable importance in the formation of the blue amylose–iodine complex. Early claims that iodide was unnecessary for complex formation failed to take account of the possible equilibrium reactions of the type

$$I_2 + H_2O \rightleftharpoons H_2OI^+ + I^- \qquad (3.9)$$

$$I_2 + H_2O \rightleftharpoons HOI + H^+ + I^- \qquad (3.10)$$

$$3I_2 + 3H_2O \rightleftharpoons IO_3^- + 5I^- + 6H^+ \qquad (3.11)$$

A complete discussion and evaluation of the various equilibrium constants is given by Allen and Keefer (1955). The addition of acid to aqueous iodine solution will displace each of these equilibria to the left, so reducing the concentration of iodide ion formed by hydrolysis of the iodine. Under conditions in which the equilibrium is displaced in this manner, complex formation is prevented.

According to Higginbotham (1949a), the iodine-binding capacity of amylose decreases with decreasing iodide concentration at potassium iodide concentrations of $10^{-3}$M to zero, but Holló and Szejtli (1968) suggested the converse, i.e. that the iodine-binding capacity decreases with increasing potassium iodide concentration, for the range 0–3 M potassium iodide. Our own results are summarized in table 3.3. In this case, the iodine-binding capacity of amylose shows a maximum value at an intermediate potassium iodide concentration (*ca.* $10^{-2}$M), and decreases at both higher or lower iodide concentrations. Kuge and Ono (1960a) reported that the absorbance at the wavelength of maximum absorption $\lambda_{max}$ of the amylose–iodine complex also passed through a maximum at approximately $10^{-2}$M potassium iodide (although the value of $\lambda_{max}$ itself remained virtually constant from $10^{-5}$–$10^{-3}$M potassium iodide, and then decreased at high concentrations). It is noteworthy that both potentiometric and spectrophotometric techniques suggest that $10^{-2}$M potassium iodide provides optimal conditions for the amylose–iodine interaction, although no explanation can at present be offered for the phenomenon.

*Table* 3.3. The iodine-binding capacity of amylose at 20.4°C as a function of potassium iodide concentration (Banks, Greenwood and Muir 1971b).

| molarity of KI | IBC (mg iodine bound/ 100 mg amylose) |
|---|---|
| 0.0001 | 18.1 |
| 0.001 | 18.6 |
| 0.005 | 19.0 |
| 0.01 | 19.5 |
| 0.05 | 19.1 |
| 0.10 | 17.8 |

Other salts normally present during the differential potentiometric titration are those comprising the buffer (potassium dihydrogen phosphate and dipotassium hydrogen phosphate). We have found that as the concentration of buffer increases in the range $4.95 \times 10^{-4}$–$9.90 \times 10^{-3}$M, the iodine-binding capacity does decrease, but the effect is quite small—only *ca.* 4.5 per cent. However, with increasing concentration of phosphate buffer, the uptake occurred at progressively lower concentrations of free iodine.

**3.2.5.** Equilibrium Time.

The presence of the digital voltmeter in the apparatus allows an accurate estimate to be made of the time necessary to attain equilibrium after adding iodine to the half-cell containing polysaccharide. We have observed that, after the initial addition of iodine, several hours may be necessary to achieve a true equilibrium potential, the exact time depending on the temperature of the titration. Figure 3.3 shows the relative changes in potential as a function of time for titrations carried out at three temperatures.

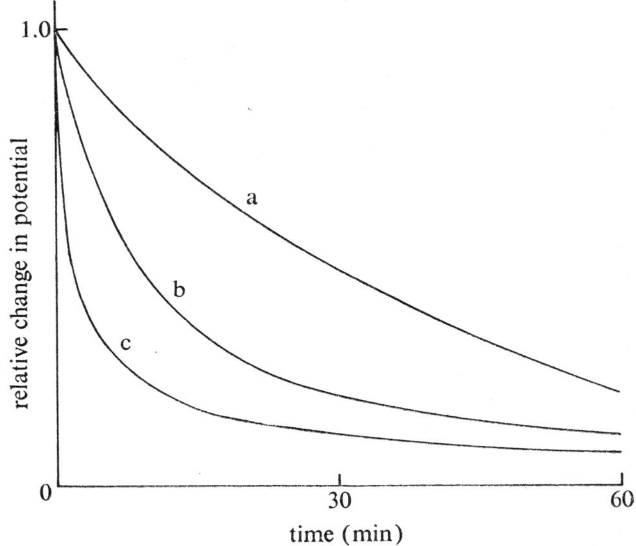


*Figure* 3.3. The relative change in potential during the initial stages of the iodine titration as a function of time: (a) 30°C, (b) 20°C, and (c) 10°C (Banks and Greenwood, unpublished work).

The experiment was carried out by adding 0.40 ml of iodine solution to the half-cell containing polysaccharide, and half that to the other. Potential was then noted as a function of time; the limiting value was obtained after fifteen hours. In each case, the change in potential is most marked in the initial stages, and as the temperature is lowered this initial change becomes more pronounced. However, it is quite obvious that several hours would be necessary in all cases before true equilibrium is attained. In the absence of polysaccharide, equilibrium is attained within thirty seconds.

Two further observations are of interest. First, if sufficient time is allowed after the initial addition of iodine for equilibrium to be attained, then equilibrium is reached within thirty minutes on subsequent additions; second, when the marked binding of iodine ceases, equilibrium is reached within seconds of completing the addition of the aliquot of iodine. However, no advantage is to be gained by adding sufficient iodine in a single aliquot to reach this portion of the titration curve—equilibrium is then achieved only after several hours.

Only the *relative* change in potential as a function of time is recorded in figure 3.3; the *absolute* change is perhaps best defined with respect to the concentration of free iodine $[I_2]_v$ at which the amylose is half-saturated with iodine. At 20°C, and allowing up to thirty minutes for equilibration, the value of $[I_2]_v$ is $1.5 \times 10^{-6}$M, whereas if fifteen hours are allowed for equilibration the value of $[I_2]_v$ is $1.2 \times 10^{-6}$M. Thus the difference in equilibration time does cause a considerable variation in the concentration of free iodine at which binding takes place.

These studies indicate that the conditions normally employed in iodine titration are far removed from equilibrium. This observation in no way affects the actual value of the iodine-binding capacity, but it does influence the concentration of free iodine at which binding occurs, moving this parameter to lower values. The above results refer to an amylose sample with $M_w$ approximately $1 \times 10^6$; as the degree of polymerization decreases below 200 glucose residues, the time required for equilibrium to be attained may be considerably increased.

**3.2.6.** Amylose Concentration.

In constructing the iodine-binding curve, the bound iodine (in mg/100 mg of polysaccharide) is graphed as a function of free iodine. Thus, one would suppose that, using various concentrations of amylose, a single iodine-binding curve should result. Within the range normally employed, $(2.5-6.0) \times 10^{-3}$ g amylose/litre of electrolyte solution, the results are superimposable,

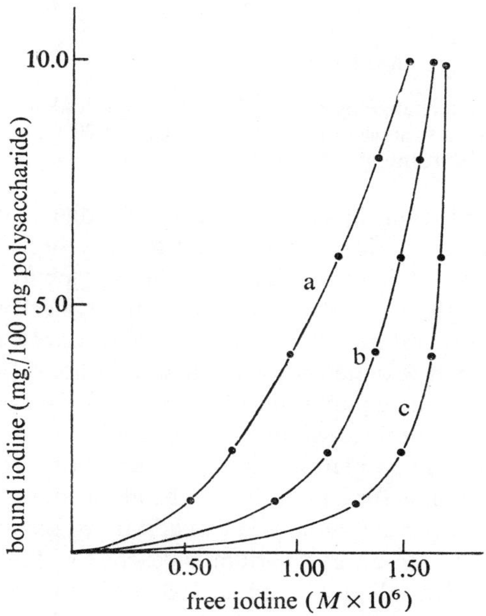


*Figure* 3.4. The initial stages of the iodine-binding curves for the same amylose sample at three different concentrations: (a) $5.9 \times 10^{-3}$ g/l; (b) $2.0 \times 10^{-2}$ g/l; and (c) $4.0 \times 10^{-2}$ g/l (Banks and Greenwood, unpublished work).

but as the amylose concentration increases the iodine-binding curve moves to higher values of free iodine, as shown in figure 3.4.

Only the initial stages of the iodine-binding process are shown in figure 3.4, as the curves merged at high values of bound iodine; i.e. the iodine-binding capacity was the same, irrespective of the weight of amylose. The titrations were carried out at 20°C, and thirty minutes were allowed to elapse after the addition of iodine to the half-cell containing amylose before the system was brought to zero potential. Thus some, if not all, of the differences observed in figure 3.4 may be due to equilibrium not having been attained. The results do emphasize that if reproducibility of the iodine binding curves is to be obtained, then the experimental conditions (concentration of amylose and the time between adding aliquots of iodine) must be strictly defined. Szejtli, Richter and Augustat (1967b) also noted that, in the photometric titration procedure, the concentration of free iodine at which iodine binding occurred varied with the amount of amylose present.

**3.2.7.** Degree of Polymerization of Amylose.

It has long been recognized that the iodine stain of amylose alters with the degree of polymerization. Swanson (1948) first investigated the qualitative relation involved by preparing a series of synthetic amylose samples, by the condensation of glucose-1-phosphate with a primer, the reaction being catalysed by phosphorylase. Later work by Bailey and Whelan (1961) refined Swanson's approach by using maltohexaose as the primer, and employing experimental conditions that would minimize any trace contamination of the phosphorylase preparation by phosphatase or *alpha*-amylase, both of which would introduce errors into the calculated degrees of polymerization. Subsequently, Pfannemüller, Mayerhöffer and Schulz (1969) and Banks, Greenwood and Khan (1971a) also prepared synthetic amylose samples, but measured the degrees of polymerization by an enzymic technique (Banks and Greenwood 1968b).

*Table* 3.4. The wavelength of maximum absorbance ($\lambda_{max}$) of the iodine complex, and iodine-binding capacity at 20.4°C for a series of enzymically-synthesized amylose oligomers of various number-average degrees of polymerization $\overline{DP}_n$ (Banks, Greenwood and Khan 1971a).

| sample | 1 | 2 | 3 | 4 | 5 | 6 | 7 | 8 | 9 | amylose |
|---|---|---|---|---|---|---|---|---|---|---|
| $\lambda_{max}$ at 20°C (nm) | 496 | 524 | 530 | 546 | 574 | 588 | 595 | 606 | 610 | 642 |
| IBC at 20.4°C[1] | —[2] | —[2] | 1.3 | 3.6 | 11.1 | n.d.[3] | 16.2 | 16.4 | 17.3 | 19.5 |
| $\overline{DP}_n$ | 22.2 | 28.9 | 31.3 | 36.4 | 50.7 | 71 | 93 | 105 | 134 | 1500 |

[1] IBC in mg iodine bound/100 mg polysaccharide.
[2] Experimental values could not be extrapolated to positive iodine-binding capacities.
[3] Not determined.

Our results for measurements of the wavelength of maximum absorption ($\lambda_{max}$), and the iodine-binding capacity as a function of the degree of polymerization, are given in table 3.4, and the relation between the two represented in figure 3.5. The form of this relation is, of course, generally accepted, but some rather important differences of detail do exist between the results of different workers. For example, Bailey and Whelan (1961) claimed that there was a fairly sharp break in the curve when the degree of polymerization reached 72 glucose residues. A similar break, but at a higher degree of polymerization ($\overline{DP}_n \sim 120$ residues), was reported by Szejtli and Augustat (1966) and by Szejtli, Richter and Augustat (1967a). However, in this latter work the degree of polymerization of the amylose was varied by subjecting the polysaccharide to different periods of mechano-chemical degradation.

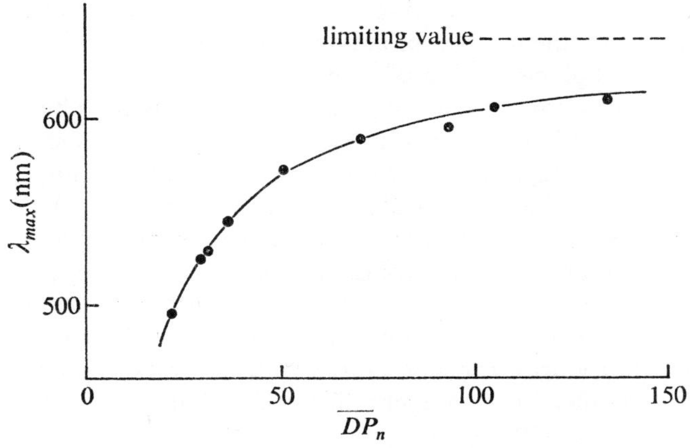


*Figure* 3.5. The wavelength of maximum absorbance $\lambda_{max}$ as a function of degree of polymerization $\overline{DP}_n$ for a series of enzymically-synthesized amylose oligomers (Banks, Greenwood and Khan 1971a).

Hence the breadth of the molecular weight distribution would be much greater than in the case of the synthetic amylose samples, making any comparison between the two difficult. [It is unlikely that the distribution obtained by Szejtli and co-workers (1966, 1967a) was narrower than the exponential, and therefore $\overline{M}_w/\overline{M}_n \sim 2$, whereas for the synthetic preparations $\overline{M}_w/\overline{M}_n < 1.1$, the exact value of the ratio decreasing with increasing $\overline{M}_n$ (see section 5.2.1).] Pfannemüller, Mayerhöffer and Schulz (1969) suggested that there was a critical chain length in the region 60–100 glucose residues at which the ordered helix was stabilized, as shown by the formation of a blue complex. In support of this contention, they also offered results of optical rotatory dispersion and circular dichroism measurements, subjects that will be discussed later (see section 3.3.3).

There can be no doubt that the major change in $\lambda_{max}$ occurs as the degree of polymerization increases towards 100 glucose residues. The range we chose to examine, $22<\overline{DP}_n<135$, covers the region in which any sharp change in $\lambda_{max}$ as a function of molecular size should be readily detectable, but in fact a smooth curve is obtained, as is clearly shown in figure 3.5. The measurements carried out by Pfannemüller, Mayerhöffer and Schulz (1969) also show that there is no discontinuity: these authors reported that the $\overline{DP}_n/\lambda_{max}$ relation, and also the corresponding $\lambda_{max}$ (circular dichroism)/$\overline{DP}_n$ relation, takes the form of a Langmuir adsorption isotherm

$$\lambda_{max}=(A\times\overline{DP}_n)/(\overline{DP}_n+B) \qquad (3.12)$$

where $A$ is the asymptotic limit of $\lambda_{max}$ at high degrees of polymerization, and $B$ is a constant. Inversion of (3.12) yields

$$1/\lambda_{max}=(1/A)+[(B/A)/\overline{DP}_n] \qquad (3.13)$$

The results of Banks, Greenwood and Khan (1971a) are shown in figure 3.6 graphed according to (3.13). Within experimental error, a linear relation is obtained with $1/A=1.558\times10^{-3}\text{nm}^{-1}$, from which $A=642$ nm. This predicted value is exactly that obtained experimentally with an amylose sample having a degree of polymerization of *ca.* 1500 glucose residues (see figure 3.6). The results of Bailey and Whelan (1961) also approximate to the form of (3.13); the parameters $A$ and $B$ derived from that work, together with those quoted by Pfannemüller and co-workers and our own values, are summarized in table 3.5. The only two values in agreement are those of Bailey and

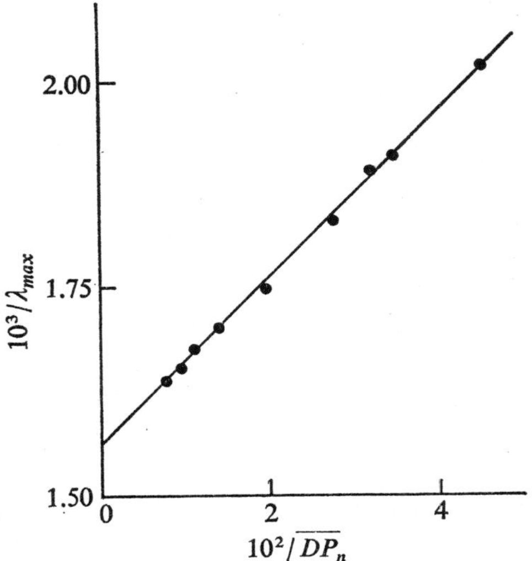


*Figure* 3.6. The reciprocal of the wavelength of maximum absorbance $1/\lambda_{max}$ as a function of the reciprocal number-average degree of polymerization $1/\overline{DP}_n$ for a series of enzymically-synthesized amylose oligomers (Banks, Greenwood and Khan 1971a).

Whelan (1961) and our own for the constant $A$; the values of $B$, however, are quite different. In fact, we used the same conditions as did Bailey and Whelan for staining the polysaccharide, hence one would expect reasonable correlation between the two sets of data. It is therefore somewhat surprising that the values of $B$ are quite so divergent, and no explanation can at present be given for the discrepancy. The German workers used radically different conditions for developing the colour, and therefore their results are not directly comparable.

*Table* 3.5. A comparison of the parameters $A$ and $B$ of equation 3.12 obtained by various workers.

| | | parameter: | |
|---|---|---|---|
| | $(1/A)\times 10^3$ | $(B/A)\times 10^4$ | $B$ |
| Bailey and Whelan (1961) | 1.56 | 88 | 5.66 |
| Pfannemüller, Mayerhöffer and Schulz (1969) | 1.637 | 92.5 | 5.65 |
| Banks, Greenwood and Khan (1971a) | 1.558 | 102.5 | 6.60 |

Iodine-binding curves (measured at 20.4°C), from which the iodine-binding capacities of the oligomeric amylose samples given in table 3.4 were obtained, are shown in figure 3.7. It is impossible to extrapolate the results for samples 1 and 2 to give a positive iodine-binding capacity; even for samples 3 and 4, the extrapolation is so lengthy that the resultant values should be regarded with caution. However, for samples containing fifty or more glucose residues, the extrapolation is fairly straightforward. The points of interest are: (a) the iodine-binding capacity is dependent on degree of polymerization, but reaches approximately 89 per cent of its asymptotic limiting value when the degree of polymerization is only 134 glucose units; and (b) the concentration of free iodine at which uptake occurs moves to higher values, and the range broadens, as the degree of polymerization decreases. It is somewhat surprising that the asymptotic iodine-binding capacity is reached so rapidly, but we have recently synthesized an amylose with $\overline{DP}_n$ of 260 glucose units, and found that its iodine-binding capacity is 19.3 mg iodine/100 mg polysaccharide, a value not very different from that of natural amylose ($\overline{DP}_n \sim 1500$ glucose residues), measured at 20.4°C. This natural amylose has an exponential distribution of molecular weights, therefore $\overline{DP}_w$ is 3000 glucose residues; only when the distribution is very narrow is it legitimate to equate the iodine stain with the number-average degree of polymerization. We may, however, conclude that the property which causes a very long amylose molecule to bind one molecule of iodine for each eight glucose residues is already present when the degree of polymerization reaches approximately 250 glucose residues. This conclusion is radically different from that of Szejtli, Richter and Augustat (1967a), who claimed that the iodine-binding capacity changed throughout the range of molecular sizes they investigated. For example, for an amylose having $\overline{DP}_n \sim 280$ glucose units, the iodine-binding capacity was only 72 per cent of the asymptotic value. We ascribe the divergence between this observation and our own to the fact that Szejtli and his co-workers used fractions having broad molecular weight distributions, which to a large extent nullify their conclusions.

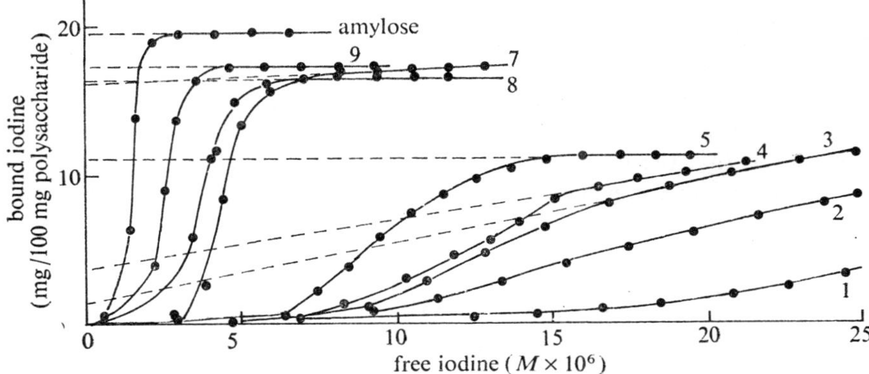


*Figure* 3.7. Iodine-binding curves at 20.4°C for amylose oligomers: sample numbers are shown, from which the characteristics of the different oligomers may be obtained using table 3.4 (Banks, Greenwood and Khan 1971a).

The increase in the concentration of free iodine at which absorption occurs, as the degree of polymerization decreases, has long been recognized. Indeed, it has been claimed (Holló and Szejtli 1958a, 1962) that amylose may be sub-fractionated according to its molecular weight by the stepwise formation of the iodine complex. The curves in figure 3.7 for amylose and oligomer sample 9 show that the concentrations of free iodine at which absorption occurs are $1.50 \times 10^{-6}$M and $2.60 \times 10^{-6}$M respectively. The difference is comparatively small, considering that the weight-average molecular weights of the samples differ by a factor of 22. We have attempted to use the technique to fractionate amylose, but have found that to obtain fractions of reasonably narrow distribution (i.e. $\overline{M}_w/\overline{M}_n < 1.7$), it is necessary to dissolve the amylose–iodine complex by heating and then re-form it by slow cooling. However, these manipulations result in extensive degradation, and the technique is not successful.

Thoma and French (1960, 1961) examined, by potentiometric titration, the reaction between iodine and various pure malto-oligosaccharides in the range maltotetraose–maltopentadecaose. Up to and including maltononaose, one molecule of iodine was present in the complex. [At the molarity of potassium iodide employed (0.25 M), the authors suggest that the iodine would be bound as $I_3^-$.] The higher oligosaccharides showed additional binding, ascribed to the initiation of a poly-iodine chain. When the formation constants of the complexes were graphed as a function of the number of glucose residues constituting the dextrins, the results showed a distinct break between maltohexaose and maltoheptaose which was ascribed to a change in the conformation of the complex from a loop to a helix. Interestingly enough, Thoma and French (1961) also reported that iodine binding by maltose and maltotriose could be detected at 0.2°C, although the points were too scattered to allow the equilibrium constants to be extricated. These reactions occur at concentrations of free iodine two orders of magnitude greater than those encountered in titrating amylose.

**3.2.8.** Temperature.

The blue colour of the amylose–iodine complex slowly disappears as the solution is heated, and reappears on cooling (see figure 4.14); the necessary corollary is that the iodine-binding capacity of the polysaccharide also decreases with increasing temperature (Higginbotham 1949b; Higginbotham

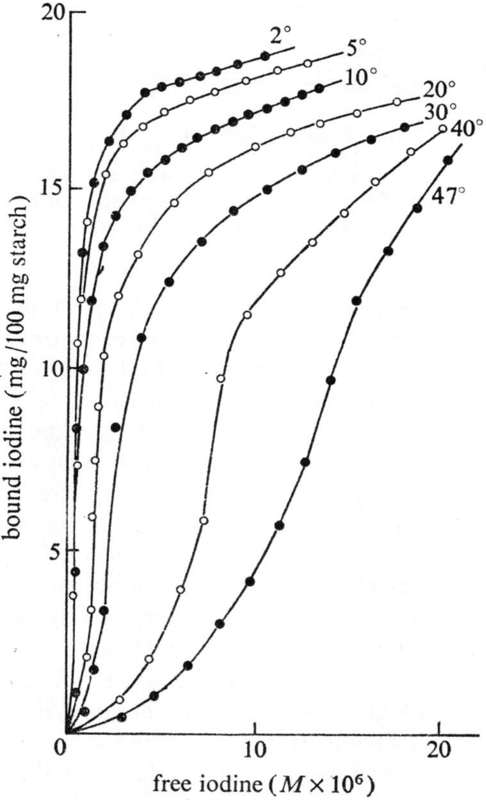


*Figure* 3.8. Iodine-binding curves for an amylomaize starch at various temperatures (Adkins and Greenwood 1966c).

and Morrison 1949; Holló and Szejtli 1957). These workers also recognized that as the temperature increased, the concentration of free iodine at which uptake occurred also moved to higher values. Strangely, few attempts were made to measure iodine-binding capacities at low temperatures, despite the clear statement by Thoma and French (1961) that at sufficiently low temperatures there was evidence that even maltose could bind iodine.

It remained for Adkins and Greenwood (1966e) to demonstrate that definite advantages could accrue from using low temperatures for the iodine titration of amylomaize starches. Figure 3.8 shows the iodine-binding curves of an amylomaize starch as a function of temperature. When the titration of this type of starch is carried out at 20°C, the curves are inflexionless, making extrapolation to zero concentration of free iodine meaningless (see figure 3.8). In contrast, on cooling the solution, the titration behaviour becomes more normal, until at 2°C it is possible to obtain results that enable an extrapolation to zero concentration of free iodine to be made (see section 2.5.3 for a full discussion).

*Table* 3.6. Values of iodine-binding capacities for amylose oligomers at various temperatures (Banks, Greenwood and Khan 1971a).

| sample[(1)] | IBC (mg of iodine/ 100 mg polysaccharide) at: | | |
|---|---|---|---|
| | 1.4°C | 7.7°C | 14.2°C |
| 1 | 7.8 | n.d.[(2)] | n.d. |
| 2 | 9.4 | n.d. | n.d. |
| 3 | 12.5 | n.d. | n.d. |
| 4 | 19.2 | n.d. | n.d. |
| 5 | 19.6 | 18.8 | 14.8 |
| 6 | 20.7 | n.d. | n.d. |
| 7 | 21.0 | 20.2 | 18.9 |
| 8 | 21.4 | 20.5 | 19.5 |
| 9 | 21.8 | 20.9 | 20.1 |
| amylose | 22.2 | 21.7 | 20.7 |

[1] As in table 3.4. [2] n.d. = not determined.

In relation to this problem, Banks, Greenwood and Khan (1971a) prepared a series of synthetic amylose oligomers as model compounds for the study of the starch–iodine interaction at various temperatures. The iodine-binding capacities of the oligomers at various temperatures are given in table 3.6, and figure 3.9 shows the titration curves obtained at 1.4°C. At this temperature, it is possible to extrapolate the results for all the oligomers to obtain the iodine-binding capacities. Moreover, a chain length of 36 glucose residues is sufficient to give an iodine-binding capacity of 85 per cent of that obtained for an amylose having a degree of polymerization of 1500 units. Technical difficulties regarding the measurement of absorbance at low temperature prevented the wavelength of maximum absorption being determined with a high degree of accuracy, but nevertheless the semi-quantitative values obtained showed very large increases in the values of $\lambda_{max}$, especially at low degrees of polymerization. For example, oligomer sample 1 with $\overline{DP}_n = 22.2$ glucose units had a red-purple colour at 1.4°C and $\lambda_{max}$ of approximately 545 nm, whilst at 20.4°C the oligomer–iodine stain was red in colour, with $\lambda_{max}$ of 496 nm. The same general observation was true for all the shorter oligomers, i.e. the wavelength of maximum absorbance of the amylose–iodine complexes moved to shorter wavelengths as the solutions were warmed from 1.4°C to 20.4°C. However, in the case of amylose of high molecular weight, there did not appear to be any marked change in the value of $\lambda_{max}$.

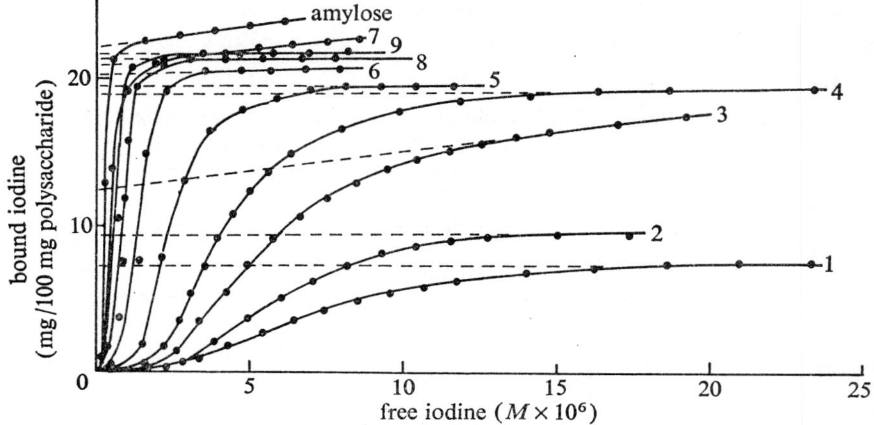


*Figure* 3.9. Iodine-binding curves at 1.4°C for amylose oligomers: sample numbers are shown, from which the characteristics of the different oligomers may be obtained using tables 3.4 and 3.6 (Banks, Greenwood and Khan 1971a).

The results shown in table 3.7 demonstrate quite clearly that for any amylose molecule containing 31 or more glucose residues, the iodine-binding capacity increases with decreasing temperature. Moreover, the effect of temperature on iodine-binding capacity becomes very much more pronounced as the degree of polymerization decreases, as shown by the variation in the function $F(N)$, which is defined as

$$F(N) = \frac{(\text{iodine-binding capacity at } 1.4°\text{C})}{(\text{iodine-binding capacity at } 20.4°\text{C})} \tag{3.14}$$

This function has a value of 9.62 for the oligomer of degree of polymerization 31.3, decreasing to 1.14 for natural amylose. The pronounced dependence of

$F(N)$ on degree of polymerization provides an excellent test for the presence of short-chain linear material in a sample, i.e. if the value of $F(N)$ is greater than 1.14, short-chain material is present.

*Table* 3.7. The ratio $F(N)$ of iodine-binding capacity at 1.4°C to that at 20.4°C for amylose molecules of different sizes (Banks, Greenwood and Khan 1971a).

| sample[1] | $\overline{DP}_n$ | $F(N)$ |
|---|---|---|
| 3 | 31.3 | 9.62 |
| 4 | 36.4 | 5.35 |
| 5 | 50.7 | 1.77 |
| 7 | 93 | 1.30 |
| 9 | 134 | 1.26 |
| amylose | 1500 | 1.14 |

[1] As in table 3.4.

Figure 3.9 shows that not only is the iodine sorption moved to lower values of free iodine as a result of using low temperatures (*cf.* figure 3.7), but the vertical portions of the curves are very much closer together. Consequently, any sub-fractionation of amylose by the stepwise addition of iodine would be even less successful than at room temperature. (Comparison of figures 3.7 and 3.9 suggests that such a fractionation process would be most efficient at high temperatures were it not for the concomitant degradation that occurs under these conditions.)

**3.2.9.** Development of Colour.

The absorbance of the amylose–iodine solution during potentiometric titration increases linearly with the amount of iodine bound by the polysaccharide, as shown in figure 3.10. Conversely, the curve of absorbance as a function of the concentration of free iodine is exactly the same as that between the latter parameter and the amount of iodine bound by the polysaccharide. Thus, irrespective of whether or not the secondary binding of iodine is present from the start (see section 3.2.2), the *total* amount of iodine bound by the polysaccharide is measured quantitatively by both the potentiometric and colorimetric methods.

In the region in which rapid binding occurs, however, there is a gradual change in the wavelength of maximum absorption. In figure 3.10, the value of $\lambda_{max}$ for the point at which 3.7 mg iodine/100 mg amylose had been bound was 605 nm; the limiting value of 624 nm for $\lambda_{max}$ was reached only when the amylose had bound 19.5 mg iodine/100 mg polysaccharide. [The conditions used for absorbance measurements were those operative during the standard differential potentiometric titration procedure, which accounts for the limiting value of $\lambda_{max}$ being different from that quoted in table 3.4.] It is interesting to note that in this case the amylose continued to bind small amounts of iodine above the value of 19.5 mg/100 mg amylose, and this was faithfully reflected in the increased absorbance of the complex, but the wavelength at which maximum absorption occurred did not increase once the iodine-binding capacity of the material had been reached.

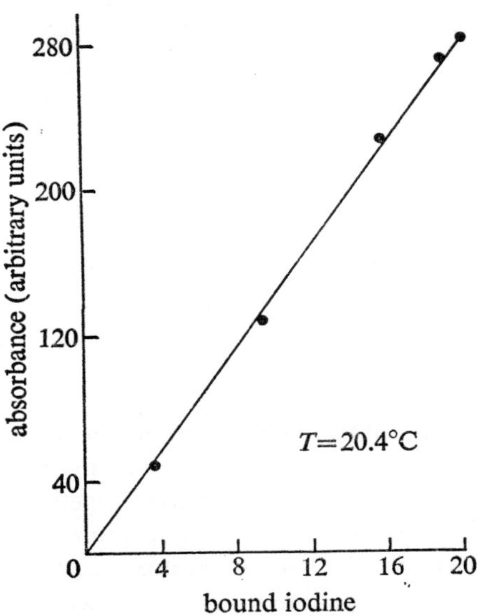


*Figure* 3.10. Absorbance (in arbitrary units) of the amylose–iodine complex as a function of the weight of iodine bound by the polysaccharide at 20.4°C (Banks, Greenwood and Muir 1974b).

The absorption spectrum of the amylose at the point at which it had bound 19.5 mg iodine/100 mg amylose is shown in figure 3.11. It shows, in addition to the major absorption peak at 624 nm, two minor peaks at approximately 350 nm and 485 nm. The former absorption band is due to the presence of the tri-iodide, but the origin of the latter band is obscure; Pfannemüller, Mayerhöffer and Schulz (1969) have suggested that it represents an interaction between iodine and the hydroxyl groups of the amylose.

As we stressed earlier, colorimetric methods for determining the amylose content of a starch are much more rapid than the differential potentiometric technique, and require the minimum of apparatus. A critical comparison of the colorimetric and potentiometric techniques (Banks, Greenwood and Muir 1974b) shows that for most purposes the former method is adequate, but difficulties arise in the case of amylomaize starch. Whilst there must be some debate as to whether the amylose content obtained by iodine titration of these starches is particularly accurate, the form of the iodine-binding curve at least signifies that the material is anomalous, an observation that could not be made in the case of a single-point colorimetric assay. This objection could be overcome by constructing a photometric titration curve as advocated by Richter and Szejtli (1966). At present, a great deal of interest is being shown by plant breeders in genetic manipulation to produce starches of high amylose content, and if new strains possess the short-chain material found in amylomaize, considerable error could result in the amylose content determined by a single-point method.

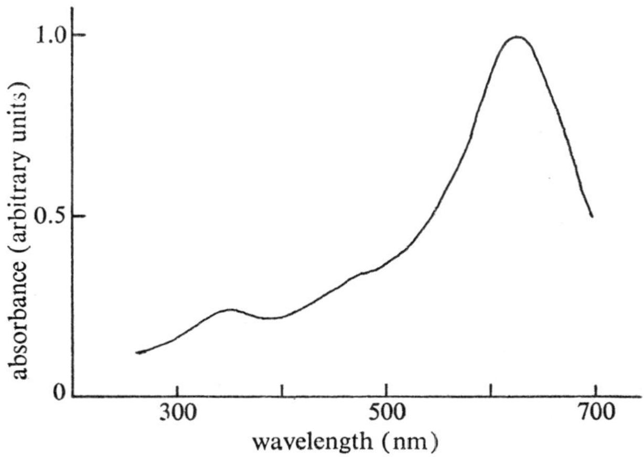


*Figure* 3.11. Absorbance of the amylose–iodine complex as a function of wavelength.

**3.2.10.** Summary.

We have dealt in some detail with the factors affecting the technique of semimicro differential potentiometric titration, because this technique is, in our view, one of the most valuable research tools in starch chemistry. Many of the factors that we have detailed as being of particular importance in affecting the titration procedure are applicable to studies of the starch–iodine interaction in general, and this theme is expanded in the following section.

## 3.3. The Nature of the Amylose–Iodine Interaction.

The interaction of amylose with iodine generates a helical inclusion complex in which the iodine molecules occupy the central cavity of the helical polysaccharide. As iodide ions must be present before the complex can form, the complexing agent may be ionic in nature, and the stoichiometry of the reaction is of considerable interest.

**3.3.1.** Stoichiometry of the Interaction.

The first study of the stoichiometry of the amylose–iodine interaction was due to Gilbert and Marriott (1948). They based their analysis on the assumption that, in solution, the complexing species is of the form $(xI_2.yI^-)$, and then attempted to measure $x$ and $y$ experimentally. At low degrees of saturation of the amylose, the activity of the complex will be proportional to its amount per unit quantity of amylose, and therefore its chemical potential $\mu_{xy}$ is given by

$$\mu_{xy} = \mu_{xy}^0 + RT \ln a \tag{3.15}$$

where $a$ is the number of moles of complex per gram of amylose. In a thermodynamically ideal system, the concentrations of the ions and molecules which are in equilibrium with the complex are given by

$$\mu_{I^-} = \mu_{I^-}^0 + RT \ln I^- \tag{3.16}$$

and

$$\mu_{I_2} = \mu_{I_2}^0 + RT \ln I_2 \tag{3.17}$$

From the relation

$$\mu_{xy} = x\mu_{I_2} + y\mu_{I^-} \tag{3.18}$$

it may be deduced for a given temperature that

$$[I_2]^x[I^-]^y = K_{xy}[a] \tag{3.19}$$

where $K_{xy}$ is an equilibrium constant. The value of $x$ may be obtained by varying the iodine concentration at a constant concentration of iodide, so reducing (3.19) to

$$\partial \log[a]/\partial \log[I_2] = x \tag{3.20}$$

Similarly, by varying the iodide concentration at constant adsorption, (3.19) reduces to

$$\partial \log[I_2]/\partial \log[I^-] = -y/x. \tag{3.21}$$

The above treatment assumes that if the values of $x$ and $y$ are not constant throughout complex formation, then at least one composition is dominant at each stage of the interaction.

Experimentally, it is of course quite easy to evaluate $x$ from (3.20); the first portion of the iodine-binding curve is examined by graphing the logarithm of the bound iodine as a function of the logarithm of the free iodine. Equation 3.21 is more difficult to use; the most satisfactory method is to construct a series of iodine-binding curves for different concentrations of iodide ion, and then determine the iodine concentrations required to produce a fixed value for iodine bound at each of these iodide concentrations. As (3.17) refers to molecular iodine rather than the total amount of iodine not bound by the amylose, one must take account of the reaction

$$I_2 + I^- \rightleftharpoons I_3^- \tag{3.22}$$

The concentration of molecular iodine may be calculated from data given by Jones and Kaplan (1928), or Davies and Gwynne (1952).

According to Gilbert and Marriott (1948), there is some indication that in the initial stages of the interaction the value of $x$ is 2, and that this subsequently increases to 3. These observations were made over a region representing *only some two per cent of the total capacity for iodine binding*. Gilbert and Marriott (1948) concluded that the structure of the complexing agent was $3I_2.2I^-$, although they also noted that as the extent of interaction increased there was a tendency for the value of $x$ to increase.

We have repeatedly attempted to reproduce these results and, although we agree with the earlier workers that the value of $x$ increases during the titration, we have additionally found that the value of $x$ is dependent on the weight of amylose being titrated, as is shown in table 3.8. (Gilbert and Marriott make no mention of investigating the interaction at different concentrations of polysaccharide.) At an iodide concentration of $10^{-2}$M, the values of $x$ increase with the weight of amylose used in the titration; the same phenomenon is seen when $10^{-3}$M potassium iodide is employed as the supporting electrolyte (i.e. the same conditions used by Gilbert and Marriott). Adjustment of the conditions therefore leads to wide variations in the possible values of the parameter $x$.

*Table* 3.8. Values of the parameter $x$ of equation 3.20 as a function of the concentration of amylose (Banks and Greenwood, unpublished work).

| conc. of amylose (g/l) | $x$ initial | $x$ subsequent |
|---|---|---|
| $5.9 \times 10^{-3}$ | 2 | 3 |
| $2.0 \times 10^{-2}$ | 3 | 4 |
| $4.0 \times 10^{-2}$ | 4 | 6 |

The reason that such wide variations in $x$ are observed is, we believe, due to the difficulty of ensuring equilibrium after each addition of iodine. We have previously noted (see section 3.2.6) that the portion of the iodine-binding curve at which it is most difficult to attain equilibrium is precisely the range used in the evaluation of the parameter $x$. The systematic change to the higher values of $x$ recorded in table 3.8 with increasing weights of amylose is probably related to the fact that the iodine-binding curves for different concentrations of amylose are not superimposable (see figure 3.4; in fact, the results recorded in table 3.8 and figure 3.4 refer to the same series of experiments).

Gilbert and Marriott (1948) also noted that the composition of the complex changed as the concentration of iodide ion was increased, until the ratio of iodide ions to iodine molecules was at least one. In support of this claim, they recorded that the loss in concentration of the iodide ion resulting from the precipitation and removal of the amylose from a 0.2 M potassium iodide solution by addition of excess iodine was such that the ratio of iodine molecules to iodide ions in the complex must be unity. This type of experiment does not prove that the iodide ions removed from solution are actually found in the helical cavity—they could equally well be adsorbed onto the surface of the amylose–iodine complex.

Higginbotham (1949a) measured the ratio $C_{I^-}/C_I$, defined as

$$\frac{C_{I^-}}{C_I}=\frac{\text{equivalents iodide bound by amylose}}{\text{equivalents iodine bound by amylose}} \tag{3.23}$$

and reported that it increased with increasing concentrations of iodide ion, changing from a value of 0.39 at $2\times10^{-3}$M iodide to 0.81 at $5\times10^{-2}$M iodine. At the low end of the concentration range, the values are quite compatible with the $3I_2.2I^-$ structure proposed by Gilbert and Marriott (1948) for a solution containing $10^{-3}$M potassium iodide.

Mukhergee and Bhattacharyya (1946) isolated the starch–iodine and amylose–iodine complexes, and determined the amount of iodine present. They obtained a measure of the total iodine and the available iodine, the difference between the two representing the amount of iodine present as iodide. In the range of potassium iodide concentrations that they employed, $0.046\text{M}<\text{KI}<1.21\text{M}$, they found the molar ratio of iodide to iodine of amylose complex to vary in the range 0.141 to 0.470, suggesting that the complexing species was $7I_2.I^-$ at the low concentrations of iodide ion, but tended towards the limiting form $I_3^-$ as the iodide ion concentration increased.

An attempt to provide a theoretical basis for the relation between the concentration of iodide ion and the wavelength of maximum absorbance of the amylose–iodine complex was made by Ono, Tsuchihashi and Kuge (1953). Their treatment was analogous to that of Kuhn (1949) for polyenes, giving

$$\lambda_{max}=\left[\frac{V_0}{hc}\left(1-\frac{1}{N}\right)+\frac{h}{8mc}\cdot\frac{(N+1)}{L^2}\right]^{-1} \tag{3.24}$$

where $L$ is the resonating chain-length, $N$ is the number of $\pi$ electrons, $h$ is Planck's constant, $m$ is the mass of the electron, $c$ is the velocity of light, and $V_0$ is the perturbation parameter. Table 3.9 shows the number $n$ of iodine atoms required to give experimental values of $\lambda_{max}$ for a range of iodide concentrations, from which results Ono and his co-workers (1953) concluded that Kuhn's theory was applicable to amylose–iodine. These authors also suggested that the iodide ions are probably found at the ends of the resonating poly-iodine chains, rather than forming an integral part of them. On close examination the theory is not convincing, for the authors obtained the value of $V_0$ from measurements on amylose in water, assuming that iodide ions were absent. They also assumed that under these conditions a single poly-iodine chain was formed in the amylose, which had a degree of polymerization of 480 units. Neither assumption is, in fact, valid.

An alternative treatment, in which the absorption maximum is calculated for different lengths of iodine chain using the electron gas theory, appears to be preferable. According to this treatment, a poly-iodine chain containing fourteen iodine atoms should exhibit its maximum absorbance at a wavelength of 625 nm (Cramer 1951, 1952a).

*Table* 3.9. The wavelength of maximum absorbance $\lambda_{max}$ calculated as a function of the number of iodine atoms $n$ in the polyiodine chain at different concentrations of potassium iodide (Ono, Tsuchihashi and Kuge 1953).

| molarity of KI | $n$ | $\lambda_{max}$ (nm) |
|---|---|---|
| zero | 160 | 628 |
| $1.2 \times 10^{-4}$ | 134 | 625 |
| $6.1 \times 10^{-4}$ | 108 | 621 |
| $5.0 \times 10^{-3}$ | 77 | 613 |
| $3.0 \times 10^{-2}$ | 52 | 600 |
| $2.1 \times 10^{-1}$ | 30 | 572 |
| $5.0 \times 10^{-1}$ | 22 | 557 |

Mould and Synge (1954) used the fact that iodide ions are taken up by the polysaccharide during the amylose–iodine interaction, so giving a charged complex, as the basis of an electrophoretic separation method for the products from the α-amylolysis and acid-degradation of amylose. They obtained four fractions, which gave iodine stains of blue, red, orange, and virtually no colour; the degrees of polymerization associated with these colours are summarized in table 3.10. Perhaps the most surprising observation about these results is that electrophoresis of the amylose–iodine complex

*Table* 3.10. Fractions obtained by electrophoresis of a degraded amylose sample; the size range of the polysaccharide associated with different colours of iodine stain (Mould and Synge 1954).

| colour | degree of polymerization |
|---|---|
| none | $<10$ |
| orange | 10–25 |
| red | 25–40 |
| blue | 40–130 |

did not yield a continuous spectrum of colours, but rather separated the material into a number of distinct zones, separated by gaps in which the polysaccharide is absent. In particular, no material giving a purple stain with iodine was isolated, suggesting that when this colour is observed it is given by a mixture of chains, one portion staining blue and the other red; i.e. the purple colour does not represent a single population of molecules but rather a mixed population. The fact that the purple colour is obtained using enzymically-synthesized material is no argument against this postulate of Mould and Synge (1954), because the breadth of the molecular weight distribution, although narrow, is finite. The authors suggest that, in the blue migrating complexes, colour and mobility are determined by the presence of a $(3I_2.2I^-)_n$ grouping, whereas the colour and mobility of the red migrating complex reflects the presence of an $(I_3^-)_n$ grouping.

The potentiometric iodine titration behaviour of the fractions obtained on electrophoresis of the amylose–iodine complex was examined by Mould (1954). At the temperature used, both the blue- and red-staining fractions bound iodine—in fact, to considerably greater extents than the limiting value of 20 per cent associated with amylose. However, the uptake occurred at much higher levels of free iodine than it did with amylose, and the red-staining material bound iodine at higher levels than did the blue-staining fraction.

The separation of the various fractions by electrophoresis shows that the amylose–iodine complex is negatively charged, almost certainly as a result of the presence of iodide (or tri-iodide) ions. It does not necessarily follow, however, that these negatively-charged ions are located solely (if at all) within the helical cavity of the amylose. Continued uptake of iodine by the polysaccharide beyond the limiting value of *ca.* 20 per cent by weight signifies that a mechanism other than helical inclusion is operative. As this alternative mechanism probably involves some kind of surface adsorption of the tri-iodide ion, the charge on the amylose–iodine complex could arise predominantly from that source, with only a minor contribution from charged species within the helical cavity.

Schneider, Cronan and Podder (1968) considered the amylose–iodine interaction as an example of co-operative binding to a one-dimensional lattice. The adsorption isotherm of iodine–iodide in the helical cavity was determined by equilibrium dialysis. Spectrophotometric measurements of these isotherms made at various concentrations of iodide indicated that the complex had the general form $I_2.I_b^-$, where $b$ is dependent on the concentration of iodide ion. At $10^{-2}$M potassium iodide the value of $b$ is approximately unity, so that the dominant bound species is in fact the tri-iodide ion, $I_3^-$.

The detailed calculations by which the above result was achieved are given by Cronan and Schneider (1969), and essentially are the same as those of Kuge and Ono (1960a) with the additional refinement that a variable amount of iodide ion may be bound together with iodine. Thus there are two species that may be bound, namely $I_2$ and $I_3^-$, and the overall average formula $I_2.I_b^-$, where $b$ can vary between zero and one, is applicable. If the absorption at 640 nm is due only to the complex, then

$$OD_{640}=\varepsilon_{640}[I_2.I_b^-] \tag{3.25}$$

where $\varepsilon_{640}$ is the molar extinction coefficient of the bound species at 640 nm. Absorption at 353 nm is supposed to be due to contributions from both bound and free species, so that

$$OD_{353}=\varepsilon'_{353}[I_2 . I_b^-]+\varepsilon''_{353}[I_2]_F+\varepsilon'''_{353}[I_3^-]_F \tag{3.26}$$

where $\varepsilon'_{353}$, $\varepsilon''_{353}$, and $\varepsilon_{353}$ are the molar extinction coefficients at 353 nm of the bound species, free iodine, and free tri-iodide, respectively. Mass conservation relations give

$$[I^-]_T = [I^-]_F + [I_3^-]_F + b[I^-]_B \tag{3.27}$$

and

$$[I_2]_T = [I_2]_F + [I_3^-]_F + [I_2]_B \tag{3.28}$$

where the subscripts *T*, *F*, and *B* refer to the total, free, and bound concentrations of each species. Also necessary for the calculation is the value of $K_F$, the equilibrium constant for the iodine–iodide interaction [see equation 3.22; Cronan and Schneider (1969) used the value of $K_F$ derived by Davies and Gwynne (1952)].

Cronan and Schneider (1969) analysed their experimental results by a computer treatment that enabled $\varepsilon_{640}$ to be treated as an adjustable parameter; the value of *b* was also adjusted in steps of 0.1, and the necessary criterion to be satisfied was that $\varepsilon_{353}$ be constant over the entire titration. Theoretical isotherms generated in this manner were then fitted to the experimental data, the best fit allowing the value of the parameter *b* to be identified.

This treatment also showed that under the conditions used by Gilbert and Marriott (1948) the value of *b* was approximately 0.7, corresponding to the formula $3I_2 . 2I^-$, i.e. the same formula reported by the earlier workers. No significance should be attached to this apparently good agreement, however, because Gilbert and Marriot reported that there was evidence of a different stoichiometry beyond the range of interactions in which they were interested (only the first 2 per cent of the total interaction), whereas Cronan and Schneider (1969) obtained their *b* values when the interaction was 50 per cent complete.

Kuge and Ono (1960a) assumed that in $5 \times 10^{-2}$M potassium iodide the interaction would take the form

$$\text{amylose} + nI_2 + nI^- \rightleftharpoons \text{complex} \tag{3.29}$$

so that the tri-iodide ion was the complexing agent. In a later work (Watanabe, Ogawa and Ono 1970), this approach was modified to give

$$\text{Amylose} + nI_2 + mI^- \rightleftharpoons \text{complex} \tag{3.30}$$

Spectrophotometric measurements were then used to deduce values of *n* and *m*, starting from the relation

$$[I_2]_B = [I_2]_T - ([I_3^-]_F + [I_2]_F) \tag{3.31}$$

The concentration of tri-iodide ion was obtained from the experimental absorbance at 288 nm and the molar extinction coefficient at that wavelength given by Awtrey and Connick (1951), and that of free iodine from the equilibrium constant of the iodine–iodide interaction (see equation 3.22), again using the experimental value of Davies and Gwynne (1952). This calculation showed that a considerable discrepancy existed between the theoretical and experimental quantities of iodine bound by the amylose. Watanabe and his co-workers (1970) therefore concluded that the poly-iodine chain in the complex also contributed to the absorbance at 288 nm (and at 353 nm). To force agreement between theory and experiment, it was necessary to assume that both ends of the poly-iodine chain were terminated by tri-iodide ions. The most probable values of the factors $n$ and $m$ of (3.30) were calculated to be 11 and 2, respectively, so that the complexing species is $I_3^-.(I_2)_9.I_3^-$.

In $10^{-3}$M potassium iodide the spectral evidence is interpreted by Cronan and Schneider (1969) to indicate that the complexing species has the formula $3I_2.2I^-$, whilst according to the treatment of Watanabe, Ogawa and Ono (1970) the same evidence yields the conclusion that the complexing species has the formula $11I_2.2I^-$. The difference between the two is quite profound.

Thompson and Hamori (1969) studied the kinetics of the amylose–iodine interaction using a stopped-flow technique. Their analysis is based on the relation

$$\text{rate} = -d[I_2]/dt = k[\text{amylose}]^a[I^-]^b[I_3^-]^c[I_2]^d \tag{3.32}$$

Assuming the iodine–iodide equilibrium to be much faster than the rate of formation of the blue amylose–iodine complex, (3.22) can be used to eliminate $[I_3^-]$ from (3.32), giving

$$\text{rate} = kK^c[\text{amylose}]^a[I^-]^{b+c}[I_2]^n \tag{3.33}$$

where

$$n = c + d \tag{3.34}$$

On the basis that only the concentration of iodine changes appreciably in the initial stages of the interaction, (3.32) can be integrated to yield

$$t_{0.25} = \frac{(0.75^{1-n} - 1)(1 + K[I^-])^{n-1}[I_2]_i^{1-n}}{kK^c[\text{amylose}]^a[I^-]^{b+c}(n-1)} \tag{3.35}$$

where $t_{0\cdot25}$ is the quarter-life of the reaction, and

$$[I_2]_i = [I_2]_0 + [I_3^-]_0 \tag{3.36}$$

where $[I_2]_0$ is the initial concentration of iodine. The iodide ion is present in large excess, and therefore its concentration does not alter during the experiment; thus a graph of $\log[I_2]_i$ as a function of $\log t_{0\cdot25}$ has a slope of $1-n$. This treatment yielded $n=4$, i.e. the initiation of the poly-iodine chain requires a stable nucleus of a tetramer of iodine molecules and/or tri-iodide ions.

Subsequent work (Thompson and Hamori 1971) failed to identify the component values of $n$ (i.e. $c$ and $d$ of equation 3.34), but suggested four possible forms: $4I_2+3I^-$; $3I_2+2I^-+I_3^-$; $2I_2+I^-+2I_3^-$; and $I_2+3I_3^-$. In each case, the complexing species possesses eleven iodine atoms, three negative charges and four $I_2$ and/or $I_3^-$ groups. The general formula is $4I_2.3I^-$; using the nomenclature of Cronan and Schneider (1969), the value of the parameter $b$ is 0.75. These authors recorded a $b$ value of approximately 0.7, whilst Gilbert and Marriott (1948) obtained $b=0.67$, all the values referring to $10^{-3}$M potassium iodide. It is doubtful if the experimental techniques are in fact sufficiently sensitive to distinguish between these slightly different $b$ values.

French (1973), discussing the results of Hamori and Senior (1973), focussed attention on one very obvious deficiency in the above treatment, namely that the formation of the complex involves three negatively-charged species coming together to form the nucleus—a somewhat unlikely event. The explanation that the negatively-charged species need not all be present simultaneously is far from convincing.

From the above brief review, it is still not clear what part the iodide ion plays in the formation of the complex. Certainly, if iodide ion is not present in aqueous solution the blue complex is not formed. However, in the solid state, if the amylose has the *V*-crystalline conformation, the presence of iodine vapour is sufficient to give the blue colour. Under these circumstances, one would imagine the production of the iodide ion to be very limited in extent. [Note that the *V*-crystalline conformation of the amylose is a necessary prerequisite for complex formation in the solid state: exposure of amorphous amylose to iodine vapour results in only a comparatively small amount of the iodine being bound by the polysaccharide (Katzbeck and Kerr 1950).]

One frequently employed argument in favour of the presence of substantial amounts of iodide ion within the helix is the fact that whilst crystalline amylose binds 26 per cent of its own weight of iodine, the complex in solution takes up only 19.5 per cent of its own weight. If in the aqueous system the sole complexing species were the tri-iodide ion rather than iodine, the measured iodine-binding capacity would be approximately 17.5 per cent; the experimental value of 19.5 per cent can be obtained by assuming that the complexing species has the formula $3I_2.2I^-$. The obvious danger in this argument is that it supposes that the dried solid complex has exactly the same dimensions as does the complex in aqueous solution. The difference in the amount of iodine bound in the crystalline state and in aqueous solution represents only a change from one iodine molecule per six anhydroglucose residues in the former case, to one iodine molecule per eight anhydroglucose residues in the latter. It is recognized that the number of residues comprising one complete helical turn can vary considerably, depending on external constraints, but if the amylose–iodine complex forms a folded-chain type of structure (see section 3.3.3), the monomer units involved in the actual folds may not be able to satisfy the geometrical requirements for complex formation. Hence, although the helical sections could have six anhydroglucose residues per turn, the residues at the fold would be unable to bind iodine; overall, one molecule of iodine would be bound by rather more than six glucose units, the actual number depending on the ratio of the number of glucose units in the helical regions to that in the disordered region. For a molecule of a given degree of polymerization, this ratio will increase as the fold-length of the chain increases; i.e. as this ratio increases, the number of glucose units required to bind one iodine molecule will apparently decrease.

We conclude that there is no definitive evidence for the composition of the poly-(iodine–iodide) chain. Whilst in solution, at least, a small amount of iodide is required for the formation of the blue complex, there must remain some doubt as to whether the ion is incorporated in the helix, or is adsorbed by some other mechanism.

**3.3.2.** Enthalpy of the Interaction.

Gilbert and Marriott (1948) obtained the enthalpy $\Delta H$ of the amylose–iodine interaction from (3.19) and the general relation between the variation of an equilibrium constant with temperature and the heat of reaction

$$\frac{\Delta H}{2.30R} = -\left(\frac{\partial \log K_{xy}}{\partial(1/T)}\right) = -x\left(\frac{\partial \log [I_2]}{\partial(1/T)}\right) \tag{3.37}$$

These authors found $\Delta H = -11.2$ kcal, i.e. the formation of the complex from molecular iodine and iodide ions is accompanied by the emission of 11.2 kcal per mole of iodine bound.

Using an amperometric method and starch (wheat and potato) rather than amylose, Holló and Szejtli (1957) obtained a value of $\Delta H = -17.2$ kcal per mole of iodine bound.

Kuge and Ono (1960b) used (3.29) to define the equilibrium constant $K$ so that

$$K = (\text{complex}]/([\text{amylose}][I_2]^n[I^-]^n) \tag{3.38}$$

or

$$K^{1/n} = [\text{complex}]^{1/n}/([\text{amylose}]^{1/n}[I_2][I^-]) \tag{3.39}$$

The concentrations of the complex and the amylose are regarded as unity, so that (3.39) reduces to

$$K^{1/n} = 1/[I_2][I^-] \tag{3.40}$$

The parameter $K^{1/n}$ is thus available from experimental measurements; its temperature dependence enables the heat of formation to be calculated. In this way, Kuge and Ono (1960b) obtained a value of $-15.5$ kcal per mole of iodine bound at 16°C.

Szejtli, Richter and Augustat (1967b) considered the reaction

$$I_2 + \text{amylose}^* \rightleftharpoons I_2 . \text{amylose}^* \tag{3.41}$$

with an equilibrium constant $K$ defined by

$$K=[I_2.\text{amylose*}]/[I_2][\text{amylose*}] \tag{3.42}$$

where [amylose*] and [$I_2$.amylose*] represent not the concentration of free and complexed amylose respectively, but rather the concentration of iodine-binding sites, i.e. the concentration of helical turns capable of binding iodine, expressed in terms of iodine equivalents. At the point of half-saturation, denoted by the subscript $v$, then $[\text{amylose*}]_v=[I_2.\text{amylose*}]_v$ from which

$$K=1/[I_2]_v \tag{3.43}$$

Relating this equilibrium constant to temperature by the usual exponential relation gave a value for the heat of formation of −9.7 kcal per mole of iodine bound.

The same theoretical treatment was used by Banks, Greenwood and Khan (1971a) on amylose and a series of enzymically-synthesized oligomers with degrees of polymerization of up to 134 glucose units. Values of the parameter $[I_2]_v$ for the different oligomers are given as a function of temperature in table 3.11 together with the derived enthalpies.

*Table* 3.11. Values of the free iodine concentration at half-saturation, $[I]_v$, as a function of temperature and the number-average degree of polymerization, $\overline{DP}_n$; and the derived enthalpies, $\Delta H$ (Banks, Greenwood and Khan 1971a).

| sample[(1)] | $\overline{DP_n}$ | $[I]_v \times 10^6$ | | | | $-\Delta H$ |
|---|---|---|---|---|---|---|
| | | 1.4°C | 7.7°C | 14.2°C | 20.4°C | (kcal/mol) |
| 5 | 50.7 | 2.44 | 3.70 | 5.86 | 9.08 | 11.4 |
| 7 | 93 | 0.90 | 1.61 | 2.61 | 4.40 | 13.4 |
| 8 | 105 | 0.70 | 1.35 | 2.36 | 3.75 | 14.4 |
| 9 | 134 | 0.50 | 0.91 | 1.73 | 2.60 | 14.8 |
| amylose | 1500 | 0.25 | 0.50 | 0.92 | 1.50 | 15.7 |

[1] As in table 3.4.

In the case of amylose, the heat of formation of the complex is −15.7 kcal per mole of iodine bound, a value in reasonable agreement with those derived by Holló and Szejtli (1957) and Kuge and Ono (1960b). Of more interest than the actual value, however, is the fact that the heat of formation of the complex is a function of the degree of polymerization of the amylose, i.e. the amount of heat evolved decreases as the molecular size decreases. This conclusion is apparently a contradiction of the work of Kuge and Ono (1960b), who stated that the heat of formation of the complex was independent of the degree of polymerization of the amylose. In fact, no contradiction exists, for although Kuge and Ono used samples of varying molecular size, the smallest molecule had a number-average degree of polymerization of at least 500 glucose residues. The values for the heats of formation given in table 3.11 suggest that when $\overline{DP}_n$ exceeds 200–500 monomer units, the effect of molecular size will be negligibly small.

All the reported values for the heat of formation of the amylose–iodine complex fall within the range of $-9.7$ to $-17.2$ kcal per mole of iodine bound. The derived values are subject to two major sources of error, namely the experimental difficulty of ensuring that each point on the titration curve represents essentially an equilibrium state, and secondly the choice of a suitable set of equations by which the reaction may be described. Restricting comments to our own results, we found that the approach to equilibrium became very much slower as the degree of polymerization decreased, and the temperature increased; for the smallest oligomer examined in table 3.11, some thirty minutes were allowed for equilibration at each point on the titration curve, but we were aware of a subsequent slow uptake of iodine over a period of many hours. The theoretical treatment we chose to use makes use of the concentration of unbound iodine (i.e. both molecular iodine and tri-iodide in solution) whereas others prefer to use the concentration of free iodine. On this latter basis, the heat of formation of the complex would be approximately $-20$ kcal per mole of iodine bound. We have used total free iodine rather than molecular iodine in the above calculation, because we think that neither the concentration of iodide ion nor that of molecular iodine are the factors controlling the equilibrium of the reaction. Two observations support this statement: (a) the iodine-binding capacity of amylose passes through a maximum on varying the iodide concentration (see table 3.3); (b) although the use of low temperatures causes the amylose to bind iodine at low concentrations of total free iodine (as would be expected of a reaction in which the concentration of tri-iodide was a dominant factor), the amount of iodine bound also increases (see tables 3.4 and 3.6), a fact which is not compatible with the tri-iodide ion being the bound species. The difficulty of deciding which is the correct theoretical treatment to use stems directly from our ignorance of the stoichiometry of the nucleation of the amylose–iodine interaction.

### **3.3.3.** The Amylose–Iodine Complex.

In their development of the concept of the amylose–iodine complex as a helical inclusion structure, Freudenberg, Schaaf, Dumpert and Ploetz (1939) constructed a model of amylose with the constituent glucose units in a boat form, and showed the internal surface of the helix to be hydrocarbon in nature, with the hydroxyl groups and the glycosidic oxygen atoms directed towards the outer surface of the helix. (Using the convention of Reeves (1949), explained in section 4.3.1, this boat form would be described as type B1.) These authors then drew a comparison with the behaviour of iodine in certain hydrocarbon solvents, to give a blue colour similar to that resulting from the amylose–iodine interaction. The similarity is by no means valid, for the blue colour developed by iodine in a non-polar solvent is not so deep as that resulting from the amylose–iodine interaction. Furthermore, the amylose–iodine interaction produces a spectrum of colours ranging from red to blue depending on the degree of polymerization of the polysaccharide, so that only over a relatively narrow range of molecular size does the blue colour of the complex correspond with that developed by iodine in a hydrocarbon solvent. And the molar extinction coefficient of iodine in the amylose–iodine complex is more than an order of magnitude greater than that for iodine in a non-polar environment (Rundle, Foster and Baldwin 1944). Final confirmation that the model was unreal came with the realization that the glucose units adopted a chair conformation (see section 4.3.1), imparting considerable *hydrophilic* character to the inside surface of the helix.

The very large differences between the extinction coefficients of iodine in amylose and in hydrocarbon solvents enabled Rundle, Foster and Baldwin (1944) to deduce that in the former material the iodine possessed dipolar character. Any species within the helix must be subjected to a considerable dipole moment from the surrounding glucose units; in the case of iodine, it was suggested, polarization occurs and an induced dipole moment results. As the number of iodine molecules entering the helix increases, there is a corresponding increase in the value of the induced dipole moment, and in the stability of the complex. This theory explained the phenomenon that had been observed by Bates, French and Rundle (1943), namely that long amylose chains were saturated with iodine before the shorter ones bound any.

The model of dipolar interaction between the iodine molecules and the glucose units constituting the helix was further developed by Stein and Rundle (1948), who calculated the interaction energy as a function of a parameter $x$, and showed that there was a critical value of $x$ at which a 'condensation' of the iodine into a poly-iodine chain would be expected to occur. Resonance of the poly-iodine chain was suggested as an additional stabilizing factor.

However, Murakami (1954, 1955) suggested that the model based on dipolar interaction was inconsistent with the X-ray measurements of West (1947) for the periodicity of iodine in the amylose–iodine complex. West's results confirmed that the iodine atoms were arranged linearly within the helix with an average interatomic spacing of approximately 3.10 Å, whilst the corresponding value for molecular iodine is 2.67 Å (Herzberg 1950). Murakami postulated that an electron transfer took place from the oxygen atoms of the amylose to the iodine molecule. Discussing the resultant structure from the viewpoint of the work of Mulliken (1950) on the intermolecular charge-transfer spectra for complexes of iodine with alcohols and others, this model requires that the glucose units adopt a chair conformation in which one hydroxyl group and the glycosidic oxygen atoms are found on the internal surface of the helix. Application of the free electron model to the species $I_4^{2-}$ (considered by Murakami to be the elemental unit of the poly-iodide chain) gave reasonable agreement between the predicted and experimental absorption spectra, whilst the charge on the complexing species accounted for the rather large interatomic distances in the poly-iodine chain. Again, the complex was supposed to be stabilized by resonance energy.

Greenwood and Rossotti (1958) attempted to establish the existence of iodine–oxygen interactions in the amylose–iodine complex by infra-red spectroscopy. Their results are shown in figure 3.12. After equilibration with water vapour, the infra-red spectra of the complex and the parent polysaccharide are identical (curve (b) of figure 3.12). However, in the case of

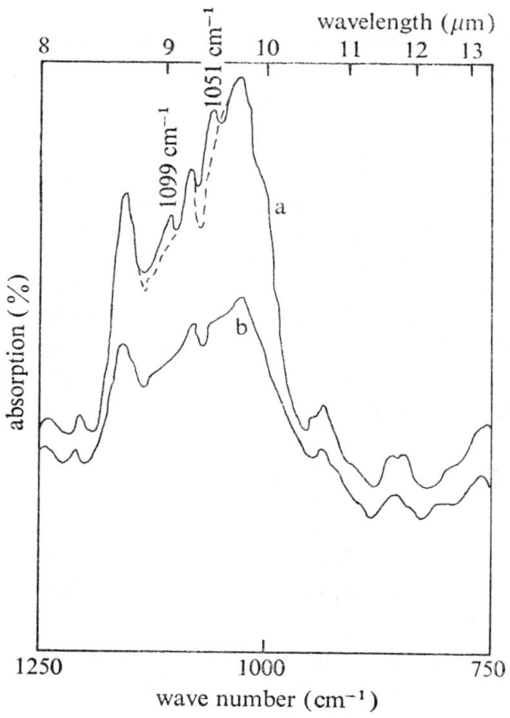


*Figure* 3.12. Infra-red absorption spectra of amylose–iodine complexes (full lines) and of amylose (broken lines), displaced vertically for convenience: (a) Samples dried over phosphorus pentoxide at 80°C *in vacuo*; (b) Samples equilibrated with water vapour—the spectra of the complex and the parent amylose are identical (Greenwood and Rossotti 1958).

amylose dried over phosphorus pentoxide at 80°C *in vacuo*, slight differences are observed between the spectra of the amylose and of the amylose–iodine complex. In particular, the complex shows small peaks at 1051 $cm^{-1}$ and 1099 $cm^{-1}$ that are absent in dried amylose. As peaks in this range were also observed in solutions of iodine in various ethers, and attributed to complex formation between iodine and the ethereal oxygen atom (Haszeldine 1954;

Glusker and Thompson 1955), it was concluded that some interaction did occur between iodine and the oxygen atoms of the constituent glucose units of the helix when the amylose–iodine complex was formed. Similar results were obtained using bromine or iodine cyanide as the complexing agent, i.e. new peaks appeared at 1051 $cm^{-1}$ and at 1099 $cm^{-1}$ in the case of the dried material, but were absent when small amounts of water were present. The ability of traces of water to obscure small changes in the spectra of polysaccharides is, of course, recognized (Cramer 1956). The above observations are consistent with the theory of Murakami (1954, 1955) based on electron transfer.

Holló and Szejtli (1957, 1958b) suggested that the force stabilizing the amylose–iodine complex was hydrogen-bonding between the halogen atoms and the secondary hydroxyl groups at $C_2$ and $C_3$. The suggestion was based on their measured value (−17.2 kcal per mole of iodine bound) for the enthalpy of complex formation in aqueous solution, which corresponds to the heat of formation of twelve weak hydrogen bonds; i.e. they postulated the formation of hydrogen bonds between the iodine molecule and two hydroxyl groups on each of the six glucose residues constituting a single helical turn. Rossotti (1959) tested this hypothesis by examining the infra-red spectra (in the range 1750–750 $cm^{-1}$) of saturated solutions of iodine in various primary and secondary alcohols. She found them to be identical with those of the pure solvents, indicating that hydrogen bonding was not likely to be of importance in the amylose–iodine complex. Further, Rossotti (1959) drew attention to the fact that the most probable conformation of the glucose residues was the C1 chair form (see section 4.3.1), a model in which it is impossible for both hydroxyl groups ($C_2$ and $C_3$) to be found on the inner surface of the helix. Thus if hydrogen bonding were operative in the formation of the helix only six such bonds could be formed, and this would account for only half of the observed heat of reaction.

Bersohn and Isenberg (1961) developed a theory for the structure of the amylose–iodine complex based on the assumption that the tri-iodide ion rather than iodine was the complexing agent. However, our brief review of the literature shows that there is no unanimity regarding the composition of the complexing species, and the fact that iodine vapour will enter the helix of amylose in the crystalline *V*-form to give a blue complex indicates that the iodide ion is not necessary for the development of colour. The fact that this ion is required for the formation of the complex in aqueous solution may suggest that its primary purpose is to bring the polysaccharide chain into a conformation in which iodine can be bound. Nevertheless, on the basis of tri-iodide being the complexing species, Bersohn and Isenberg (1961) presented a model of amylose as a one-dimensional metal (the second assumption built into their model was that the length of the helical complex was that of the extended polymer molecule—an assumption we now know to be incorrect (Bittiger and Husemann 1968)), and predicted that electrical

8

conductivity would occur through the complex. Peticolas (1963) showed that such conductivity could be detected only when the solid complex contained adsorbed water, and this was later substantiated by Paulson (1965), who ascribed the conductivity to an ionic conduction in the wet solid.

Support for the tri-iodide ion as the complexing species in the amylose–iodine interaction came from a study of the benzamide–tri-iodide complex (Robin 1964; Reddy, Knox and Robin 1964). As the electronic spectra of the two types of complex are fairly similar, the crystal structure of the benzamide–tri-iodide complex may be used as a model for the solid amylose–iodine complex. Accordingly, the latter complex may be regarded as a linear chain of tri-iodide ions, each of which maintains its individual identity. The internal distance of the atoms in the tri-iodide ion is 2.92Å, and the separation of adjacent ions is 3.80Å, so that the average separation of iodine atoms in the complex is 3.10Å, in agreement with the X-ray measurements of West (1947) for the amylose–iodine complex. The van der Waals distance for tri-iodide ions is 4.6Å; their closer approach within the helix requires that the electrostatic repulsion between the negatively-charged species be overcome, and it is suggested that close-range dispersion forces provide the necessary stabilizing influence. The chromophoric group in the benzamide–tri-iodide complex is, according to Robin (1964), an infinite tri-iodide chain in which the individual tri-iodide ions are strongly coupled, the excitation energy resonating between them. In the amylose–iodine complex, however, the poly-iodine chain need not be very long (see below); whether the length would then be sufficient to give a stable blue-coloured complex, if tri-iodide were the complexing species, is questionable.

Cronan and Schneider (1969) concluded that the composition of the complexing species was $I_2.I_b^-$, where $b$ may take any value from zero to one depending on the concentration of iodide ion, and they postulate that each helical turn of amylose is capable of binding quite strongly either an iodine molecule or a tri-iodide ion. In fact, their calculations show that the degree of cooperativity required to form the amylose–iodine complex is only moderate, and it is the strong intrinsic binding of the complexing species to each helical turn that is responsible for the stability of the chain. This conclusion is somewhat surprising for, from the earliest work of Bates, French and Rundle (1943), successive investigations have shown, at least in quantitative terms, that the amylose–iodine interaction involves a high degree of co-operativity. Rundle, Foster and Baldwin (1944) demonstrated that if the complex was precipitated by the addition of potassium iodide to an amylose solution containing half the amount of iodine required to saturate the polysaccharide, then *all* the iodine was found in the portion of amylose precipitated, and the polysaccharide fraction remaining in solution (50 per cent of the total) was so devoid of iodine as to be colourless. Whilst we have criticized the use of iodine for fractionating amylose, we accept that there is a relation between degree of polymerization and the concentration of iodine at which binding takes place—a phenomenon to be expected from the original observation of Rundle, Foster and Baldwin (1944).

Cronan and Schneider (1969) also claim a gradual increase in the chain length of bound species as the extent of interaction increases from 0.1 to 1.0. At the former level they envisage a chain length of two to four units, rising to a value of fifteen at saturation. Thus, they conclude that there is an appreciable probability of finding short chain-lengths of the complexing species even at relatively advanced stages of the interaction. In view of the fact that the wavelength of maximum absorbance changes hardly at all as the amylose–iodine interaction proceeds, it is astonishing that such a claim could be made. The results of Cronan and Schneider (1969) so differ from other studies that it seems pertinent to inquire into the reason for the discrepancy. Fortunately, they provide detailed values of a titration carried out in $10^{-2}$M potassium iodide at 20°C, i.e. the conditions under which we use the semi-micro differential potentiometric technique. We have therefore graphed their values in the manner we normally employ for our own data, and show the result in figure 3.13. For comparison, we show a graph obtained by ourselves for a

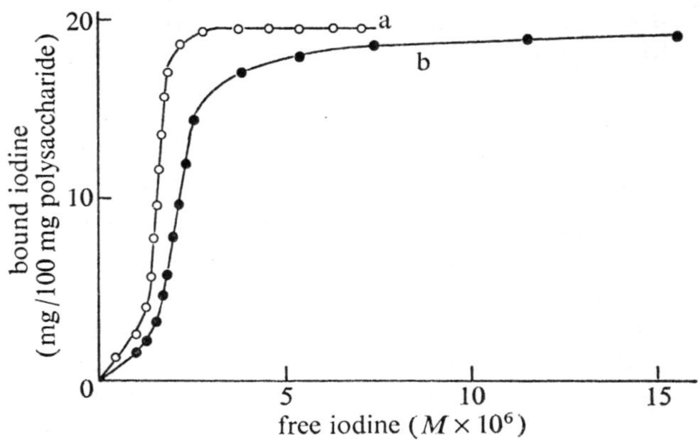


*Figure* 3.13. Iodine titration curves: (a) amylose, determined in our own laboratory; (b) amylose, curve calculated from data given by Cronan and Schneider (1969).

linear amylose ($\overline{DP}_n \sim 1500$ glucose units), and the origins of the unique conclusions of Cronan and Schneider (1969) become apparent. The concentration of free iodine (i.e. molecular iodine plus tri-iodide) at which binding occurs indicates that the *average* degree of polymerization of the material they use is probably only 200–300 anhydroglucose residues; the continued slow uptake of iodine at concentrations of free iodine greater than $3 \times 10^{-6}$M indicates the presence of short-chain material ($50 < \overline{DP}_n < 100$). Cronan and Schneider (1969) chose to use a commercial amylose sample, reputed to have a molecular weight of 150 000 (whether this is the number- or weight-average is unspecified), and dissolved a fraction of it by stirring the solid in water at room temperature for 24 hours. Such conditions must favour the preferential dissolution of the material at the low end of the molecular-weight distribution. We have already drawn attention to the anomalous amylose–iodine interactions encountered at low degrees of polymerization (see section 3.2.8), and suggest that Cronan and Schneider (1969) have unwittingly involved themselves with a range of molecular weights in which factors other than those they treat from a theoretical standpoint influence the stability of the complex.

The amylose–iodine interaction has also been studied using optical rotatory dispersion and circular dichroism. Using the former technique, Wolf and Schulz (1967) established that the interaction of amylose with iodine caused the appearance of a positive induced Cotton effect, the wavelength of maximum rotation being 610 nm whilst that for maximum absorbance was 600 nm, i.e. the Cotton effect was moved to a slightly higher wavelength than the absorption peak. The optical rotatory dispersion curve for amylose in aqueous solution shows no peaks, the value of the optical rotation decreasing smoothly from a value of $+1500°$ at 250 nm to approximately $+200°$ at 300 nm (Neely 1961; Kuge and Ono 1961); in the presence of iodine and iodide, the optical rotation of amylose was found to be $+2000°$ at 610 nm (Wolf and Schulz 1967). The same authors (Wolf and Schulz 1968) examined the optical rotatory dispersion curves for starch–iodine solutions rather than amylose–iodine solutions for technical reasons, and established two other facts: that the optical rotation of the Cotton peak reached a limiting value when the molar ratio of iodine to anhydroglucose was approximately 1:6, and that the value of the optical rotation (but not the absorbance) at the Cotton peak was grossly dependent on time, increasing by a factor of 2.5 over forty hours. This latter effect was quite reversible in that, after heating the solution for two minutes at 70°C to destroy the complex, the optical rotation returned to approximately its initial value, and subsequently increased progressively again on storage at 25°C.

A detailed study of optical rotatory dispersion and circular dichroism of members of the amylose oligomer series in the presence of iodine–iodide was carried out by Pfannemüller, Mayerhöffer and Schulz (1969). The wavelength at which the complex displayed the maximum rotation increased with increasing degree of polymerization, the values in each case being displaced to wavelengths slightly higher than these obtained by measuring the wavelength of maximum absorbance. Also, instead of the single Cotton effect obtained by previous workers, two other minor peaks were noted at 350–360 nm and 485 nm. In both cases, the appearance of these peaks was independent of the degree of polymerization of the amylose (both peaks are obvious in absorption spectra, too; see figure 3.11). The positive maximum of the circular dichroism curve was found to be strongly dependent on the degree of polymerization, the wavelength being in all cases slightly lower than that of maximum absorbance. The relation between the degree of polymerization and the wavelength of the maximum of the circular dichroism curve was governed by (3.13); in fact, the value of the ratio $B/A$ was the same for the circular dichroism and absorbance measurements. In each technique (absorption, optical rotatory dispersion and circular dichroism) the wavelength at which the maximum effect is achieved changes smoothly with the degree of polymerization, which indicates strongly that different aspects of the same phenomenon are being examined—namely the helical amylose–iodine inclusion complex. Because of the existence of a multiple Cotton effect, Pfannemüller, Mayerhöffer and Schulz (1969) reject the theoretical model devised by Cramer (1952b) in favour of that of Robin (1964). They suggest that the maximum at 485 nm may be due to an interaction between iodine and the hydroxyl groups of the amylose, since it is also observed in solutions of iodine in water and ethanol (the appearance of a Cotton effect shows that the absorbance is related to the presence of the helical structure). Against this postulate must be set the fact that infra-red studies (Rossotti 1959) failed to detect any hydrogen bonding between iodine and alcohols. Similarly, the peak at 350–360 nm, usually ascribed to the presence of the tri-iodide ion, must represent a property of the helical complex, but whether that property is present on the surface of the complex, or is an integral part of the complexing species, cannot be decided.

In their studies, Pfannemüller and her co-workers (1969) found two parameters that were time-dependent, namely the rotation of the complex and the coefficient of dichroic absorption $\Delta\varepsilon$, defined by

$$\Delta\varepsilon = \Delta D / cl \tag{3.44}$$

where $\Delta D$ is the observed value of the difference in absorbance between left and right circularly-polarized light, $c$ is the concentration of the monomer unit in moles per litre, and $l$ is the path length in cm. Both parameters increased as the complex aged, and for this reason they were measured ten minutes after adding iodine to the aqueous solution of the polysaccharide. Relating values of the molar rotation per residue, and the coefficient of dichroic absorption obtained in this somewhat arbitrary manner, to the degree of polymerization showed that both parameters passed through a maximum when the amylose contained 70–90 residues. Furthermore, in the same region, the relation of the molar extinction coefficient to molecular weight showed a dramatic change, decreasing very rapidly as the degree of polymerization decreased but increasing only comparatively slowly at higher molecular weights. Two alternative explanations for the time-dependent phenomena were suggested, namely (a) some type of restructuring of the complex involving a partial orientation, or (b) a secondary process in which disordered portions of the complex gradually take up iodine. It is difficult to envisage the latter explanation being applicable to the range of degrees of polymerization below 100 units, in which the authors agree that only one helical segment is present and yet which show abnormally high values for both the molar rotation per glucose residue and the coefficient of dichroic absorption, even after ten minutes. Since the authors also report that in the range $65 < \overline{DP}_n < 100$ association effects are improbable on the basis that no visible turbidity develops even after 140 hours, it is unlikely that any type of restructuring can occur either.

The interpretation of these results appears to be influenced by the model of amylose developed by Szejtli and his co-workers (Szejtli and Augustat 1966; Szejtli, Richter and Augustat 1967a). According to their concept, amylose exists in aqueous solution as a number of helical segments separated from each other by regions deficient in order (this model is considered in detail in section 4.3.3). The length of the helical segments when conditions (concentration of iodide ion, pH, and temperature) are held constant is then a function only of the degree of polymerization of the macromolecule. For a degree of polymerization less than approximately 120 glucose units, only a single segment exists; as the degree of polymerization increases, the number of helical segments and their lengths also increase. Their conclusions are summarized in table 3.12. According to Szejtli and Augustat (1966), the

*Table* 3.12. The helical segment-length as a function of the degree of polymerization of the amylose (after Szejtli and Augustat 1966; Szejtli, Richter and Augustat 1967a).

| $\overline{DP}_n$ | glucose residues per helical segment | helical turns per segment |
|---|---|---|
| 60 | 60 | 7 |
| 100 | 100 | 12 |
| 360 | 115 | 14 |
| 1200 | 122 | 15 |

blue amylose–iodine complex is formed in aqueous solution without any detectable change in the molecular shape of the polysaccharide, which must therefore exist as a helix prior to the addition of the complexing agent. However, as the hydrodynamic behaviour of the macromolecule as a whole conforms to that of a random coil (see section 4.2.1), it is necessary to invoke intervening non-ordered regions which act as hinge points between the rigid helical sections, conferring flexibility on the macromolecule. Accepting that the largest amylose molecules are preferentially saturated with iodine requires that the length of the helical segments increase with increasing degree of polymerization; the molecular size at which only a single helical segment can exist is taken to be the point at which there is a discontinuity in the graph of maximum absorbance wavelength of the amylose–iodine complex as a function of degree of polymerization. At first sight, this apparently constitutes a reasonable model. However, as we have documented in some detail in section 4.3.7, it is based on a mistaken observation, for the viscosity number of amylose (a measure of the molecular shape of the polysaccharide) does *not* remain constant as the complex is formed. The model is not valid, in fact, and cannot be used as a basis for reasoning.

A great deal of attention has been focussed on the length of the amylose chain required to produce a stable blue-coloured complex. For this purpose, synthetic amylose oligomers or degraded amylose fractions have been employed, and efforts made to detect the point in the molecular-size–iodine-stain relation at which a discontinuity appears: a degree of polymerization in the range of 70–120 glucose residues has been suggested. This approach appears to us to be a gross over-simplification, in that it assumes the amylose–iodine interaction to constitute the sole stabilizing force for the complex. Theories to date have attempted to treat the amylose–iodine complex either as a single linear array of iodine molecules (i.e. the complex may be regarded as a rod), or as a number of independent helical segments separated by regions of disorder. We suggest that both of these approaches are wrong, the former because such a model is unlikely on energetic considerations, and the latter because no consideration has been given to the temperature-dependence of the reaction with iodine of amylose oligomers of $\overline{DP}_n < 150$ glucose units.

The results in table 3.7 are of quite profound significance in this respect, as they show that the temperature-dependence of the amylose–iodine interaction depends grossly on the degree of polymerization of the polysaccharide at values of $\overline{DP}_n < 150$ glucose units. We have already remarked on the fact that, at these low degrees of polymerization, the wavelength of maximum absorbance of the complex moves towards higher values as the temperature is decreased. Although technical difficulties prevented an accurate assessment of the change in wavelength, the iodine-binding capacities recorded in table 3.6 leave no doubt that sample 4, having a $\lambda_{max}$ of 546 nm at 20°C, should exhibit a wavelength of maximum absorbance in excess of 600 nm at 1.4°C. We base this assessment on two factors, namely: (a) the measured value of the iodine-binding capacity is 86.5 per cent of that obtained for amylose at that temperature and almost the same as the value recorded for amylose at 20.4°C (see table 3.4); and (b) it is most unlikely that the glucose residues at chain ends can enter into iodine binding, and, by allowing a value of five residues per chain for this purpose, samples 4 to 9 exhibit an iodine-binding capacity at 1.4°C that is virtually identical to that of amylose at the same temperature. If, therefore, we were to look for some discontinuity in the behaviour of the iodine complex at low temperature, it would apparently occur at a degree of polymerization of approximately 30 glucose units. On this basis, the helical segment length in amylose would be deduced to be 30 residues; i.e. approximately 25 per cent of the value postulated for room temperature. Because the spectral characteristics of high molecular weight amylose do not show much variation with temperature, one would then conclude that a segment length of 30 residues accounted for the known properties of the complex. Assuming that the helical segment length is 30 residues at 1.4°C and 120 at 20.4°C, makes it difficult to explain why the former type of segment binds iodine at a concentration of free iodine some six times lower than does the latter (see table 3.11).

We suggest that these ambiguities are eliminated if the helical regions are regarded not as separate entities, but rather as interacting ones, and, furthermore, that this interaction constitutes a stabilizing force. The segmental interaction that we envisage is crystalline in nature, and specifically arises because the helical molecule folds back on itself, forming ultimately a molecular crystallite. We have noted above that the wavelength of maximum absorbance of amylose is virtually independent of temperature over a wide range, and therefore suggest that there is no great variation in the fold length of the amylose–iodine complex with temperature. Accepting this argument, the value of 30 glucose units that we find necessary to give the theoretical binding of iodine by internal residues must represent a *maximum* value for the fold length; it is quite probable that if measurements could be made at lower temperatures, the degree of polymerization necessary for binding of iodine by the internal residues would be further decreased. Also, the fact that the iodine-binding capacity of amylose increases with decreasing temperature (see table 3.7) suggests that a higher proportion of glucose units are found in the helical form at low temperature, i.e. the fold length is the greater the lower the temperature.

In summary, we suggest that the length of the resonating species giving the characteristic blue colour is probably less than 40Å; the resonating species is stable only if the helical segment in which it is incorporated is part of a crystalline array. The stability of the crystallite will increase with the number of folds, so amylose molecules of highest molecular weight are complexed preferentially and the reaction goes to completion with one molecule before others are started. Such a model explains why certain parameters of the amylose–iodine interaction are independent of time, whilst others are not. We envisage that the former group would constitute those related to the length of the resonating species, which is trapped within the fold length and therefore constant, whilst those parameters varying with time would be measures of the overall order of the crystalline assembly (relaxation processes would be expected to increase the three-dimensional order of the microcrystallite).

On this hypothesis, low temperatures enable aggregation to occur, so stabilizing the shorter chains. Because no visible aggregates are formed even after a number of days (Pfannemüller, Mayerhöffer and Schulz 1969), some factor must limit the size of such aggregates. It is possible that the stability of the aggregate is governed by an ionic charge, and that a charge/volume ratio is the size-limiting factor. In view of the fact that the amylose–iodine complex may be salted from solution, the colloidal-type crystallite that we suggest is not improbable.

According to our model, one cannot envisage the purple-staining complexes as being anything other than a mixture of red- and blue-coloured species, as was indeed suggested by Mould and Synge (1954). In a population of amylose molecules of a given size, a certain fraction will in the presence of iodine be found in the crystalline form, and will therefore give a blue-coloured solution; the remainder will exist in a form having no order, and therefore absorb at a much lower wavelength. The dynamic equilibrium set up between the two species will depend on factors such as concentration of free iodine, degree of polymerization, concentration of iodide ion, and temperature.

We would interpret the results of Bittiger and Husemann (1968) to be consistent with this hypothesis. These authors showed electron micrographs of amylose–iodine complexes in which the degree of polymerization of the macromolecule ranged from 1950–11 500 glucose units. In each case, rods were obtained, some 40Å in diameter and with lengths varying from 530Å to 2720Å. It should be noted that whilst accepting that the chains had to be folded, the authors rejected the concept that the fold length could lie along the 40Å dimension, but, they suggested, had to lie at right angles to it. The former explanation is rejected because, they say, a minimum length of 100Å is required for the ordered helix necessary to give the required absorbance characteristics—a value again based on studies using model compounds. Electron micrographs of the amylose–iodine complex precipitated from relatively concentrated solutions are also available (Bittiger and Husemann 1968). These show the appearance of fibrils, which are indicative of the rods having associated end-to-end rather than laterally. Since, according to their model, the axes of helices lie parallel to the major dimension of the rod, the association of the rods must occur by interactions between the least ordered segments at the actual chain folds. One would expect an association to occur between the best-ordered surfaces—in that case, the formation of fibrils could be explained if the axes of the helices were parallel to the short dimension of the rod.

It should also be noted that single crystals have been obtained of the butan-1-ol complex (Manley 1964; Yamashita and Hirai 1965; Bittiger and Husemann 1968). Although very much larger than the amylose–iodine crystals, their thickness (70–80Å) is not very different from that of the iodine complex. In most polymers, the molecular axis or helix axis is found to lie along the short dimension of the crystal. Thus we would expect *a priori* that the helical chains in the amylose–iodine chain lie parallel to the short dimension, as we have suggested.

From the work carried out on the nature of the complexing species, there is no reason to suppose that the helical segments need be very long in order to accommodate the chromophore. According to Cramer's calculations, fourteen iodine atoms (*ca.* 40Å) should give a complex having its absorption peak at 620 nm, whilst most attempts to elucidate the stoichiometry of the reaction are agreed that a fairly small species is involved.

Our proposed model does not agree with the flow dichroism measurements of Rundle and Baldwin (1943) who showed that the major absorption occurred in a molecular axis parallel to the flow lines. These authors deduced that the linear array of iodine atoms responsible for the blue colour also lay parallel to the flow lines. In our model, the resonating poly-iodine chain would lie at right angles to the flow lines, and constitute the short axis of the microcrystal; the major axis, lying parallel to the flow lines, would contain a considerably longer array of regularly-ordered, but unconnected, iodine atoms. It is not known whether such a model is compatible with the measurements of flow dichroism.

The nature of the iodine species constituting the chromophore group remains obscure. We would comment only on the proposal that the stoichiometry is very dependent on the stage of the interaction or the iodide concentration. In such a case, one might expect the molar coefficient of absorption, or the wavelength of maximum absorption, to change. In fact, both are remarkably insensitive to the stage of the interaction, and only at very high concentrations of iodide ion is the wavelength of maximum absorption markedly moved to lower values.

### 3.4. The Reaction of Iodine with Amylopectin and Glycogen.

Both these branched glucans stain with iodine, the interaction with glycogen producing a wavelength of maximum absorbance in the region 430–490 nm, whilst for amylopectin the corresponding value occurs in the region 530–540 nm (Manners and Wright 1960; Archibald, Fleming, Liddle, Manners, Mercer and Wright 1961). Under normal potentiometric titration conditions the iodine-binding capacity of glycogen is zero, and that of amylopectin is very small. However, by increasing the concentration of free iodine, potato amylopectin will take up more than 19.5 mg iodine/100 mg polysaccharide (Higginbotham 1949b). Again at low temperatures the uptake of iodine by the branched material is increased (e.g. the iodine-binding capacity of waxy-maize amylopectin is 0.10 mg/100 mg polysaccharide at 20°C, and 0.15 mg/100 mg polysaccharide at 1.5°C).

Anomalous amylopectin fractions, containing abnormally long chains, give appreciable iodine-binding capacities, and this effect can be demonstrated by synthesis of a series of amylopectin samples differing in the length of outer chain by using phosphorylase to extend the chains of waxy-maize amylopectin (Banks, Greenwood and Khan 1970a). The properties of such samples are shown in table 3.13. As the external chain-length increases, the $\beta$-amylolysis limit and the wavelength of maximum absorbance of the iodine complex increase; *beta*-amylase reduces each sample to the same high molecular weight residue, characterized by a $\lambda_{max}$ of 525 nm. The most notable aspect of the iodine-binding capacities is the profound increase at low temperatures. However, in certain respects the iodine-binding curves were anomalous. Figure 3.14(a) shows the iodine-binding curves for the experiments carried out at 1.5°C. The curve for sample Ap1 is decidedly

*Table* 3.13. The properties of waxy-maize amylopectin, and of synthetic amylopectin with long external chains prepared from it (Banks, Greenwood and Khan 1970a).

| sample | waxy maize amylopectin Ap0 | synthetic amylopectin Ap1 | Ap2 | Ap3 | Ap4 |
|---|---|---|---|---|---|
| $\lambda_{max}$ (nm) | 529 | 544 | 556 | 575 | 587 |
| $\beta$-amylolysis limit (%) | 58 | 71.5 | 75 | 85 | 93 |
| chain length | 19.7 | 29.2 | 36.0 | 52 | 115 |
| external chain length | 13.7 | 23.4 | 29.9 | 46.5 | 109.5 |
| internal chain length | 6.0 | 5.8 | 6.1 | 5.5 | 5.5 |
| IBC at 20°C (%) | 0.10 | 0.3 | 7.1 | 10.0 | 14.3 |
| IBC at 1.5°C (%) | 0.15 | 7.3 | 9.6 | 16.1 | 17.1 |
| $\lambda_{max}$ of $\beta$-amylolysis limit dextrin (nm) | 525 | 525 | 525 | 525 | 525 |

anomalous, in that it crosses the curve for sample Ap2, and also because the slope of the 'linear' portion between $(5\text{–}10) \times 10^{-6}$M iodine is much larger than could reasonably be expected. In fact, this polysaccharide sample continued to take up iodine as the concentration of free iodine was further increased, and at a free-iodine concentration of $25 \times 10^{-6}$M had bound 20 mg iodine/100 mg polysaccharide without reaching the limiting value. Similar remarks apply to the measurements carried out at 20°C (see figure 3.14(b)), but at this temperature it is sample Ap3 that is most noticeably abnormal in its iodine titration behaviour.

Thus, at both temperatures, there is a chain length at which decidedly odd behaviour is observed, in that the increase in iodine bound with successive additions of iodine is so large. Mould (1954) noted a similar phenomenon in

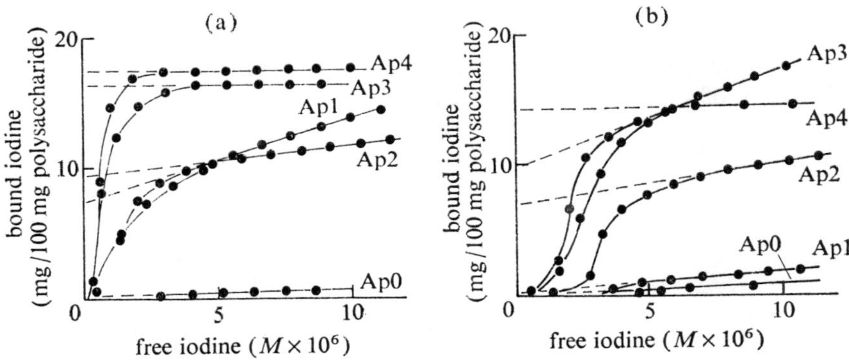


*Figure* 3.14. Iodine titration curves for amylopectin samples at (a) 1.5°C and (b) 20°C; sample numbers are shown, from which the characteristics of the different amylopectins may be obtained using table 3.13 (Banks, Greenwood and Khan, 1970a).

his study of degraded amylose fractions, and explained it on the basis of a suggestion by Dvonch, Yearian and Whistler (1950) to the effect that surface adsorption of the iodine molecules on the polysaccharide was occurring. If this explanation is correct, then surface adsorption is apparently limited to a small range of molecular sizes, the actual value of which is dependent on the temperature, and which are apparently just insufficiently long to give stable helices. Why such a structure should favour surface adsorption is not clear.

To summarize, it is apparent that when the external chain-length is less than about 15 glucose units, virtually no iodine is bound at either low or high temperature. When the external chain-length reaches 20–25 glucose units there is still little iodine bound at 20°C, but an appreciable amount is bound at 1.5°C, although there is evidence of some kind of surface effect as well as helical formation. At an external chain-length of 30 glucose residues and greater, iodine is bound in considerable amounts at both 1.5°C and 20°C.

In these experiments only the external chain-length was varied, the core of the branched macromolecule being the same in all cases. However, as the branch points may be regarded as a source of interference in the formation of the regular structure necessary to yield a stable complex, it is highly probable that changes in degree of branching involving variation of both outer and inner chain-lengths would also be reflected in changes in iodine-binding capacity.

Glycogen does not bind significant quantities of iodine in the range of free-iodine concentrations available for use in the semi-micro differential potentiometric techniques. An attempt to prepare model compounds of fixed internal structure, but having variable external chain-lengths, succeeded only in lengthening a small proportion of chain ends by a large amount (Banks, Greenwood and Khan 1970b).

The only interest of the reaction of iodine with glycogen and amylopectin is that it may provide an insight into the mechanism by which iodine can be taken up by the polysaccharides in a manner other than that involving the formation of a stable helical structure.

# 4. The Conformation of Amylose in Dilute Solution

## 4.1. Introduction.

In discussing the reaction of amylose with iodine (section 3.1), we noted that the complex presented the first instance in which a *helical structure* was postulated and proved, both in the solid state and in solution, for a biological macromolecule. It is often implied, however, that this imposing work of Rundle and his collaborators showed amylose itself to be helical in solution, rather than just the amylose–iodine complex. This implication was specifically rejected by Rundle and Baldwin (1943) in the following words: 'The authors wish to point out again that the proposed helical configuration of the starch chain in the starch–iodine complex does not imply a helical configuration for the chain in granular or retrograded starch. . . . Neither would the helical configuration of the starch chains in the iodine complex require that starch in solution have the helical configuration.'

The helical conformation in the solid state is, of course, by no means restricted to biological macromolecules. Stereoregular synthetic polymers crystallize in either an extended zig-zag or helical form (mathematically, the former is merely a degenerate helix), and the helix is favoured by those that are isotactic. Amylose is a highly stereoregular polymer—the 1:4-glucosidic linkage gives the isotactic polymer when the configuration at $C_1$ is in the $\alpha$-form (amylose), and the syndiotactic polymer when this configuration is in the $\beta$-form (cellulose).

Whilst both biological macromolecules and synthetic polymers may adopt helical conformations in the solid state, their behaviour in solution differs in a most astonishing fashion. Thus synthetic polypeptides, proteins, and DNA can, by the choice of a suitable solvent, be induced to retain their ordered conformations in the molecularly-dispersed state. No such phenomenon has been observed in the older field of classical polymer chemistry. It is therefore of interest to establish the conformation of amylose in solution, with the object of finding whether it is similar in behaviour to macromolecules such as proteins and DNA in that the molecular conformation of the crystalline state is retained in solution, or whether its behaviour is more akin to that of the synthetic polymers, which in solution (and in the amorphous state) adopt the random coil conformation: i.e. we must identify whether the conformation in solution is ordered or random.

The ordered conformation in the solid state involves co-operative interactions both between chains and within chains, but only the latter are of importance in solution. The term 'co-operative interaction' is applied to the forces which stabilize, for example, the helical structures of polypeptides and DNA. These forces are thought to depend mainly on hydrogen- and hydrophobic-bonding. Each of these secondary bonds supplies individually only a small amount of stabilization energy to the macromolecule, but the co-operative effect of many such bonds is to stabilize the conformation. Thus the shape of the polymer is fixed by the co-operation of small, localized attractive forces. As a result of the co-operative nature of these attractive forces, very small regions of helix (one or two turns) are generally unstable, whereas the longer the helical region becomes, the more stable it is. This explanation accounts for the very sharp transitions from ordered to disordered states observed on heating, or on changing the pH of solutions of biopolymers.

The random coil conformation, on the other hand, depends on the flexibility of the macromolecule. There is usually some degree of free rotation about each of the inter-unit bonds in a polymer, enabling the chain to adopt an infinite number of conformations. As a result of thermal energy there is constant rotation about bonds, and consequently the polymer molecule takes on a vast number of shapes in a given time interval, ranging from the fully-extended to the highly-coiled. The internal energy of the molecule does not change as the shape alters, but the entropy does. If the ordered state is maintained in solution, only a single fixed conformation is allowed, whereas adoption of the random coil conformation allows the polymer molecule to explore myriads of shapes. Thus there is a large gain in entropy in going from the ordered to the random conformation, and this is the factor that offsets the attractive (stabilization) forces favouring a single fixed conformation.

To summarize, we would expect the amylose macromolecule in solution to retain the helical (ordered) form typical of the crystalline state (in which state it presumably attains its minimum free energy) if the energy barrier to rotation about the glycosidic bonds is sufficiently high; i.e. if the only reasonably accessible energy minimum of the non-bonded interactions occurs when the molecule is a helix. On the other hand, if the energy barrier to rotation is fairly low, the entropy gain resulting from the increased number of accessible conformation states will cause the dissolved amylose molecule to adopt the form of a random coil.

The experimental determination of the shape of the amylose molecule in solution involves light-scattering, viscometric, and ultracentrifugal techniques. However, the interpretation of these experimental results requires some knowledge of the classical theories of polymer conformation and hydrodynamic behaviour. We therefore begin by providing a brief description of the necessary theoretical background. A much fuller account may be obtained from Flory (1953) and Tanford (1961).

## 4.2. The Linear Polymer Molecule in Solution.

### 4.2.1. Mathematical Basis of the Model.

The mathematical basis of analysing polymer behaviour in solution is the theory of the random, or drunkard's, walk. If $n$ steps, each of length $l$, are taken so that the direction of any step is independent of that of its predecessors, the mean-square end-to-end distance (the displacement length)

$\langle r_0^2 \rangle$* between the starting point and the finishing point is given by

$$\langle r_0^2 \rangle = nl^2 \tag{4.1}$$

Equation 4.1 yields a value for the mean-square displacement length that is averaged over all possible conformations of the $n$ steps. Translating this mathematical concept into the physically real terms of the polymer molecule, we may regard $n$ as the degree of polymerization and $l$ as the length of each monomer unit; the averaging procedure is carried out by the macromolecule itself, in that rotations about inter-residue bonds enable it, in a finite time, to explore a vast number of conformations.

Only comparatively recently has the direct determination of $\langle r_0^2 \rangle$ become experimentally feasible. Usually, it has been derived from the light-scattering radius of gyration $R_{G0}$ (see section 4.2.2). For the *freely-jointed chain model*

$$\langle R_{G0}^2 \rangle / nl^2 = (n+2)/6(n+1) \tag{4.2}$$

In this case, $\langle R_{G0}^2 \rangle / nl^2$ is dependent on $n$, whereas in the corresponding relation for the mean-square end-to-end distance, $\langle r_0^2 \rangle / nl^2$ is independent of $n$ (see equation 4.1). However, as $n$ increases, the ratio $\langle R_{G0}^2 \rangle / nl^2$ decreases, rapidly reaching the limiting relation

$$\langle R_{G0}^2 \rangle / nl^2 = 1/6 \tag{4.3}$$

as $n$ approaches infinity. Thus, in the limit of large $n$

$$\langle R_{G0}^2 \rangle = \langle r_0^2 \rangle / 6 \tag{4.4}$$

a relation first derived by Debye (1946) in his study of the light-scattering technique. The use of (4.4) enables $\langle r_0^2 \rangle$ to be obtained experimentally; we should, however, bear in mind that we have introduced the simplifying assumption of a large degree of polymerization in obtaining this relation.

Comparison of this model with the real polymer molecule shows that the model is deficient in that it assumes the direction of any step is independent of that of its predecessors, a state not applicable to the molecule because of the constraints introduced by fixed bond angles. This difference might at first sight suggest that the model would have to be abandoned. However, Kuhn (1936, 1939) pointed out that the correlation between bonds consequent upon the imposition of fixed bond angles would disappear as the distance along the chain between the bonds increased, assuming that there was free rotation about the bonds to introduce a randomizing influence. Kuhn therefore approximated the real polymer chain by an equivalent *freely-jointed chain* in which each statistical element consists of a sequence of $n_k$ bonds instead of an individual bond. By choosing an appropriate value for $n_k$, the direction in space of successive statistical elements can then be taken as

* Angled brackets around a parameter are used to denote an average value. A bar over a parameter may also be used.

random. The number of bonds in a sequence depends on the geometry of the chain, and the hypothetical model is assumed to have the same projected contour length $r_{max}$ as the real polymer chain. Thus the polymer may be regarded as a random coil composed of $n/n_k$ segments (referred to as Kuhn statistical segments), each of length $A_m$, so that

$$r_{max}=(n/n_k)A_m \tag{4.5}$$

and

$$\langle r_0^2\rangle=(n/n_k)A_m^2 \tag{4.6}$$

For the freely-jointed chain,

$$r_{max}=nb_0 \tag{4.7}$$

where $b_0$ is the projected bond length. Substitution of (4.5) and (4.6) in (4.7) gives

$$\langle r_0^2\rangle=nb_0A_m \tag{4.8}$$

In going from the model of the freely-jointed chain to that of the freely-rotating chain, certain constraints have been imposed. For the former model, the characteristic ratio $C_n$, defined by

$$C_n=\langle r_0^2\rangle/nl^2 \tag{4.9}$$

assumes a value of unity for all values of $n$, whereas in the case of the freely-rotating chain the characteristic ratio reaches a constant value only in the asymptotic limit of large $n$, when

$$C_\infty=(1+\cos\theta)/(1-\cos\theta) \tag{4.10}$$

where $\theta$ is the angle between the positive directions of successive bonds.

Combination of (4.9) and (4.10) gives the general relation

$$\langle r_0^2\rangle=nl^2(1+\cos\theta)/(1-\cos\theta) \tag{4.11}$$

which is valid for any freely-rotating chain of sufficient length. Inherent in the derivation of (4.11), however, is the supposition that the value of $\cos\theta$ does not approach unity.

The freely-jointed chain is itself an approximation in that it assumes the angle of rotation about the inter-unit bond to be 360°, a situation not encountered in practice because of energy barriers. It is, however, intuitively obvious that the effect of restricting the rotation about individual bonds will be merely to increase the number of bonds in a given statistical segment, i.e. the Kuhn statistical segment length $A_m$ will increase as a result of hindered rotation, but the proportionality between $\langle r_0^2\rangle$ and $n$ will be retained. Oka (1942) and Benoit (1947) have shown that for a model in which the bond angle is fixed and the rotation about the inter-unit bond is hindered, there exists the relation

$$C_\infty = [(1+\cos\theta)/(1-\cos\theta)] \times [(1+\langle\cos\phi\rangle)/(1-\langle\cos\phi\rangle)] \quad (4.12)$$

where $\theta$ is defined in (4.10) and $\langle\cos\phi\rangle$ is the average value of $\cos\phi$, the angle $\phi$ being that through which rotation is permitted. The derivation of (4.12) assumes that the hindrance is of the simplest type, namely one that acts symmetrically about $\phi=0$ so that the repulsion for angle $\phi$ should be identical to that for $-\phi$.

Combining (4.9) and (4.12) gives

$$\langle r_0^2\rangle = nl^2[(1+\cos\theta)/(1-\cos\theta)] \times [(1+\langle\cos\phi\rangle)/(1-\langle\cos\phi\rangle)] \quad (4.13)$$

The parameters $l$, $\theta$ and $\phi$ are constant for a given polymer at a fixed temperature, and (4.13) may therefore be written in the form

$$\langle r_0^2\rangle = ns \quad (4.14)$$

where $s$ is a skeletal function that measures the short-range forces operative in the polymer chain, i.e. those between adjoining monomer units.

The basic forms of (4.1) and (4.14) are similar, in that both predict a linear proportionality between the mean-square displacement length and the degree of polymerization. However, whilst (4.1) is valid for *all* values of $n$, the validity of (4.14) is restricted to high values of $n$. Therefore, if the limiting form of the theory is to be applied to the real polymer molecule, then that molecule must contain a sufficiently large number of residues so that, when they are grouped in sequences of $n_k$ units (the value of $n_k$ being large enough to make negligible the effect of one sequence on the direction of the following one), the number of statistical elements is such that the so-called Gaussian distribution will still be valid. The lower limit for the validity of this approximation is generally thought to be about ten statistical segments. The actual polymer molecule has a finite size, and it is therefore quite possible that if the hindrance to rotation is sufficiently severe, the length of the statistical segment may be so great that the asymptotic value of (4.14) is not realized; i.e. the value of $n$ at which (4.14) becomes valid increases progressively with greater restrictions on rotations, so increasing the probability that the polymer molecule will not attain this size. Polymers of this type are often described as having 'stiff' chains, or as being non-Gaussian. They are intermediate between the completely flexible macromolecules and those which are so rigid (i.e. having high energy barriers to rotation) as to be described as rod-like. These definitions are to some extent arbitrary, for as the degree of polymerization increases the macromolecule will pass successively through the stages of a rigid rod-like structure, a non-Gaussian chain, and finally a Gaussian chain.

[The term 'Gaussian (or non-Gaussian) chain' arises because the probability distribution for the ends of a random walk or flight has the mathematical form of the Gaussian error law, and, for this particular case, (4.1) is obeyed. By common usage the definition has become inverted, i.e. if the chain lengths of a homologous series of polymer fractions are sufficiently great for the ratio $\langle r_0^2 \rangle / n$ to attain constancy, the polymer is said to exhibit Gaussian behaviour. Conversely, when the chains are so short that this asymptotic relation has not been realized, they are said to be non-Gaussian. This is a misuse of the term 'Gaussian'—the real polymer chains may attain constancy for the ratio $\langle r_0^2 \rangle / n$ without the probability distribution for the chain ends being accurately Gaussian, or the distribution may approximate closely to the Gaussian form without attaining the asymptotic value of the ratio. Therefore, our use of the term 'Gaussian' for chains described by (4.14) is not strictly valid, but this usage has become so common in the literature that we persist in its use.]

*Excluded Volume Effect.* In developing the model of the random coil, we have moved through stages that have gradually brought it closer in behaviour to that of the real chain. However, there is inherent in the mathematical treatment an assumption that can have no physical counterpart. In the random walk, the path taken may be self-intersecting. This observation would imply that in the case of the actual polymer molecule, two segments could simultaneously occupy the same point in space—an obvious physical impossibility. The effect of this interference in space between segments of the chain must occur with highest probability in compact conformations. Such conformations therefore cannot exist in the case of the real chain, and correspondingly its dimensions must increase relative to those of its mathematical analogue. The consequence of this, the *excluded volume effect*, is to modify the probability distribution of the segments so that the mean-square end-to-end separation of the real polymer is increased.

Flory and Fox (1951) introduced to the model a semi-empirical parameter $\alpha$, such that

$$\langle r^2 \rangle = \alpha^2 \langle r_0^2 \rangle \tag{4.15}$$

where $\langle r^2 \rangle$ is the mean-square end-to-end separation of the actual chain, and $\langle r_0^2 \rangle$ is, as before, the corresponding value in the absence of the excluded volume effect. The proportion of forbidden conformations increases rapidly with the number of steps in the random walk. The physical counterpart to this observation is that the parameter $\alpha$ should be dependent on the number of the segments in the chain, i.e. on the degree of polymerization. Thus the net result of the excluded volume effect is to abolish the proportionality between the mean-square end-to-end separation of the polymer chain and the number of segments it contains, a proportionality which has survived the limitations imposed by fixing the bond angle and restricting the rotation about inter-unit bonds.

*Solvent Interactions.* A second factor has also been neglected in the theoretical development. We have treated only the polymer molecule, almost as though it could exist in a vacuum. Such, of course, is not the case. The physical measurements carried out on the polymer are performed in dilute solution, and therefore there is an interplay between solute and solvent. This previously neglected factor affects the calculations, because solute–solvent interactions will lead to certain conformations being favoured at the expense of others. In good solvents, in which the interaction between solute and solvent is greater than that between segments of the polymer chain (solute–solute interaction), conformations allowing the maximum amount of solvent–solute interaction will be favoured, whereas those allowing maximum contact between the polymer segments will be reduced in frequency. Obviously, the favoured state in this case must correspond to extended conformations, so that the mean-square end-to-end separation of the polymer coil is greatly increased with respect to the value calculated using (4.14), for example. Similarly, in poor solvents, solute–solute contacts will be preferred, and therefore compact conformations which allow such contacts will be favoured. In thermodynamic terms, compact conformations will be avoided if segmental interaction increases the free energy of the polymer, as would be the case in a good solvent.

Thus the mathematical model is deficient in two respects, in that it ignores the finite volume of the real chain segments, and also the effect of solvent interaction with that chain. These two effects are related: according to Flory (1949) the excluded volume is defined as the effective covolume of a pair of segments present in the given solvent, and not merely as the physical volume of the chain segment. In a given solvent, therefore, the parameter $\alpha$ defined by (4.15) will depend markedly on temperature; at constant temperature, $\alpha$ will vary with the solvent employed. We must therefore conclude that a knowledge of the structural geometry of the chain will be of no assistance in evaluating $\alpha$.

*Theta-Temperature.* In a good solvent, the effect of solute–solute interaction is to increase the free energy of the system, and for this reason conformations leading to such interactions are suppressed. The converse is also valid: in a poor solvent, the requirement of minimum free energy will favour compact conformations. Thus the effect of the poor solvent will be to decrease the mean-square end-to-end distance of the polymer molecule, whilst the finite volume of the chain will tend to increase it. Because these two effects are opposed, the judicious choice of solvent and temperature enables a balance to be attained, in which the mutual attractions between chain segments resulting from the use of a poor solvent exactly compensates for the finite volume of the segment. In such a solvent, the covolume is zero, and $\alpha=1$ so that (4.15) reduces to

$$\langle r^2\rangle=\langle r_0^2\rangle \tag{4.16}$$

Equation 4.16 shows that, by appropriate choice of solvent, the perturbations of the conformation consequent upon the finite volume of the chain elements can be made to vanish. Under these conditions, the proportionality between the mean-square end-to-end distance of the actual polymer molecule and its

degree of polymerization will again be valid. For a given solvent, the temperature at which the excluded volume effect vanishes is referred to as the Flory temperature, or *theta-temperature*. At the *theta*-temperature, the polymer adopts its unperturbed dimensions, denoted by the subscript zero; i.e. $\langle r_0^2 \rangle$ and $\langle R_{G0}^2 \rangle$ are the mean-square end-to-end distance and radius of gyration, respectively, of the polymer molecule in a *theta*-solvent.

The above treatment considers only the segmental interactions within an isolated polymer molecule (*intramolecular* interactions), and shows that they disappear at the *theta*-temperature. This same disappearance of segmental interaction might be expected to occur in the case of a pair of polymer molecules. This assertion assumes that there is no difference in the interactions between a pair of segments belonging to the same macromolecule and a pair belonging to different molecules—an assumption that could scarcely be challenged. At the *theta*-temperature, therefore, *intermolecular* interactions between the polymer molecules vanish. The *theta*-solvent is thus thermodynamically ideal, in that the second virial coefficient obtained in the virial expansion of the osmotic pressure or light-scattering relations is zero.

*Summary.* In (4.14) the complex effects of constancy of bond angle and restrictions to bond rotations were simplified by the introduction of a skeletal parameter $s$ which is dependent solely on the geometry of the chain, and represents the accumulation of short-range interaction. The excluded volume effect, on the other hand, is due to the mutual interactions of polymer segments that are remote from each other along the chain, but are brought into juxtaposition in space as the molecule explores all the available conformations open to it. This latter type of interaction may therefore be attributed to the operation of long-range forces.

Polymer dimensions obtained from measurements carried out in dilute solution will be in general compounded from both short- and long-range forces; only in the absence of the latter type of interaction is the proportionality between the mean-square end-to-end separation of the polymer chain and the degree of polymerization maintained. Thus, if the theoretical treatments developed earlier for describing polymer dimensions are to be used, either the measurements must be carried out at the *theta*-temperature of the solvent, or the relation between the expansion factor $\alpha$ and the degree of polymerization must be determined. Unfortunately, no single theoretical treatment of the exceedingly complex relation existing between these two latter parameters has achieved general acceptance for the entire range of $\alpha$ values; some of those in current use are detailed later (see section 4.3.2). In the main, the theories fall into two main classes, governed by either a fifth- or a third-power law. This terminology refers to the form of the asymptotic relation between $\alpha$ and $n$ at very high values of $\alpha$; according to the fifth-power law

$$\alpha^5 \propto n^{1/2} \qquad (4.17)$$

whereas according to the third-power law

$$\alpha^3 \propto n^{1/2} \tag{4.18}$$

Substitution of (4.15), (4.17) and (4.18) in (4.14) gives the general relation

$$\langle r^2 \rangle \propto n^{1+\delta} \tag{4.19}$$

where $\delta$ is a function of molecular weight reaching a limiting value of 0.2 if the theory from which (4.17) is derived is valid; on the other hand, the asymptotic value is 0.33 if the derivation of the third-power law (4.18) is valid.

Although it is a simple matter to distinguish between a random coil in the unperturbed state and a rod-like molecule (for the former model, the mean-square radius of gyration is proportional to the degree of polymerization; for the rod, it is proportional to the square of this parameter), there is an intermediate region in which some doubt must exist as to which model is appropriate. For example, if the parameter $\delta$ of (4.19) is found to have a value of 0.30, we may be dealing with a molecule which would conform to the Gaussian model but for the excluded volume effect, or a rather inflexible molecule that has not yet reached the asymptotic limit of Gaussian behaviour, or even a compound effect involving both excluded volume and skeletal inflexibility. Needless to say, the unambiguous interpretation of experimental results in such circumstances is difficult.

**4.2.2.** Interpretation of Measurements on Dilute Solutions of Polymers.

*Light Scattering.* Until the fairly recent development of the technique of measuring the dipole moments of polymers, which yields directly the mean-square end-to-end separation of the chain, the only direct method of obtaining an unequivocal measurement of the dimensions of the dissolved macromolecule was by means of light scattering. In essence, the light-scattering method is quite straightforward: the intensity of light scattered by the polymer molecules is measured as a function of (a) the angle between the incident beam and the direction of observation, and (b) the concentration of the scattering molecules, and related to the intensity of the incident beam. No information regarding the shape of molecules having a dimension less than approximately one-twentieth that of the wavelength of light in the medium can be derived directly from light-scattering measurements. However, when a polymer molecule has a dimension in excess of the approximate limit given above, it no longer acts as a point source of radiation, and, as a result, there is destructive interference in the light scattered from different parts of the molecule. This effect is a function of the angle of viewing $\theta$, the interference increasing with increasing angle. It is therefore necessary to introduce a particle scattering factor $P_\theta$, defined as the ratio of the observed intensity ($R_\theta$ observed) to that expected in the absence of interference ($R_\theta$ ideal), that is

$$P_\theta = R_\theta \text{ observed} / R_\theta \text{ ideal} \tag{4.20}$$

The particle scattering factor can be related to the dimensions of the molecule by employing simple models, and the relation

$$Kc/R_\theta = 1/MP_\theta + 2A_2c + \ldots \tag{4.21}$$

is obtained, where $K$ is a constant, and $A_2$ is the second virial coefficient. At $\theta = 0°$, $P_\theta = 1$ for any model—sphere, rod or coil. The molecular weight is therefore obtained by extrapolation of experimental results, obtained over a range of $\theta$, to $\theta = 0$. Another series of extrapolations, to zero concentration of solute, is then necessary if the second virial coefficient is finite. The extrapolations may be made simultaneously by using the method of Zimm (1948), according to which the experimental data are graphed as a function of $[\sin^2(\theta/2) + k'c]$, where $c$ is the concentration of the polymer, and $k'$ is an arbitrary constant chosen to give the term in $c$ the same magnitude as that of $\sin^2(\theta/2)$.

Both the lines of zero angle and zero concentration extrapolate to a common point on the $Kc/R_\theta$ axis. The numerical value of this intercept is the reciprocal of the molecular weight. The mean-square radius of gyration is obtained from the slope of the line of zero concentration as

$$\langle R_G^2 \rangle = (3\lambda_0^2/16\pi^2 n_0^2)(\text{initial slope/intercept}) \tag{4.22}$$

where $\lambda_0$ is the wavelength of the incident light *in vacuo*, and $n_0$ is the refractive index of the solvent. The slope of the line of zero angle is directly proportional to the second virial coefficient.

The theoretical basis of the light-scattering technique is thoroughly developed and quite rigorous. Difficulties frequently arise, however, in the actual experimentation. Particularly important in this respect are optical artefacts in the apparatus and, more especially, the presence of dust particles in the polymer solution. These particles, being similar in size to the polymer molecules themselves, are consequently exceedingly difficult to remove from solution without simultaneously withdrawing a fractional amount of the polymer. The problem, unfortunately, is particularly severe in the aqueous systems with which we are concerned in the case of amylose. However, the addition of *alpha*-amylase to the amylose solution after completion of the scattering measurements enables the contribution of dust to the scatter to be taken into account (Banks, Greenwood and Sloss 1969).

*Hydrodynamic Techniques.* Although light scattering and the measurement of dipole moment present the only methods by which the dimensions of the polymer molecule may be obtained unambiguously, it is also possible to relate certain hydrodynamic characteristics of the macromolecule to its dimensions. It must be stressed, however, that the theoretical interpretation of hydrodynamic behaviour is far from rigorous, and hence the information obtained in this way may be of uncertain significance.

When a particle suspended in a viscous fluid is subjected to an applied force, it accelerates until that force is neutralized by the frictional forces generated by its movement. Information regarding the shape of the particle may then be obtained by (a) measuring its resultant steady-state velocity, or (b) determining the increase in the rate at which energy is dissipated in that fluid, when subjected to shear forces, as a result of the presence of the suspended particle. The former type of measurement can be made by means of the ultracentrifuge; the latter by viscometric techniques. The way in which the shape of a particle may be deduced from such measurements will now be outlined.

*Sedimentation and Frictional Coefficients.* The ultracentrifuge provides the means by which a high gravitational field can be applied to the dissolved macromolecule, and the resultant steady-state velocity measured. Denoting the mass of the molecule by $m$, with $\overline{V}$ being its partial specific volume in a solution of density $\rho$,

$$m(1-\overline{V}\rho)\omega^2 x = f\,dx/dt \tag{4.23}$$

where $\omega$ is the applied angular velocity, $x$ is the distance of the molecule from the centre of rotation, $dx/dt$ is the steady-state velocity and $f$ is the frictional coefficient. Defining the sedimentation coefficient $S$ as

$$S = (dx/dt)/\omega^2 x \tag{4.24}$$

reduces (4.23) to

$$m(1-\overline{V}\rho) = fS \tag{4.25}$$

or

$$M(1-\overline{V}\rho) = fSN_A \tag{4.26}$$

where $N_A$ is Avogadro's number, and $M$ is the molecular weight of the polymer.

The frictional coefficient $f$ may be obtained for simple models by the application of Stoke's Law. Thus for a spherical molecule

$$f_{sphere} = 6\pi\eta_0 R_s \tag{4.27}$$

where $\eta_0$ is the solvent viscosity and $R_s$ is the radius of the sphere. For a spherical molecule, the molecular weight is proportional to the volume occupied by the sphere. Consequently, $R_s$ is proportional to $M^{1/3}$, and we have the general relation

$$f_{sphere} \propto M^{1/3} \propto n^{1/3} \tag{4.28}$$

The relation between the frictional coefficient of a rod-like molecule and its degree of polymerization may be obtained from considering the limiting case of a prolate ellipsoid of very high axial ratio. This treatment yields

$$f_{rod} \propto n^{2/3} \tag{4.29}$$

To treat the hydrodynamic problem of a randomly-coiled molecule in mathematical form it is usual to take as the model the so-called 'pearl necklace', independently derived by Brinkman (1947) and by Debye and Bueche (1948). In this model, the polymer molecule is represented as a swarm of rigid beads joined by infinitely thin bonds, the whole having spherical symmetry with the beads distributed uniformly throughout the volume of the sphere; only the beads offer any resistance to solvent flow. The hydrodynamic resistance to flow through the swarm can then be visualized in terms of one of two extreme situations. First, if the beads are so far apart that the flow of solvent is disturbed only locally in the vicinity of each bead, and these disturbances do not interact, the overall frictional coefficient will be proportional to the number of beads. For this, the *free-draining coil*, the following relations must be valid:

$$f_{coil} \propto M \propto n \tag{4.30}$$

At the other extreme, the disturbances produced in the solvent flow by the beads being close together may interact to such an extent that solvent is entrapped between the beads. Hydrodynamically, the swarm and the entrapped solvent constitute a single entity, known as the *impermeable*, or *matted, coil.* This concept of solvent being entrapped within the domain of the polymer coil is rather analogous to the situation pertaining in an aqueous gel—even at macromolecular concentrations as low as one per cent, the water can be completely immobilized by the polymer. For this impermeable model, the real polymer coil can be replaced by a fictitious *equivalent sphere*, the radius $R_e$ of which is so chosen as to give the correct frictional coefficient of the polymer coil, so that

$$f_{coil} \propto R_e \propto \langle R_{GO}^2 \rangle^{1/2} \tag{4.31}$$

Equation 4.31 assumes that the radius of the hydrodynamically-equivalent sphere is proportional to a linear dimension of the real coil, such as the root-mean-square radius of gyration $\langle R_{GO}^2 \rangle^{1/2}$. This latter parameter is, of course, proportional to the square-root of the molecular weight in the case of the random coil, so that

$$f_{coil} \propto M^{1/2} \propto n^{1/2} \tag{4.32}$$

It is instructive at this stage to correlate the relations between the frictional coefficient and the molecular weight obtained using the various models described above. This is done in table 4.1, and since the experimental quantity in which we are interested is the sedimentation coefficient, this is also given as a function of molecular weight; from (4.26)

$$S \propto M/f \propto n/f \tag{4.33}$$

The proportionalities obtained for the frictional coefficient in the above models are substituted in (4.33) to yield the relation between $S$ and $M$ for spheres, rods, free-draining coils, and impermeable coils.

The results in table 4.1 clearly demonstrate that the relation between $S$

and $M$ is very dependent on the model used and, in the case of coiled molecules, on the assumptions made regarding the flow of solvent through the swarm of beads. Thus, for the free-draining model $S$ is independent of $M$, whereas for the impermeable coil $S$ is proportional to $M^{1/2}$.

*Table* 4.1. The relation of the frictional coefficient $f$ and the sedimentation constant $S$ to the molecular weight for the sphere, the rod, the free-draining coil, and the impermeable coil, models.

| | sphere | rod[1] | coil (free-draining) | coil (impermeable) |
|---|---|---|---|---|
| $f$ | $M^{1/3}$ | $M^{2/3}$ | $M$ | $M^{1/2}$ |
| $S$ | $M^{2/3}$ | $M^{1/3}$ | $M^0$ | $M^{1/2}$ |

[1] A rather more accurate treatment of the rod-like model, carried out by Riseman and Kirkwood (1950), suggests that $f$ is proportional to $L/\log L$, so that $S$ varies only with $\log L$ (or $\log \bar{M}$).

In practice, it is found that most synthetic polymers obey a relation of the form

$$S_0 = K_c M^c \qquad (4.34)$$

where $S_0$ is the sedimentation coefficient extrapolated to infinite dilution, and the exponent $c$ has a value slightly less than 0.50. This low value suggests that the polymer coil may be intermediate in its behaviour between the extremes of the free-draining and the impermeable coil. To represent its behaviour, it is necessary to invoke the model of *partial free-draining*, i.e. as the polymer coil gradually increases in size, it passes successively through the stages of free-draining, partial free-draining, and impermeability. Thus the effect of free draining will be most apparent at low molecular weights, and will become steadily less important with increasing molecular weight. The exponent $c$ of (4.34) would therefore be expected to approximate to a constant over a limited range of molecular weights, intermediate in value between the zero expected for a free-draining coil and the value of 0.50 predicted for the impermeable model. Over a more extended series of molecular weights, $c$ might be expected to range from zero at low values to the asymptotic limit of 0.50 at high molecular weight.

In the limiting case of the impermeable coil

$$f_{coil}/\eta_0 = P_0 \langle R_{G0}^2 \rangle^{1/2} \qquad (4.35)$$

where the quantity $P_0$ has the value $9\pi^{3/2}/4 = 12.6$, and $\eta_0$ is the viscosity of the solvent. The frictional coefficient depends, therefore, only on the magnitude of the root-mean-square radius of gyration, and not on the chemical composition of the polymer. The parameter $P_0$ may thus be regarded as a constant, universal for all coiled polymers in any solvent.

All mathematical treatments of free draining show that, in the limit of very high molecular weight, the hydrodynamic behaviour of the coil approaches that of the impermeable model. Flory and Fox (1951), however, argued that this assumption is actually valid over a wide range of molecular weights, and Flory (1953) has shown, in fact, that in any randomly-coiled linear macromolecule, the contribution of free draining must be negligibly small over most of the molecular weight range of interest.

This conclusion, if valid, effectively removes the model of the partially free-draining coil from discussion, and an alternative explanation must be found for values of $c$ greater than 0.50. In the theories considered above, the effect of excluded volume has been ignored. If we use (4.35) in the more general form

$$f_{coil}/\eta_0 = P_0 \langle R_G^2 \rangle^{1/2} = P_0 \alpha \langle R_{G0}^2 \rangle^{1/2} \tag{4.36}$$

it will be seen that the frictional coefficient of the impermeable coil is proportional to $\alpha \langle R_{G0}^2 \rangle^{1/2}$. The expansion factor $\alpha$ is a function of molecular weight, so that at the limit of very high molecular weight and pronounced solvent–solute interaction

$$f_{coil} \propto M^{0.60} \propto n^{0.60} \tag{4.37}$$

or

$$f_{coil} \propto M^{0.67} \propto n^{0.67} \tag{4.38}$$

Equation 4.37 assumes that the relation between $\alpha$ and $n$ conforms to the fifth-power law; (4.38) is based on the third-power law. Expressing these relations in terms of the sedimentation coefficient, we see from (4.33) that the exponent $c$ of (4.34) is predicted to vary in the range $0.40 \leqslant c \leqslant 0.50$ or $0.33 \leqslant c \leqslant 0.50$, depending on the theory used for expressing $\alpha$ in a closed form. (For the majority of synthetic polymers, $c$ is found to vary in the range $0.40 \leqslant c \leqslant 0.50$; of the natural polymers, the derivatives of cellulose are peculiar in that the observed values of $c$ are found to be somewhat lower, usually about 0.35.)

It must be stressed that the theory of Flory has a completely different philosophical basis to that of partial free-draining. In order to obtain an exponent of 0.50 in (4.34) two conditions must be satisfied, namely (a) the radius of the equivalent hydrodynamic sphere is proportional to a linear dimension—the root-mean-square radius of gyration, for example—of the polymer coil, and (b) the probability distribution of the chain elements in space is such that the mean-square radius of gyration of the chain is proportional to the number of elements. To explain values for the exponent other than 0.50, the theories of partial free-draining accept assumption (b), but reject (a); according to Flory, however, assumption (a) is valid over the entire range of molecular weights of interest, and it is (b) that is invalid, specifically because of its failure to recognize the existence of the excluded volume effect.

*Viscosity*. The simplest, and most widely used, technique of characterizing a polymer is to measure the increment in viscosity, the *specific viscosity* $\eta_{sp}$, caused by the addition of a small amount of polymer to the solvent. The specific viscosity is defined by

$$\eta_{sp}=(\eta-\eta_0)/\eta_0 \tag{4.39}$$

where $\eta$ is the viscosity of the polymer solution, and $\eta_0$ is the viscosity of the solvent.

The viscosity number, defined as the ratio $\eta_{sp}/c$ where $c$ is the polymer concentration, usually depends on concentration, and hence it is necessary to extrapolate the experimental results to zero concentration in order to obtain a parameter that characterizes the macromolecule; this is the limiting viscosity number $[\eta]$, defined as

$$[\eta]=\underset{c\to 0}{\mathrm{L}}\ \eta_{sp}/c \tag{4.40}$$

The necessary extrapolation may be carried out using the relation due to Huggins (1942)

$$\eta_{sp}/c=[\eta]+k[\eta]^2c \tag{4.41}$$

where $k$ is a constant (the Huggins constant).

Einstein, in his classical treatment of viscosity, showed that for a sphere the limiting viscosity number is independent of the molecular weight, that is:

$$[\eta]\propto n^0 \tag{4.42}$$

The theoretical treatment of asymmetrical particles subjected to a velocity gradient in a fluid is rather more difficult, because of their tendency to orientate so that their major axis is parallel to the flow lines. Consequently, the viscous contribution that they make is dependent on their orientation. However, if the shearing effect is so small that it may be overcome by the Brownian motion of the particles, it is possible to treat such asymmetric molecules theoretically. For example, with this simplifying assumption, the relation

$$[\eta]\propto n^{1.8} \tag{4.43}$$

is obtained for rod-like molecules.

For coiled macromolecules, there are again the alternatives of the free-draining and the impermeable model to be considered. In the case of the former model, if the distribution of chain segments about the centre of mass of the 'pearl necklace' is Gaussian in form, it can be shown that

$$[\eta]\propto n \tag{4.44}$$

At the other extreme of hydrodynamic behaviour, when the coil may conform to the impermeable model, the actual polymer coil can, as in the case of the frictional coefficient, be replaced by a hypothetical equivalent hydrodynamic sphere, the radius $R_e$ of which is so chosen as to give the same viscosity increment as the real molecule. This treatment yields the relation

$$[\eta]=(10/3)N_A\pi R_e^3/M \quad (4.45)$$

If the radius of the equivalent sphere is again assumed to be proportional to a linear dimension of the polymer molecule, such as the root-mean-square radius of gyration,

$$R_e=\xi\langle R_{GO}^2\rangle^{1/2} \quad (4.46)$$

where the parameter $\xi$ is a proportionality constant. Substituting (4.46) in (4.45) yields

$$[\eta]=(10/3)N_A\pi\xi^3(\langle R_{GO}^2\rangle/M)^{3/2}M^{1/2} \quad (4.47)$$

For a polymer coil, the segments of which are distributed in space according to the Gaussian approximation, the ratio $\langle R_{GO}^2\rangle/M$ is independent of molecular weight, so that (4.47) reduces to

$$[\eta]_\theta=K_\theta M^{1/2} \quad (4.48)$$

where

$$K_\theta=(10/3)N_A\pi\xi^3(\langle R_{GO}^2\rangle/M)^{3/2} \quad (4.49)$$

The subscript $\theta$ in (4.48) and (4.49) emphasizes that these relations are valid only in the absence of the excluded volume effect, i.e. in a thermodynamically ideal solvent; it should be noted that the numerical value of $K_\theta$ is dependent upon the type of polymer, and the temperature.

In practice, solutions of linear high polymers are found to conform to relations of the type

$$[\eta]=K_a M^a \quad (4.50)$$

where in most instances the exponent $a$ varies between values of 0.50 and 0.80, but values as high as 1.0 have occasionally been recorded (Kurata and Stockmayer 1963).

The limiting viscosity number is proportional to the molecular weight in the case of the free-draining coil, and to the square-root of the molecular weight for the model of the impermeable coil. Since the experimental values of $a$ are found to lie in a region intermediate between those predicted for the free-draining and impermeable models, it is tempting to postulate that, in the case of the real polymer molecule, the concept of partial free-draining is operative. Many theoretical treatments of this intermediate model have been attempted. For example, the model of Kirkwood and Riseman (1948) leads to the relation

$$[\eta]=\pi^{3/2}N_A XF(X)\langle R_{GO}^2\rangle^{3/2}/M \quad (4.51)$$

where $F(X)$ is a function of the drainage parameter $X$. The quantity $XF(X)$ varies with molecular weight in such a way as to account for values of the exponent $a$ in (4.50) intermediate between 0.50 and 1.0. However, in the limit of sufficiently large $n$, this quantity tends to an asymptotic limit. Using this limiting value gives, in the notation of Flory, the viscosity constant $\Phi_0$, defined as*

$$\Phi_0 = (10/3)\mathrm{N}_A \pi \xi^3 = 4.22 \times 10^{24} \quad (4.52)$$

and therefore

$$[\eta]_\theta = \Phi_0(\langle R_{G0}^2 \rangle / M)^{3/2} M^{1/2} \quad (4.53)$$

so that

$$K_\theta = \Phi_0(\langle R_{G0}^2 \rangle / M)^{3/2} \quad (4.54)$$

The viscosity constant, like the frictional constant $P_0$, is independent of both polymer and solvent, according to this treatment. Whilst all theoretical treatments give the proportionality between limiting viscosity number and the square-root of molecular weight in the asymptotic limit of high molecular weight, Flory claims that (4.53) is valid over a very wide range of molecular weights. The departure of the exponent $a$ of (4.50) from the expected value of 0.50 is then due to the excluded volume effect, and this is reflected in the solution no longer behaving in an ideal fashion, that is, the second virial coefficient becomes finite. In such a solution

$$[\eta] = \Phi(\langle R_G^2 \rangle / M)^{3/2} M^{1/2} \quad (4.55)$$

$$= \Phi\alpha^3(\langle R_{G0}^2 \rangle / M)^{3/2} M^{1/2} \quad (4.56)$$

where $\alpha$ is the expansion factor.

The theories considered here fall into two quite distinct categories, namely those dealing with partial free-draining, and those taking cognisance

*The units of limiting viscosity number are those of reciprocal concentration; unfortunately, concentrations are sometimes quoted in terms of g/ml, and sometimes as g/dl. The numerical value of the viscosity constant $\Phi_0$ will then vary by a factor of 100, depending on which units of concentration are employed. Also, the numerical value of $\Phi_0$ is related to whether the radius of gyration or the end-to-end distance of the chain is employed; that is, if the factor $\langle R_{G0}^2 \rangle / M$ of (4.51) is replaced by $\langle r_0^2 \rangle / M$, the numerical value of $\Phi_0$ must be adjusted. To avoid confusion, we give the numerical values of $\Phi_0$ for the various systems. They are:

| | |
|---|---|
| $[\eta]$ in ml/g, $\langle R_{G0}^2 \rangle / M$ | $\Phi_0 = 4.22 \times 10^{24}$ |
| $[\eta]$ in dl/g, $\langle R_{G0}^2 \rangle / M$ | $\Phi_0 = 4.22 \times 10^{22}$ |
| $[\eta]$ in ml/g, $\langle r_0^2 \rangle / M$ | $\Phi_0 = 2.86 \times 10^{23}$ |
| $[\eta]$ in dl/g, $\langle r_0^2 \rangle / M$ | $\Phi_0 = 2.86 \times 10^{21}$ |

The units of limiting viscosity number used throughout this book are ml/g. Various theoretical treatments give slightly different numerical values for the parameter $XF(X)$ and therefore for those of $\xi$ and $\Phi_0$. The above values represent the upper limit for $\Phi_0$; at the other extreme, the values are only about 60 per cent of those quoted here.

of the excluded volume effect. No effort has been made to deal with both effects occurring simultaneously—a not unlikely situation in practice. The mathematical manipulations necessary for such a study are rather more complex than for either of the simple extremes, but Yamakawa and Kurata (1958) carried out perturbation calculations, based on the model of Kirkwood and Riseman (1948), and obtained

$$[\eta]=[\eta]_\theta[1+p(X)z-\ldots] \tag{4.57}$$

$$[\eta]_\theta=\pi^{3/2}N_AXF(X)(\langle R_{G0}^2\rangle/M)^{3/2}M^{1/2} \tag{4.58}$$

where $p(X)$ and $F(X)$ are functions of the drainage parameter $X$, and $z$ is a parameter related to solvent–polymer interaction. Both $XF(X)$ and $p(X)$ attain their limiting values of 1.26 and 1.55, respectively, at fairly low molecular weights.

*Summary of The Relations Predicted for The Various Models.* For a series of polymer homologues, we have the general relations

$$[\eta]=K_aM^a \tag{4.59}$$

$$\langle R_G^2\rangle=K_bM^b \tag{4.60}$$

$$S_0=K_cM^c \tag{4.61}$$

The hydrodynamic shape of the macromolecule can be deduced from the experimental values of the exponents $a$, $b$ and $c$. In table 4.2, we present a summary of the values predicted for the various models we have considered in this section.

*Table* 4.2. The predicted values of the exponents in the equations relating limiting viscosity number, mean-square radius of gyration, and sedimentation coefficient to molecular weight for various models.

| | sphere | rod | coils: | | | |
|---|---|---|---|---|---|---|
| | | | free-draining | impermeable | partial free-draining | Flory model |
| $a$ | 0 | 1.8 | 1.00 | 0.50 | 1.00–0.50 | 0.50–0.80 |
| $b$ | 0.67 | 2.0 | 1.00 | 1.00 | 1.00 | 1.00–1.20 |
| $c$ | 0.67 | 0.33 | 0 | 0.50 | 0–0.50 | 0.50–0.40 |

Thus, to study the conformation of amylose in a given solvent, it is necessary to measure the limiting viscosity number, radius of gyration, sedimentation coefficient, and molecular weight for a series of samples differing in molecular weight. Since the amylose from any natural source contains molecules differing widely in molecular weight, the simplest method of obtaining a suitable range of samples is to divide the parent material into a series of fractions of different molecular weights by some technique such as fractional precipitation from solution (see section 4.3). The hydrodynamic properties of the fractions can then be used to infer the shape of the dissolved macromolecule.

*Molecular Weight Averages.* In developing the descriptive theories for the polymer molecule, we have considered the isolated molecule. The real polymer differs from our mathematical models in that it contains a series of molecules differing in molecular weight and consequently in dimensions, and therefore any attempt to measure these two quantities must inevitably yield average values. The most important of these averages, and the ways in which they can be obtained, are outlined below.

The simplest is the *number average*. In the determination of molecular weight, any method (osmometry, determination of reducing power, cryoscopy and ebulliometry) that depends upon the colligative properties of the solution yields the number-average molecular weight $\bar{M}_n$, defined as

$$\bar{M}_n = \Sigma N_i M_i / \Sigma N_i \tag{4.62}$$

where $N_i$ is the number of moles of species $i$ in the population, and $M_i$ is the molecular weight of that species. Alternatively, the number-average molecular weight may also be defined as

$$\bar{M}_n = c/\Sigma(c_i/M_i) = 1/\Sigma(w_i/M_i) \tag{4.63}$$

where $c_i$ is the concentration of species $i$, $c = \Sigma c_i$ and $w_i = c_i/c$.

Correspondingly, the weight-average molecular weight $\bar{M}_w$, obtained from the light-scattering technique, is defined by any of the following relations:

$$\bar{M}_w = \Sigma N_i M_i^2 / \Sigma N_i M_i = \Sigma c_i M_i / c = \Sigma w_i M_i \tag{4.64}$$

The other molecular-weight average accessible to experimental evaluation, by means of the ultracentrifuge, is the $Z$-average. This quantity, designated $\bar{M}_z$, is characterized by any of the following expressions:

$$M_z = \Sigma N_i M_i^3 / \Sigma N_i M_i^2 = \Sigma c_i M_i^2 / \Sigma c_i M_i = \Sigma w_i M_i^2 / \Sigma w_i M_i \tag{4.65}$$

The $Z$-average is of particular interest in that the polymer dimensions obtained from the light-scattering technique correspond to this average. Thus the mean-square radius of gyration calculated from such measurements is defined by

$$\langle R_{G0}^2 \rangle_z = \Sigma N_i M_i^2 \langle R_G^2 \rangle_i / \Sigma N_i M_i^2 \tag{4.66}$$

where $\langle R_G^2 \rangle_i$ is the mean-square radius of gyration of the $i$th species, the averaging procedure in this case being carried out over all possible conformations. In the case of a spherical or a rod-like molecule $R_G^2$ is not time-dependent because the shape of the molecule is fixed, but a $Z$-average value is obtained when the sample is polydisperse; for a flexible coil, however, with its constantly changing shape, we obtain a quantity which is averaged over all possible conformations for a given species $i$, and also averaged according to (4.66).

The same type of averaging procedure when applied to the limiting viscosity number yields a Mark–Houwink–Sakurada relation of the form:

$$[\eta]=K_a\bar{M}_v^a \tag{4.67}$$

where $\bar{M}_v$ is the viscosity-average molecular weight, and

$$\bar{M}_v=[\Sigma N_iM_i^{(1+a)}/\Sigma N_iM_i]^{1/a} \tag{4.68}$$

When $a=1$, $\bar{M}_v$ and $\bar{M}_w$ are identical; in the case of the exponential or most probable distribution, in which $\bar{M}_w/\bar{M}_n=2$, $0.84\bar{M}_w\leqslant\bar{M}_v\leqslant0.92\bar{M}_w$ for $0.50\leqslant a\leqslant0.80$. For fractions characterized by a narrower distribution, the difference between $\bar{M}_v$ and $\bar{M}_w$ becomes even less, so that the use of light-scattering measurements to establish the molecular-weight–limiting-viscosity-number relation is justified. If the fractions used in order to establish this relation have widely different polydispersities (the ratio $\bar{M}_w/\bar{M}_n$), the derived value of $a$ has little or no theoretical significance.

**4.2.3.** Polymer Dimensions from Computer Model-Building.

The advent of computer techniques has considerably advanced the study of the short-range forces operative within the polymer molecule. By indexing the positions of the various atoms of the dimer, it is possible to compute the potential energy as rotation takes place about the inter-residue bond. Where there is only one type of monomer, and interactions between anything other than nearest neighbours are negligible, this simple approach allows the dimensions of the macromolecule to be predicted; a more complex treatment is required for those macromolecules in which there are specific interactions between residues some distance apart along the chain but brought close together in space by the adoption of a fixed conformation, e.g. a helix. The use of this technique may be illustrated by the work carried out on amylose, and summarized below.

The repeat unit of amylose is α-D-glucose, and we will pre-empt our later discussion (see section 4.3.1) by assuming that it exists in the Cl conformation, using the nomenclature of Reeves (1950). Conformational analysis of saccharides in the pyranose form usually assumes that all bond lengths, valence angles, and torsion angles of the carbon and oxygen atoms forming the ring are constant. This assumption is equivalent to considering the pyranose rings as rigid. No serious error is introduced by this particular assumption since no evidence has been found of any conformation other than Cl in amylose, and calculations show that this form has the lowest energy (see section 4.3.1). A simplifying consequence of considering the rings as rigid is that the distance $l_v$ between the glycosidic oxygen atoms is fixed; hence amylose may be considered as a chain of virtual bonds of length $l_v$, and the flexibility of the macromolecule results solely from the rotation about the glycosidic bond, i.e. by the angles $\phi$ and $\psi$ (see figure 4.1), the torsions about the bonds $C_1$–$O_1$ and $O_1$–$C_4$, respectively. Although the atoms comprising the pyranose rings are regarded as rigidly fixed, rotation can still occur about the bonds $C_2{-}O_2$, $C_3{-}O_3$, $C_5{-}C_6$ and $C_6{-}O_6$, and energies are calculated allowing such rotations. In practice, the various bond angles and bond distances are obtained from appropriate crystallographic studies, and are used to fix the coordinates of the atoms in a Cartesian co-ordinate system. Although the origin of this system may be arbitrary, it is usually taken as the glycosidic oxygen atom. The angle $\phi$ is then varied through 10° increments, the new coordinates of the atoms computed and the interactions between them calculated. Similar calculations are performed for incremental values of $\psi$. The initial orientations of the two residues $(\phi, \psi)=(0, 0)$ are taken to be those when $C_1$, $O_1$ and $O_4$ of the first residue lie in the same plane as $C_4'$, $O_4'$ and $O_1'$ of the second glucose residue. Positive rotation of $\phi$ and $\psi$ occurs when both operations are carried out in the right-handed sense, i.e. when viewing along the $C_1{-}O_1$ or $C_4'{-}O_4'$ towards the origin of the Cartesian coordinate system at the glycosidic oxygen atom, rotation in the anti-clockwise direction gives positive values of $\phi$ and $\psi$.

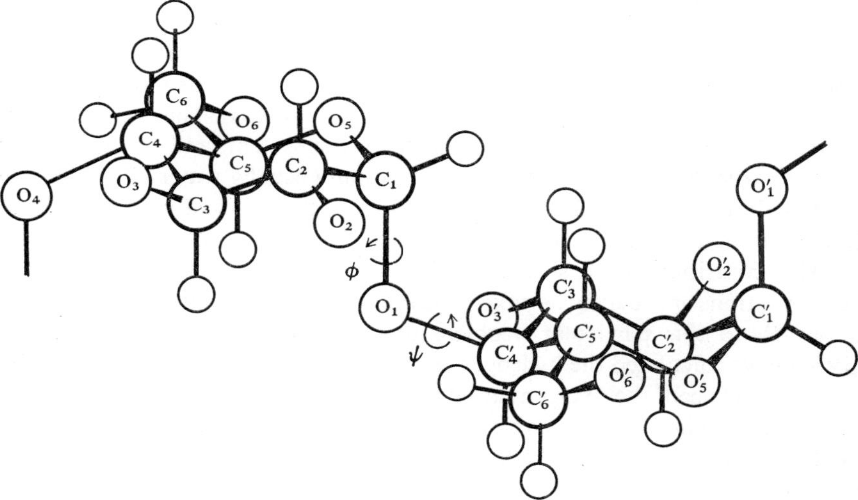


*Figure* 4.1. Dimeric skeletal segment of the amylose chain (after Goebel, Dimpfl and Brant 1970).

The computations are carried out by calculating the non-bonded interactions between the two glucose residues as the angles $\phi$ and $\psi$ are varied. These interactions contain torsional, steric and electrostatic contributions. The torsional energies of the $C_1{-}O_1$ and $C_4'{-}O_4'$ bonds are evaluated using a three-fold cosine term, first employed by Scott and Scheraga (1966a, b) for the *n*-alkanes. This relation takes the form

$$V_{rot}(\phi, \psi)=0.5V_{rot}(1+\cos 3\phi)+0.5V_{rot}(1+\cos 3\psi) \qquad (4.69)$$

where $V_{rot}(\phi, \psi)$ is the torsional potential energy resulting from rotating the residues through angles $\phi$ and $\psi$, and $V_{rot}$ is the energy barrier to rotation about the C–O bond. (The value of $V_{rot}$ is generally taken to be 0.8–1.0 kcal/mol.)

The steric contribution may be evaluated by use of a function due to Kitaigorodsky (1961), according to which

$$V_{nb}(i, j) = 3.5[8600 \exp(-13Z) - 0.04/Z^6] \tag{4.70}$$

where $V_{nb}$ is the non-bonded interaction energy between a pair of atoms $i, j$ and

$$Z = r_{ij}/r_0 \tag{4.71}$$

with $r_{ij}$ being the distance between them, and $r_0$ the equilibrium distance between the atoms $i$ and $j$, i.e. $r_0$ is the sum of the van der Waals radii for the atoms concerned.

The electrostatic term takes the form

$$V_{es}(i, j) = 332 Q_i Q_j / \varepsilon' r_{ij} \tag{4.72}$$

where $V_{es}(i, j)$ is the potential energy of the interaction between atoms $i$ and $j$ and $Q_i$ and $Q_j$ are the fractional charges on the atoms, as defined by Del Re (1958) and Del Re, Pullman and Yonezawa (1963). The parameter $\varepsilon'$ represents the effective dielectric constant of the system, and is much lower than the true dielectric constant of the medium (usually water) in which the carbohydrate is dissolved. An average value of 3.5 is usually ascribed to $\varepsilon'$ in the computations. The numerical factor 332 gives the result in kcal.

The contribution of specific hydrogen bonding in stabilizing certain conformations must also be taken into account. In the most sophisticated approach to this particular problem to date, Goebel, Dimpfl and Brant (1970) have calculated the hydrogen-bond energy by taking cognisance of the distance between the oxygen atoms, the angle subtended by the donor oxygen atom (in the most stable hydrogen bond the atoms involved –O . . . H–O– are co-linear), and whether the acceptor oxygen atom is in an acetal or hydroxyl group.

The potential energy contours for maltose have been calculated independently by a number of groups (Rao, Sundararajan, Ramakrishnan and Ramachandran 1967; Goebel, Dimpfl and Brant 1970; Whittington 1971). The various calculations were made in somewhat different ways by each group of workers—only Goebel *et al.* (1970), for example, allowed for electrostatic interactions in deriving their potential energy diagram, but they ignored the contribution to the total energy made by the torsional bond energies on the ground that it is difficult to assign a meaningful value to $V_{rot}$ (see equation 4.69) and that this inherent potential for C–O bond rotation would be expected to contribute only a minor contribution to the energy map. The results of these independent calculations are similar, however, and a typical diagram is shown in figure 4.2. As the omission of electrostatic contributions to the total potential energy does not greatly affect the result, the dominant factor in determining the molecular conformation of maltose must be steric in nature. Inspection of a molecular model confirms the very high degree of steric interaction as rotation about the glycosidic bond takes place. For example, starting from $(\phi, \psi)=(0, 0)$ and rotating $\phi$ through 180° leads to one ring virtually lying on top of the other, so that the van der Waals repulsion is very large.

The most important point to emerge from this study is the comparatively small area of conformational space available to maltose. In figure 4.2, in fact,

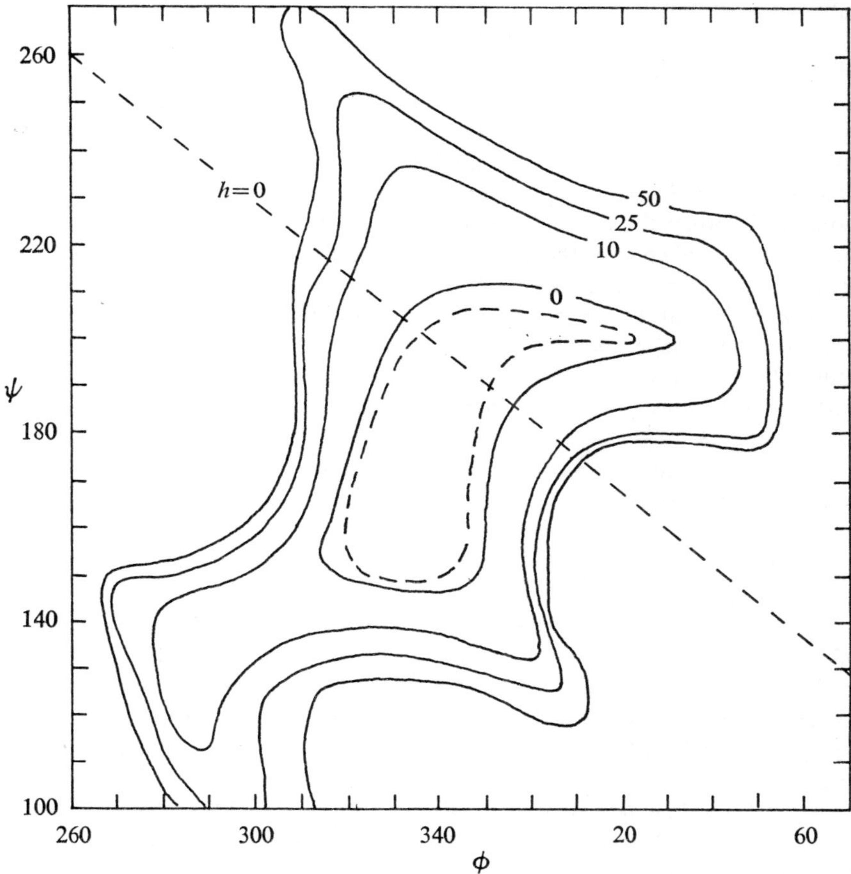


*Figure* 4.2. Conformational energy diagram for maltose, calculated for a glycosidic bond angle of 115°. Contours are labelled in kcal per mole; the dashed contour lies 1.5 kcal per mole above the energy minimum. The broken line ($h=0$) defines the locus of conformations for which the translation $h$ per glucosidic unit along the axis of a helical chain is zero (after Brant and Dimpfl 1970).

only some one per cent of the entire conformational area can be occupied by maltose. This very low value reflects the high degree of steric hindrance occurring across the glycosidic bond—very little movement is possible before interactions of $H_1$ with $H_3'$, $H_5'$, $(OH)_3'$ and $(CH_2OH)_6'$ produce impossibly high steric barriers.

Once the conformational map of the dimer has been obtained, predictions can be made with respect to the polymer for (a) the crystal structure of amylose, and (b) the dimensions of the molecule in solution, particularly the mean-square end-to-end separation of the molecule. A different approach is necessary for each of these calculations, inasmuch as the helical model applicable to amylose in the crystalline state demands that all maltose units are equivalent, and therefore that the angles $(\phi, \psi)$ assume the same values at all glycosidic linkages. The helix has an additional stabilizing interaction, namely hydrogen bonding between contiguous residues and also between those residues which are fairly distant along the chain, but which are brought together in space by the formation of the helix. For example, if there are six glucose residues in one complete helical turn, one might expect some hydrogen bonding between the $i$th and the $(i-6)$th residue. In generating the amylose random coil, of course, the constrictions on the constancy of $\phi$ and $\psi$ are relaxed, and the approach adopted by Brant and Dimpfl (1970) to the problem of intramolecular hydrogen bonding in solution was to ignore it, on the grounds that the potential of the hydroxyl groups of the polysaccharide molecule for hydrogen bond formation will be satisfied by their interaction with solvent molecules, which in dilute solutions of amylose are present in vast excess.

The mean-square end-to-end distance of the amylose chain of degree of polymerization $n$ in the random coil form is given by

$$\langle r_0^2 \rangle = nl_v^2 + 2 \sum_{i<j} \langle \mathbf{l}_i . \mathbf{l}_j \rangle \tag{4.73}$$

where $l_v$ is the length of the virtual bond, and $\mathbf{l}_i$ is the $i$th virtual bond vector of magnitude $\mathbf{l}$. The vector $\mathbf{l}_i$ starts at $O_4$ of the non-reducing end of the molecule, and $\mathbf{l}_n$ terminates at $O_1$ of the reducing end. The quantity in (4.73) to be averaged is the statistical mechanical average over all conformations, i.e. the energy for all sets of $(\phi, \psi)$ in the maltose unit is calculated, and then averaged according to the Boltzmann distribution.

*General criticisms of the technique.* First, the values for the various bond angles and lengths by which the pyranose ring is defined are obtained from X-ray crystallographic analysis of various low molecular-weight compounds such as $\alpha$-D-glucose (McDonald and Beevers 1952), $\alpha$-D-glucose monohydrate (Killean, Ferrier and Young 1962), and $\beta$-methyl maltoside (Chu and Jeffrey 1967). Although the bond lengths obtained in these different studies are quite consistent, the same is not true of the bond angles—the latter show quite significant variations between structures. These results are to be expected—it has been calculated that the energy required for a stretching of ten per cent from the equilibrium length (1.54Å) of single C–C bonds is of the order of 10–100 kcal/mol. By contrast, assuming that the difference in the heats of combustion of cyclohexane and cyclobutane arise only from the change in bond angle, it can be shown that a 10° deviation from the normal tetrahedral angle of 109.5° corresponds to an energy difference of only approximately 1.5 kcal/mol. Thus a molecule may well relieve strain by adjustment of one or more of the bond angles. This is particularly noticeable in the case of isotactic polymers having a carbon–carbon skeleton—the bond angle in polymethylene is 109.5°, but is 145.5° in polypropylene and 116.5° in polystyrene. Of particular importance in the case of amylose is the bond angle to be assigned to $C_1$–O–$C'_4$. The value is believed to be in the region of 110°–120°, but at these two extremes the conformational space available is radically different, as shown by Goebel, Dimpfl and Brant (1970). Construction of even a molecular model for maltose will show that as the angle at the glucosidic oxygen atom increases, the ease of rotation increases.

A second criticism may be directed against the relations by which the various interaction energies are obtained. They are at best semi-empirical in nature, and slightly different forms of the various equations are in use; the differences are of sufficient magnitude to generate quite large variations in, for example, the characteristic ratio of the polymer (see section 4.2.1).

There is even some doubt about the values of the van der Waals radii which should be used for $r_0$ in (4.71). Indeed, in the procedure of Goebel, Dimpfl and Brant (1970), $r_0$ is treated as an adjustable parameter. Augmenting the normally accepted values of the van der Waals radii by as little as 0.15Å, was found by Brant and Dimpfl (1970) to make quite dramatic changes in the dimensions of the amylose random coil.

*Summary*. With so many ill-defined and adjustable parameters available in the conformational analysis of polysaccharides, the predicted dimensions for random coil conformations are not unambiguous: an elegant technique is not of itself sufficient to justify the results it produces. As Rees (1973) has so rightly said: 'Unfortunately, computer calculations can too easily give a veneer of sophistication and spurious rigour to investigations that start from doubtful premises.'

## 4.3. Amylose and Its Water-Soluble Derivatives in Solution.

The results and conclusions of early work in this field are of doubtful significance, because the difficulties involved in preparing samples free from contaminating amylopectin, and in devising a suitable fractionation procedure, were not always fully appreciated. The first successful, and reproducible, fractionation scheme was, in our opinion, that employed by Everett and Foster (1959a). Our survey, therefore, will start with this work.

### 4.3.1. Conformation of the Monomer Unit.

An additional complication in interpreting the solution behaviour of polysaccharides is that their constituent monomer units are, theoretically, capable of assuming different conformations.* The atoms comprising the pyranose rings are not co-planar, but may adopt either a *chair*- or a *boat*-form.

For the pyranose sugars, Reeves (1950) designated the eight strain-free conformations—two chair forms and six boat forms—as C1, 1C, B1, 1B, B2, 2B, B3 and 3B. The orientations of the hydroxyl groups for α-D-glucose in these conformations are detailed in table 4.3.

*Table* 4.3. The orientations of the hydroxyl groups in α-D-glucose for different conformations (axial orientation ax, equatorial orientation eq).

| conformation | hydroxyl group at | | | | |
|---|---|---|---|---|---|
| | $C_1$ | $C_2$ | $C_3$ | $C_4$ | $C_6$ |
| C1 | ax | eq | eq | eq | eq |
| 1C | eq | ax | ax | ax | ax |
| B1 | eq | ax | eq | eq | eq |
| 1B | ax | eq | ax | ax | ax |
| B2 | ax | eq | eq | ax | ax |
| 2B | eq | ax | ax | eq | eq |
| B3 | ax | ax | ax | ax | eq |
| 3B | eq | eq | eq | eq | ax |

Early studies suggested that a major source of instability in the ring was the presence of axially-oriented hydroxyl groups. Also, according to the so-called Hassel–Ottar principle, conformations in which the $-CH_2OH$ group is oriented axially on the same side of the ring as another axially-oriented –OH group will be particularly unstable. Reeves (1950) added a third cause of instability—the Δ2 effect. Again, a conformation in which the –OH group on carbon atom 2 is oriented axially in such a way that the C–O bond bisected the angle of the two C–O bonds of carbon atom 1 is supposed to be a marked cause of instability.

Using the above criteria, Reeves (1950) postulated that only three conformations of α-D-glucose could be expected to be stable, namely the C1, B1

* The use of the single term 'conformation' to describe the shapes of both the monomer units and the macromolecules is rather unfortunate. Szejtli and Augustat (1966) suggested that 'conformation' be applied only to the monomer units, and that 'configuration' be retained for the description of the shape of the macromolecule. To add further confusion the polymer physicist uses the two terms interchangeably, whereas to the organic chemist the two words define quite distinct phenomena. Whilst Flory (1969) has rightly pointed out that the specific stereochemical meanings attached to the terms 'conformation' and 'configuration' are of comparatively recent origin, and that consequently the interchangeable usage is strictly correct, we have decided to retain the convention of the organic chemist.

and 3B forms, each of which has only a single axial –OH group. From studies of the comparative solubilities of amylose and cellulose in cuprammonium, Reeves (1954) concluded that in the former polysaccharide both B1 and 3B forms were present. In fact, he went further and suggested that from a study of models it was possible to conclude that the C1 conformation for both glucose units in maltose was most unlikely, because of steric effects.

The importance of determining the conformation of the individual monomer lies in the fact that this parameter greatly affects the dimensions of the macromolecule in solution. Burchard (1960) calculated the following values of the ratio $(\langle R_{G0}^2\rangle/n)^{1/2}$ for amylose chains composed of glucose units in each of the allowed conformations, assuming free rotation about the glycosidic bonds:

$$\text{C1 conformation } (\langle R_{G0}^2\rangle/n)^{1/2} = 2040\times10^{-11}\text{ cm} \qquad (4.74)$$

$$\text{B1 conformation } (\langle R_{G0}^2\rangle/n)^{1/2} = 3260\times10^{-11}\text{ cm} \qquad (4.75)$$

$$\text{3B conformation } (\langle R_{G0}^2\rangle/n)^{1/2} = 3060\times10^{-11}\text{ cm} \qquad (4.76)$$

Thus, merely by altering the ring form of the monomer residues, changes of approximately 60 per cent are caused in the unperturbed dimensions of amylose.

From a study based on model building, Holló, Szejtli and Toth (1961) pointed out that an amylose chain composed of units in the 3B conformation was incapable of forming a helical structure. They further suggested that, in alkali, amylose was composed of an equilibrium mixture of glucose residues in the C1 and 3B conformations.

Experimentally, however, it has not been possible to show the presence of glucose units having a conformation other than C1. This conformation is favoured in the solid state (see, for example, Senti and Witnauer 1948; Greenwood and Rossotti 1958; Hybl, Rundle and Williams 1965); the use of n.m.r. has similarly shown that the C1 conformation is retained in solution (Rao and Foster 1963a, 1965; Geddes 1965).

Postulating the existence of the B1 and 3B conformations ignores the fact that the boat form itself has a high energy relative to the chair form. Thus, rather more than thirty years ago, Gorin, Kauzmann and Walter (1939) suggested that the boat conformation was inherently unstable, and therefore unlikely to be found in pyranose sugars. Their suggestion was indeed correct—an energy difference of approximately 5.5 kcal/mol between the chair and boat forms of cyclohexane has been detected experimentally (Johnson, Margrave, Bauer, Frisch, Dreger and Hubbard 1961). Computer calculations carried out by Rao, Sundararajan, Ramakrishnan and Ramachandran (1967) confirm that the energy of the C1 conformation is much lower than that of any other form.

The available evidence indicates, therefore, that the solution behaviour of amylose must be explained on the basis of a model in which the constituent glucose residues are always to be found in the C1 conformation.

**4.3.2.** Reproducibility of the Viscosity–Molecular Weight Relation. Most information regarding the conformation of amylose in solution has been deduced from the relation between limiting viscosity number and molecular weight, that is, the Mark–Houwink–Sakurada equation (4.59). It is therefore of some interest to compare the values of the parameters $K_a$ and $a$ obtained by different investigators and shown in table 4.4.

The most important inference to be drawn from table 4.4 is that the model of amylose as a rod-like molecule is not applicable. As we have seen, theory predicts a value of 1.8 for the exponent $a$ if this model is applicable; of the 22 relations recorded in table 4.4, only four (equations 4.85, 86, 89, and 90) are greater than the limiting value of 0.80 ascribed by Flory to a coil with excluded volume. We must therefore explain the hydrodynamic behaviour of amylose in terms of a coiled molecule. However, the interpretation is made rather more complex by the existence of $a$ values of greater than 0.8, suggesting, if the Flory model is accepted, that the chain has not reached a sufficient size for either (a) the Gaussian approximation, or (b) the model of the impermeable coil, or both, to be valid.

*In aqueous* KCl *solution*, the consensus would suggest that a value for the exponent $a$ of approximately 0.50 is appropriate. This observation is in agreement with the fact that the light-scattering virial coefficient of amylose is zero at about 25°C in aqueous KCl solution (Everett and Foster 1959b; Banks and Greenwood 1963a, b). However, it is difficult to reconcile these findings with the result obtained by Burchard (1963) for amylose in water. The high value of $a$ recorded in this instance (see equation 4.77) would suggest either a basically inflexible molecule, or one swollen by solvent–solute interaction, or both. Burchard (1963) also records a high value for the second virial coefficient of the amylose–water system, suggesting that water is a thermodynamically good solvent for the polysaccharide. This appears to be unlikely because of the marked instability of amylose in water, a subject we have touched upon earlier. We have shown (Banks and Greenwood 1963a) that the amylose samples employed in our work exhibit exactly the same limiting viscosity number in water as they do in aqueous salt solution. Griffin, Erlander and Senti (1967) also found that same value of the limiting viscosity number in both water and aqueous salt solution.

In our studies, we have invariably used the amylose in either the form of the wet butan-1-ol complex, which is readily soluble in water, or in the dehydrated form obtained by repeated stirring of the complex with butan-1-ol and vacuum drying. The latter material is not completely soluble in water, even after prolonged boiling, and for this reason aqueous solutions are prepared by dissolving the dried amylose in cold dilute caustic alkali, and subsequently neutralizing with acid (hence the presence of KCl in the solution). To obtain solutions of amylose in water, we employ the wet butan-1-ol complex, which is easily soluble in warm water, and subsequently remove excess butan-1-ol by steam distillation. On the other hand, Burchard

*Table* 4.4. The parameters $K_a$ and $a$ of the Mark–Houwink–Sakurada equation for amylose, and water-soluble amylose derivatives, in various solvents.

| solvent | $K_a$ ($[\eta]$ in ml/g) | $a$ | equation number | reference |
|---|---|---|---|---|
| | | AMYLOSE | | |
| water | $1.32\times10^{-2}$ | 0.68 | 4.77 | Burchard (1963) |
| 0.33M KCl | $1.13\times10^{-1}$ | 0.50 | 4.78 | Everett and Foster (1959b) |
| | $1.12\times10^{-1}$ | 0.50 | 4.79 | Banks and Greenwood (1963a) |
| | $1.15\times10^{-1}$ | 0.50 | 4.80 | Banks and Greenwood (1968c) |
| 0.50M KCl | $0.55\times10^{-1}$ | 0.53 | 4.81 | Cowie (1961a) |
| 0.15M KOH | $8.36\times10^{-3}$ | 0.77 | 4.82 | Banks and Greenwood (1968d) |
| 0.20M KOH | $6.92\times10^{-3}$ | 0.78 | 4.83 | Banks and Greenwood (1969a) |
| 0.50M KOH | $8.50\times10^{-3}$ | 0.76 | 4.84 | Everett and Foster (1959b) |
| 1.00M KOH | $1.18\times10^{-3}$ | 0.89 | 4.85 | Cowie (1961a) |
| 0.50M NaOH | $3.64\times10^{-3}$ | 0.85 | 4.86 | Burchard (1963) |
| dimethylsulphoxide | $3.06\times10^{-2}$ | 0.64 | 4.87 | Everett and Foster (1959b) |
| | $1.51\times10^{-2}$ | 0.70 | 4.88 | Banks and Greenwood (1968d) |
| | $3.95\times10^{-3}$ | 0.82 | 4.89 | Burchard (1963) |
| | $1.25\times10^{-3}$ | 0.87 | 4.90 | Cowie (1961a) |
| dimethylsulphoxide–acetone mixture | $8.3\times10^{-2}$ | 0.51 | 4.91 | Burchard (1963) |
| ethylene-diamine | $1.55\times10^{-2}$ | 0.70 | 4.92 | Cowie (1961a) |
| formamide | $2.26\times10^{-2}$ | 0.67 | 4.93 | Burchard (1963) |
| | $3.05\times10^{-2}$ | 0.62 | 4.94 | Banks and Greenwood (1968d) |
| | | SODIUM CARBOXYMETHYL AMYLOSE(1) | | |
| 0.65M NaCl | $1.35\times10^{-1}$ | 0.50 | 4.95 | Patel, Patel and Patel (1967) |
| 0.50M NaCl | $2.09\times10^{-1}$ | 0.53 | 4.96 | Brant and Min (1969) |
| 0.78M NaCl | $3.70\times10^{-2}$ | 0.61 | 4.97 | Goebel and Brant (1970) |
| | | DIMETHYLAMINOETHYLAMYLOSE HYDROCHLORIDE(1) | | |
| 0.78M NaCl | $8.34\times10^{-2}$ | 0.55 | 4.98 | Goebel and Brant (1970) |

[1] These ethers of amylose vary in degree of substitution *DS*, which is defined as the proportion of glucose residues substituted. The relevant values of *DS* are: equation (4.95), $DS=0.80$; (4.96), $DS=0.55$; (4.97), $DS=0.29$; (4.98), $DS=0.20$.

(1963) uses a drying procedure that results in water-soluble material. It may be that the difference in the drying procedures accounts for the apparently anomalous behaviour of amylose in water recorded by that author. Our observation, that the light-scattering technique of Banks, Greenwood and Sloss (1969) cannot be successfully applied to amylose samples prepared by the method of Buchard, gives some support to our suggestion—the addition of *alpha*-amylase, which should reduce the polymer to fragments too small to scatter light, causes the expected initial rapid decrease in turbidity, but this is followed by an increase, a phenomenon we have witnessed only when using amylose dried in this specific way.

Whilst there is pleasing agreement in the values of the exponents recorded by those who have examined the amylose–aqueous KCl system, it must be pointed out that the results of Cowie (1961a) can only be brought into coincidence with those of Everett and Foster (1959b) and Banks and Greenwood (1963a, 1968c) if the molecular weights he records are divided by a factor of two. Burchard (1968) also suggested that Cowie's results for amylose in dimethylsulphoxide would be virtually identical with his own, if this supposition were made. Although such an operation would indeed improve the overall consistency of the results in table 4.4, it is difficult to conceive how an error of this magnitude could be made in the first place.

*For the results in aqueous alkali*, there is a high measure of agreement between equations 4.82, 83 and 84 and also between equations 4.85 and 86. It is most unfortunate that there is so little similarity between the two sets of results.

*In the case of dimethylsulphoxide*, four investigations have yielded four quite distinct results—the values of the exponent $a$ are to be found in the range 0.60–0.87. Again, it is very difficult to present any rational explanation as to why such wide variations should be encountered. Dimethylsulphoxide is a good solvent for starch components, and in our experience such solutions are stable for prolonged periods; we would expect a reasonable degree of correlation between different laboratories.

*Other Solvents.* The remainder of the solvents for amylose have not yet been sufficiently widely used to make a useful inter-laboratory comparison, and, in the case of the amylose ethers, we note only that the exponent $a$ decreases with increasing degree of substitution.

*Retrogradation.* In dealing with aqueous solutions of amylose one must constantly be aware of the possibility of retrogradation, for this phenomenon would invalidate any physical measurements carried out on the solution. The molecular weights on which (4.79) is based were obtained from light-scattering measurements on the native amylose in aqueous KCl solution; (4.80), on the other hand, made use of molecular weights obtained from measurements on amylose triacetate, in which there is no possibility of retrogradation. The excellent agreement between (4.79) and (4.80) shows quite clearly that retrogradation does not take place in solutions of the concentrations necessary for physical measurements, within the time-scale of the experiment.

*Linearity of Amylose Fractions.* Husemann, Burchard, Pfannemüller and Werner (1961) and Burchard (1963) have criticized the use of fractions of natural amylose in hydrodynamic studies, and they prefer to use enzymically-synthesized material in order to guarantee its linearity. We agree with the principle of their criticism, if not with its direct expression. Husemann and Burchard were concerned with the possibility of the presence of contaminating amylopectin—a situation that we consider unlikely because of the ease of establishing the purity of the amylose samples by their reaction with iodine. On the other hand, it is our contention that even in a population of molecules which bind the amount of iodine associated with the term 'amylose', a considerable fraction possess long-chain branching (see section 2.3.5). The presence of such molecules invalidates the various theoretical relations established in earlier sections, and it is therefore essential that the fractions employed in hydrodynamic studies be shown to be linear. Thus a necessary pre-condition to such a study is the enzymic characterization of the material. Many of the results quoted in table 4.4 were obtained using samples of amylose that had been inadequately characterized by enzymic techniques, and this makes their interpretation rather uncertain. For this reason, we shall concentrate mainly on the results obtained in our own laboratories.

*Shear Dependence of* $[\eta]$. In dealing with polymers in good solvents, an effect is encountered that is liable to be much more serious than it is in poor solvents, namely the shear dependence of the limiting viscosity number. The situation with respect to flexible macromolecules is rather confused; there is no unanimity of opinion concerning the origin of the shear dependence of the limiting viscosity number. In fact, the theories of Kirkwood and Riseman (1956), Rouse (1955), and of Zimm (1956) do not even predict the existence of such an effect; a number do predict its existence, for example, those of Kuhn and Kuhn (1943), Peterlin and Čopič (1956), and of Cerf (1948), but are not agreed as to its source.

The importance of shear effects is that they depend on the limiting viscosity number, becoming greater as this quantity increases. Thus when the limiting viscosity numbers of a series of fractions of a linear polymer measured at a given shear rate, there will be a progressive error in the values obtained with increasing molecular weight—the limiting viscosity numbers will progressively become lower than they would if measured at zero rate of shear. As a consequence of this error, the value recorded for the exponent of the Mark–Houwink–Sakurada relation will be reduced, and the deductions made from such a value, using the various theoretical treatments considered earlier, may well be suspect. For this reason, the viscosity behaviour of amylose in a very good solvent (0.15 M potassium hydroxide) was measured as a function of the rate of shear, and the limiting viscosity number at zero rate of shear obtained by an extrapolation procedure (Greenwood, Hourston and Procter

1970). A Couette-type viscometer enabled viscosities to be measured at very low shear rates (in the range 0.5–50 $s^{-1}$), so that the extrapolation to zero shear rate could be carried out with high accuracy. The values so obtained may then be compared with those measured in the standard Ubbelohde suspended-level viscometer, with a shear rate of approximately 1200 $s^{-1}$— this latter instrument has been employed in all our studies of the solution behaviour of amylose, and viscometers having similar shear rates have also been used by most other workers. The results of this study in table 4.5 show that, over the range of shear-rates of 0–1200 $s^{-1}$, there is no change in the limiting viscosity numbers, within experimental error, of any of the amylose fractions in aqueous alkali. If there were a significant effect of shear on limiting viscosity number, one would expect to detect quite profound differences in table 4.5. Even for the samples of largest molecular size (A1 and P1), no difference was evident. We may conclude, therefore, that limiting viscosity numbers measured in suspended-level viscometers will not be in serious error due to the neglect of shear corrections.

*Table* 4.5. The effect of shear on the limiting viscosity numbers (in ml/g) of amylose in 0.15 M KOH (Greenwood, Hourston and Procter 1970).

| fraction | $[\eta]_0$ (1) | $[\eta]_{1200}$ (2) |
|---|---|---|
| P1 | 1040 | 1010 |
| P2 | 855 | 855 |
| P3 | 760 | 770 |
| P5 | 510 | 510 |
| A1 | 1060 | 1090 |
| A2 | 625 | 625 |
| A3 | 430 | 430 |

[1] $[\eta]_0$ is the limiting viscosity number extrapolated to zero rate of shear.
[2] $[\eta]_{1200}$ is the limiting viscosity number measured at a shear rate of 1200 $s^{-1}$.

This conclusion is, in fact, the concensus of all who have investigated the hydrodynamic behaviour of amylose with the exception of Cowie (1961a, b) who has employed very large corrections for shear effects, both for amylose and its derivatives in a variety of solvents. If, as seems very likely, the corrections are erroneous, the values of the exponent *a* obtained by Cowie for amylose in various solvents and recorded in table 4.4 will be too high. Without shear corrections all of the values of *a* obtained by Cowie would be less than 0.80, the upper limit predicted by the Flory theory for a non-draining coil in a good solvent. Only the results of Burchard, who agrees that shear corrections are unnecessary, would then appear anomalously high.

Whilst the limiting viscosity number of amylose is independent of the rate of shear, the same is not true of the viscosity number. Figure 4.3 shows typical graphs of viscosity number as a function of concentration. The two sets of data, referring to values obtained at shear rates of 0 $s^{-1}$ and 1200 $s^{-1}$ respectively, can be extrapolated to a common origin, but the straight lines are quite distinct, i.e. although the limiting viscosity number does not depend on rate of shear, the Huggins' constant does. There is as yet no complete theoretical understanding of the parameters that determine the magnitude of this constant. Various hydrodynamic theories have been developed that make use of the Huggins' constant; in view of the fact that this parameter is very dependent on the rate of shear, we have avoided discussing them here.

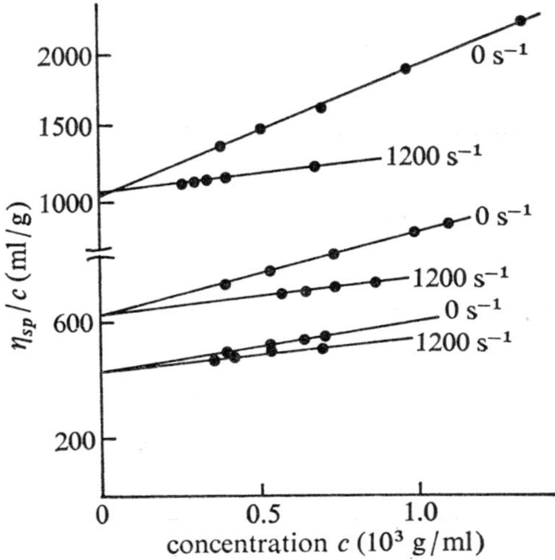


*Figure* 4.3. The effect of rate of shear on the viscosity number of amylose in 0.15 M potassium hydroxide (Greenwood, Hourston and Procter 1970).

### 4.3.3. Amylose in Neutral Aqueous Solution.

The following relations have been obtained for amylose in neutral, aqueous solution at 25°C:

$$[\eta] = 1.15 \times 10^{-1} \bar{M}_w^{0.50} \tag{4.80}$$

$$\langle R_{GO}^2 \rangle = 9.42 \times 10^{-18} \bar{M}_w^{1.00} \tag{4.99}$$

$$S_0 = 1.11 \times 10^{-15} \bar{M}_w^{0.50} \tag{4.100}$$

The experimental results (Banks and Greenwood 1963a, 1968c, 1969a) on which these relations are based are given in tables 4.6(a) and (b). Those in table 4.6(a) were obtained by applying the light-scattering technique to solutions of amylose in aqueous KCl, whereas in table 4.6(b) the molecular weights were obtained from light-scattering measurements on the triacetate in nitromethane.

All of the exponents in equations 4.80, 99 and 100 are those expected for an impermeable coil in the absence of excluded volume, i.e. at the *theta*-temperature. We have already noted that the light-scattering virial coefficient is zero in aqueous KCl solution, as indeed it is in water (Banks, Greenwood and Sloss 1969). The conclusion to be drawn from these observations is that, hydrodynamically, amylose conforms to the model of the impermeable random coil.

*Table* 4.6(a). Results of viscosity sedimentation and light-scattering measurements on amylose in 0.33 M potassium chloride solution (Banks and Greenwood 1963a,b).

| fraction | $[\eta]$ (ml/g) | $S_0 \times 10^{13}$ (s) | $\bar{M}_w \times 10^{-6}$ | $\langle R_G^2 \rangle^{1/2}$ (Å) |
|---|---|---|---|---|
| 1b | 187 | 16.3 | 2.29 | 600 |
| 2c | 158 | 14.8 | 1.70 | 505 |
| 5 | 97 | 9.83 | 0.80 | 345 |
| 6 | 62 | 6.58 | 0.35 | 230 |
| 7 | 42 | 4.55 | 0.16 | 160 |

*Table* 4.6(b). Viscosity and molecular weight data for amylose in 0.33 M potassium chloride solution (Banks and Greenwood 1968c).

| fraction | $[\eta]$ (ml/g) | $\bar{M}_w \times 10^{-6}$ | fraction | $[\eta]$ (ml/g) | $\bar{M}_w \times 10^{-6}$ |
|---|---|---|---|---|---|
| 1 | 154 | 1.75 | 7 | 78 | 0.460 |
| 2 | 126 | 1.22 | 8 | 74 | 0.424 |
| 3 | 113 | 0.945 | 9 | 71 | 0.380 |
| 4 | 100 | 0.754 | 10 | 67 | 0.320 |
| 5 | 90 | 0.613 | 11 | 52 | 0.211 |
| 6 | 81 | 0.488 | 12 | 33 | 0.083 |

The characteristic ratio $C_n$ can be calculated from the data shown in tables 4.6(a) and (b), either directly by making use of the light scattering radius of gyration, or indirectly by calculating this parameter from sedimentation and viscosity measurements, using the appropriate values for the constants $P_0$ and $\Phi_0$. However, in view of the somewhat uncertain nature of the values to be attributed to these constants, we prefer, when possible, to make use of the values of $\langle R_{G0}^2 \rangle$ obtained directly from light-scattering. The values of the characteristic ratio, defined as

$$C_n = 6\langle R_{G0}^2 \rangle_w / \bar{n}_w l^2 \qquad (4.101)$$

are shown in table 4.7, the value of $l$ being taken as 4.4Å. Because the light scattering measurements yield a weight-average molecular weight and a

$Z$-average radius of gyration, it is necessary to convert the experimental ratio of $\langle R_{GO}^2 \rangle_z / \bar{n}_w$ to $\langle R_{GO}^2 \rangle_w / \bar{n}_w$. This conversion is performed by assuming that the fractions possess a distribution of molecular weights governed by the Schulz (1939) relation

$$W(M) = [y^{h+1}/\Gamma(h+1)] M^h \exp(-yM) \quad (4.102)$$

with

$$y = h/\bar{M}_n = (h+1)/\bar{M}_w = (h+2)/\bar{M}_z \quad (4.103)$$

and where $\Gamma(h)$ is the gamma function of $h$. The factor bringing the radius of gyration to a weight average value is then given by

$$\langle R_{GO}^2 \rangle_z / \bar{n}_w = [(h+2)/(h+1)][\langle R_{GO}^2 \rangle_w / \bar{n}_w] \quad (4.104)$$

Fractions similar to those used in this work were found to have $\bar{M}_w/\bar{M}_n = 1.5$, so that $h = 2$.

*Table* 4.7. The characteristic ratio $C_n$ obtained for amylose in neutral aqueous solution from light-scattering measurements.

| $\langle R_{GO}^2 \rangle_z$ ($\times 10^{12}$ cm$^2$) | $\bar{n}_w$ ($\times 10^{-4}$) | $\dfrac{6\langle R_{GO}^2 \rangle_z}{\bar{n}_w l^2}$ | $C_n = \dfrac{6\langle R_{GO}^2 \rangle_w}{\bar{n}_w l^2}$ |
|---|---|---|---|
| 36.0 | 1.41 | 7.9 | 5.9 |
| 25.5 | 1.05 | 7.5 | 5.6 |
| 11.9 | 0.494 | 7.5 | 5.6 |
| 5.3 | 0.216 | 7.6 | 5.7 |
| 2.5 | 0.099 | 7.8 | 5.8 |

The values of the characteristic ratios, corrected for heterogeneity, of the various fractions are shown in table 4.7. They are constant, within the constraints of experimental error, over the range of molecular weights investigated, and may therefore be identified with $C_\infty$, the asmyptotic value of the characteristic ratio reached at high molecular weight. The mean value of the characteristic ratio for the fractions used in this work is thus only 5.7, a value not much removed from those tabulated (Flory 1969) for various synthetic polymers.

Another measure of the flexibility of the polymer coil is provided by the parameter $\sigma$, where

$$\sigma = (\langle R_{GO}^2 \rangle / \langle R_{GO}^2 \rangle_f)^{1/2} \quad (4.105)$$

and $\langle R_{GO}^2 \rangle_f$ is the mean-square radius of gyration calculated on the basis of free rotation. For amylose composed of glucose units in the C1 conformation, $\langle R_{GO}^2 \rangle_f$ may be obtained from (4.74). Substitution of this value in (4.105) then gives $\sigma = 2.4$. This value of the parameter $\sigma$ is quite comparable to those listed by Kurata and Stockmayer (1963) for a range of synthetic polymers, all of which are usually regarded as flexible in the hydrodynamic sense.

The values of the above relations, together with the zero value for the second virial coefficient, show that amylose has the characteristics of a random coil in its unperturbed state. However, although these studies show the macromolecule *overall* to be sufficiently flexible for it to conform to the random coil model, they do not reveal any information regarding the fine structure of the polymer. Thus, it was suggested, any of the three structures shown in figure 4.4 could conform to the relations expressed in equations 4.80, 4.99 and 4.100.

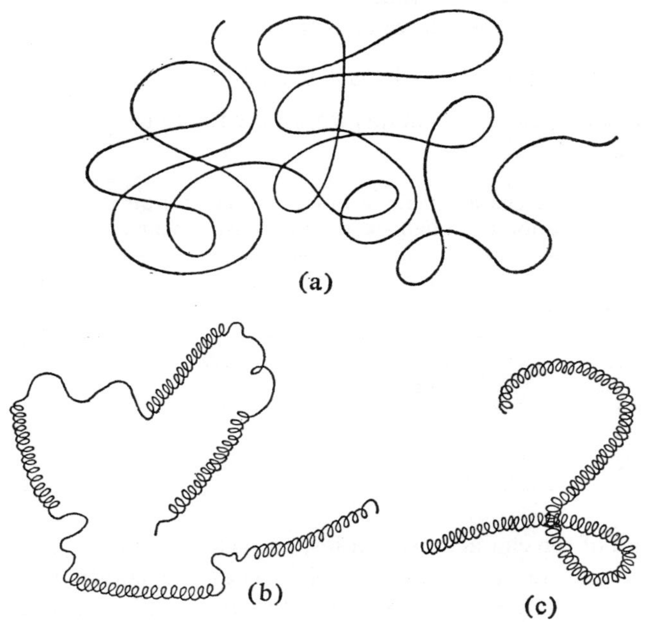


*Figure* 4.4. Models proposed for the amylose molecule in aqueous solution: (a) random coil; (b) interrupted helix; (c) deformed helix (Banks and Greenwood 1971b).

In figure 4.4(a) the polymer is completely lacking in tertiary structure, and the arrangement of successive units is *random* within the constraints placed upon the system by the skeletal parameters (fixed bond angle and restricted rotation about glycosidic bonds). Figure 4.4(b) shows the concept of amylose developed by Holló and Szejtli (1958c)—the *interrupted helix*—in which the molecule is regarded as being composed of long helical sections, each containing approximately 120 glucose residues, with short regions of random coil between them. The latter type of region confers on the macromolecule as a whole the flexibility necessary for it to conform to the model of the random coil. Figure 4.4(c) shows the model of amylose proposed by Rao and Foster (1963b), and subsequently elaborated by Foster (1965).

This *deformed helix* is rather similar to the model of the worm-like coil developed by Kratky and Porod (1949).

It is of interest to note at this point that the models of the interrupted helix and the deformed helix have also been employed in attempted explanations of the hydrodynamic behaviour of double-stranded DNA. In this case, the interrupted helix model postulates that the disordered (random coil) regions arise as a result of the mis-matching of base pairs in the two strands. Such regions are, of course, absent in the model of the deformed helix, and the overall conformation of the DNA duplex then results from the considerable degree of coiling that has occurred at the very high molecular weights characteristic of this macromolecule. We must emphasize yet again that any linear polymer would be expected to conform to the model of the random coil in the limiting case of exceedingly high molecular weight. Indeed, Kurata and Stockmayer (1963) attribute to Benoit the remark that an infinitely long steel rod would behave as a random coil.

In the case of amylose, the proponents of both the interrupted helix and deformed helix models lay great stress upon the fact that any linear polymer will tend to the random coil model, no matter how inflexible its structure. The necessary corollary, namely that the desired constancy in the ratio of $\langle R_{G0}^2 \rangle / M$ is attained only at much higher values of molecular weight that it is for a more flexible polymer, is seldom mentioned. The results of the light-scattering studies show that the asymptotic limit of the characteristic ratio is valid to at least a degree of polymerization of 1000 glucose units, corresponding to an interrupted helix of only eight segments; if this parameter is derived from the viscosity–molecular weight relation of (4.80), constancy is observed even to a lower limit ($\bar{n}_w = 500$) at which only four segments can be presented. In qualitative terms, it is difficult to comprehend how a coil of but four segments can immobilize sufficient solvent so as to be considered, hydrodynamically, impermeable.

However, various experimental facts have been offered in support of the two helical models. The most important of these are: (a) amylose has a large statistical-segment length; (b) the temperature dependence of the limiting viscosity number in neutral aqueous solution is decidedly anomalous; (c) the unperturbed dimensions of the macromolecule are significantly dependent on the solvent medium employed; (d) the dimensions of the macromolecule, determined by viscosity measurements, do not change as iodine is taken up until the point of saturation (*ca.* 20 mg iodine/100 mg amylose, at 20°C) is reached; and (e) on going from neutral to alkaline solution, amylose undergoes a conformational change which is analogous to the helix-to-coil transition observed with many proteins and synthetic polypeptides. The relevance of these observations will be considered in the following sections.

### **4.3.4.** Statistical Segment Length.

Probably the two parameters most used to measure the 'stiffness' of a polymer are the Kuhn statistical segment-length $A_m$ and the Kratky–Porod persistence length $q$. These parameters are defined by

$$\langle R_{G0}^2\rangle_w/\bar{n}_w = A_m b_0/6 \tag{4.106}$$

and

$$\langle R_{G0}^2\rangle_w/\bar{n}_w = q b_0/3 \tag{4.107}$$

and thus are obtained directly from the measured radius of gyration, if the value of the projected bond length $b_0$ is known. The parameters $A_m$ and $q$ are closely related, so that it is necessary only to consider one—we have chosen the Kuhn statistical segment length.

The quantity $A_m$ can also be obtained from measurement of the limiting viscosity number, using the Flory theory, according to which

$$[\eta] = K_\theta M^{1/2} \tag{4.108}$$

where $K_\theta = \Phi_0(\langle R_{G0}^2\rangle/M)^{3/2}$. Taking an average figure of $3.5\times10^{24}$ for the viscosity constant $\Phi_0$ gives a value for $\langle R_{G0}^2\rangle/\bar{n}_w$ of $16.0\times10^{-16}$ cm. The segment length may then be calculated from (4.106).

The statistical segment length may also be calculated directly from the viscosity and sedimentation data, using the relations of Kuhn, Kuhn and Silberberg (1954)

$$n/[\eta] = (100M_0/0.43N_A b_0^2 A_m)[-1.6+2.3\log(A_m/d_h)+ +(b_0 n/A_m)^{1/2}] \tag{4.109}$$

and

$$S_0 = [M_0(1-\bar{V}\rho)/N_A b_0 \eta_0][-0.03+0.16\log(A_m/d_h)+ +0.136(nb_0/A_m)^{1/2}] \tag{4.110}$$

where $d_h$ and $M_0$ are the hydrodynamic thickness and the molecular weight of the monomer unit. The partial specific volume $\bar{V}$ was found to have a value of 0.60 (Banks and Greenwood 1969a).

In all the calculations necessary to derive $A_m$, there is also the rather ambiguous parameter $b_0$—the projected length of the monomer unit. If amylose has no tertiary structure, $b_0$ is merely the length of the glucose residue in the C1 conformation (4.40Å), whereas if the macromolecule is assumed to be helical, and to have the same dimensions as in the crystalline state (i.e. a pitch of 8Å, with six glucose residues per turn), the projected bond length is (8/6)Å. These two extreme values of $b_0$ have been used in conjunction with equations 4.106, 4.109 and 4.110 to obtain the segment lengths shown in table 4.8.

These results demonstrate quite clearly the fallacy in using derived hydrodynamic parameters such as $A_m$ (or $q$) to make deductions regarding the stiffness of the polymer chain. The apparent stiffness is dominated by the value of $b_0$ employed in the calculations. Thus the number of monomer units per statistical segment increases by an order of magnitude on changing $b_0$ from 4.40Å to 1.33Å. Effectively, if one assumes that amylose is helical, the derived statistical segment length will be in agreement with that contention. Conversely, if one assumes that amylose is non-helical, the calculated value of $A_m$ will support that assumption.

*Table* 4.8. The Kuhn statistical segment length, $A_m$, of amylose in neutral aqueous solution (Banks and Greenwood 1968c).

| technique | $b_0$=4.40Å | | | $b_0$=1.33Å | | |
|---|---|---|---|---|---|---|
| | $A_m$ | $n_k$[1] | $M_k$[2] | $A_m$ | $n_k$ | $M_k$ |
| light-scattering | 25 | 6 | 1000 | 84 | 63 | 10200 |
| viscosity—Flory | 22 | 5 | 800 | 73 | 55 | 8900 |
| viscosity—Kuhn | 24 | 5 | 800 | 90 | 68 | 11000 |
| sedimentation—Kuhn | 26 | 6 | 1000 | 86 | 65 | 10500 |

[1] $n_k$ is the number of monomer units per statistical segment.
[2] $M_k$ is the molecular weight of the statistical segment.

The model of the interrupted helix was dealt with in some detail by Hearst and Stockmayer (1962). They specifically considered a broken chain, i.e. straight, rigid segments of equal length connected by universal joints, and suggested that the slope of a graph of $S_0$ (in Svedbergs) as a function of $M^{1/2}$ is given by

$$\text{slope} = 3.24 \times 10^{-4}[(1-\bar{V}\rho)/\eta_0](M_L/L)^{1/2} \tag{4.111}$$

where $M_L = M_0/b_0$ ($b_0$ measured in Å) and $L$ is the length of the statistical segment. From (4.100), the slope of the relevant graph is $1.11 \times 10^{-2}$, giving $L$=164Å for $b_0$=1.33Å, corresponding to 123 glucose units per segments. On the other hand, assuming $b_0$=4.40Å, $L$=50Å, and there are but eleven glucose residues per segment. Once again, the conclusion is predetermined by the chosen value of $b_0$.

A similar criticism applies to the theory of Nagai (1961), which employed the model of the interrupted helix to obtain the following relation:

$$\langle r_0^2 \rangle / M = (1-\tau)l^2 s/M_0 + \tau b_0^2 n_N/M_0 \tag{4.112}$$

where $\tau$ is the fraction of monomer units in the helical form; $s$, the skeletal factor, is $(\langle r_0^2 \rangle/M)(M_0/l^2)$; $l$ is the bond-length of the random coil sections, $b_0$ is the length per monomer unit of the helix, and $n_N$ is the weight-average number of monomer units in the helical segments. In the case of a sufficiently large helical content, (4.112) reduces to

$$\langle r_0^2 \rangle / M = b_0^2 n_N/M_0 \tag{4.113}$$

and calculation shows $n_N$ to be 55. Comparison of (4.6), (4.106) and (4.113) shows that in the extreme case of limited random-coil content, the parameters $n_k$ and $n_N$ are identical.

We conclude, therefore, that the calculation of statistical segment lengths provides no evidence either for, or against, amylose in solution possessing a tertiary structure.

### 4.3.5. Temperature-Dependence of the Limiting Viscosity Number in Neutral Aqueous Solution.

The various hydrodynamic relations, the measured dimensions of the polymer, and the thermodynamic evidence of a negligible second virial coefficient all suggest that a neutral aqueous salt solution at 25°C constitutes a *theta*-solvent for amylose. As we have stressed earlier, the *theta*-solvent must be defined with respect to temperature—increasing the temperature by a few degrees above the *theta*-point leads to the appearance of non-ideal behaviour, due to the now finite excluded volume effect, whereas lowering the temperature causes precipitation of the polymer. However, the behaviour of the amylose in water, or in aqueous salt solution, does not conform to the tenets of classical polymer theory: dilute aqueous solutions of amylose are sufficiently stable down to 2°C to allow physical measurements to be carried out, and the limiting viscosity number of amylose does not change appreciably on heating, as shown in table 4.9.

*Table* 4.9. The limiting viscosity number $[\eta]$ of an amylose fraction ($\bar{n}_w$=10500 glucose residues) in various aqueous solutions as a function of temperature (Banks and Greenwood 1963a).

| solvent | temperature (°C) | $[\eta]$ (ml/g) |
|---|---|---|
| 0.33 M KCl | 17.5 | 157 |
| | 22.5 | 158 |
| | 30.0 | 158 |
| | 45.0 | 158 |
| 0.2 M sodium acetate | 22.5 | 158 |
| | 45.0 | 158 |
| water | 22.5 | 158 |
| | 45.0 | 158 |

Cowie (1962) also reported that the temperature dependence of the limiting viscosity number of amylose in 0.5 M KCl solution was negligible. However, Rao and Foster (1963b) found a value of approximately $-0.003$ deg. $C^{-1}$ for $d\ln[\eta]/dT$, as did Burchard (1963), over a similar range of temperatures. Maywald, Leach and Schoch (1968) measured the specific viscosity of a 0.15 per cent solution of amylose in water over the temperature range 25–95°C and reported a progressive, smooth decrease in this quantity with increasing temperature. The relative decrease in reduced viscosity between these temperature extremes was not very large—only about 16 per cent. Equating the reduced viscosity with the limiting viscosity number shows that $d\ln[\eta]/dT$ is approximately $-0.0025$ deg. $C^{-1}$, a value in close agreement with those obtained by Rao and Foster, and by Burchard.

According to Rao and Foster (1963b) the above experimental observations are compatible with a model which is largely helical at room temperature. They then suggest that on raising the temperature, two mutually-opposed effects are called into play. First, the increased thermal energy of the system causes the helical segments to break down, and the resultant increase in the flexibility of the linear chain leads to a decrease in polymer dimensions. Second, the solvent power of the water is increased by raising the temperature, leading to an expansion of the coil dimensions. If the two effects are of approximately the same magnitude, the consequent change in limiting viscosity number with increasing temperature would be quite small, in agreement with experimental results.

In the above argument a quite unnecessary assumption has been introduced, in which the term 'stiffness' is correlated with the presence of helical structures. To argue that the flexibility of the macromolecule is increased by raising the temperature is justified; to suppose that the state of lower flexibility is necessarily associated with a helical model is not.

However, even if we accept the basic premise of Rao and Foster, and assume the presence of helical regions, the explanation they advance is itself not without anomaly. In particular, they envisage a gradual melting of the helical structure over the entire temperature range. By analogy with polypeptides and DNA, in which over 90 per cent of the change from helix to coil generally occurs within a temperature range of 10 deg. C, this behaviour is decidedly anomalous. The lack of any discontinuity in the relation between limiting viscosity number and temperature caused Maywald *et al.* to reject the helical models of amylose. They suggested that the effect of heat was to dissociate the hydrogen bonding between polysaccharide and water, thereby decreasing the amount of water held by the molecule and causing it to shrink.

It could, of course, be argued that the melting of the helices occurs over a fairly narrow temperature range, but that it is also in this same range that the most rapid increase in excluded volume occurs. Thus it is conceivable, if not very likely, that the change in conformation could occur in such a way that the hydrodynamic volume does not alter markedly. One would expect this transition from a helical model to a random coil to be easily detected by a study of the optical rotation of the system. Dintzis and Tobin (1969) have carried out such a study, and found no evidence to support the postulated helix–coil transformation.

Independently, we came to a somewhat similar conclusion to that of Maywald *et al.* Our explanation (Banks and Greenwood 1968c) is based upon the fact that solvent–polymer interaction depends upon the formation of hydrogen bonds, and so does polymer–polymer interaction. As the temperature is increased, both types of interaction will tend to be diminished. Again, the effects of the two phenomena will be opposed, so that if they are of the same magnitude any change in the polymer dimensions, and hence in the limiting viscosity number, would be rather small. We must emphasize a point which is too often overlooked, namely the nature of the solvent. In the present instance we are concerned with aqueous solutions, and the structure of the solvent, water, is not well understood. It must be remembered that in any highly polar solvent, and particularly in water, there is probably considerable orientation of the solvating molecules at the surface of the polymer. The effect of raising the temperature of such a system would not be easy to predict.

Our main conclusion, however, is that the temperature-dependence of the limiting viscosity number of amylose in neutral aqueous solution, despite being at odds with the theories of polymer solution behaviour, provides no evidence favouring a helical structure for the macromolecule. Although the absence of a marked temperature-dependent change in limiting viscosity number, or in optical rotation, does not completely eliminate the possibility that a helical structure exists, it is our opinion that the failure to observe evidence of such a conformation transition cannot be reconciled with the model originally suggested by Holló and Szejtli (1958c), in which hydrogen bonding was invoked as the stabilizing influence in the helical regions.

**4.3.6.** Unperturbed Dimensions in Good Solvents.

If indeed amylose adopts a helical conformation in solution as a result of inter-residue bonding (hydrogen bonding, for example) one would not expect that bonding to exist in all solvents. Therefore, when Rao and Foster (1963b) reported that the unperturbed dimensions of amylose were grossly dependent on the nature of the solvent, it appeared that at last a major piece of evidence had been provided in favour of the existence of helical structures in at least some solvents. The conclusions were based not on direct experimental evidence, but rather on the use of various theoretical treatments that were applied to the viscosity data obtained using good solvents, enabling the effect of long-range (excluded volume) forces to be separated from the short-range (skeletal) forces. The way in which polymer theory may be manipulated in order to perform this separation of long- and short-range forces will now be considered. As an example, we use the results obtained by ourselves for amylose in a series of good solvents—formamide, dimethylsulphoxide and 0.15 M KOH solution. The experimental results (Banks and Greenwood 1968d) are shown in table 4.10; the Mark–Houwink–Sakurada relations are given by (4.83) for the alkali, (4.88) for dimethylsulphoxide and (4.94) for formamide.

*Schaefgen–Flory Relation.* According to Flory

$$\alpha=([\eta]/[\eta]_\theta)^{1/3} \qquad (4.114)$$

and

$$\alpha^5-\alpha^3=Cz \qquad (4.115)$$

where $C$ is a constant and $z$ is defined by

$$z=0.33BM^{1/2}(\langle r_0^2\rangle/M)^{-3/2} \qquad (4.116)$$

*Table* 4.10. Molecular weights and limiting viscosity numbers for amylose fractions in good solvents (Banks and Greenwood 1968d).

| fraction | $\bar{M}_w \times 10^{-6}$ | $[\eta]_F$[1] (ml/g) | $[\eta]_{Me_2SO_4}$[2] (ml/g) | $[\eta]_{KOH}$[3] (ml/g) |
|---|---|---|---|---|
| 1 | 1.75 | 230 | 365 | 495 |
| 2 | 1.22 | 182 | 268 | 395 |
| 3 | 0.954 | 155 | 232 | 340 |
| 4 | 0.754 | 132 | 197 | 285 |
| 5 | 0.613 | 118 | 172 | 250 |
| 6 | 0.488 | 102 | 145 | 205 |
| 7 | 0.460 | 96 | 133 | 190 |
| 8 | 0.424 | 94 | 125 | 180 |
| 9 | 0.380 | 88 | 121 | 165 |
| 10 | 0.320 | 78 | 104 | 145 |
| 11 | 0.211 | 61 | 79 | 105 |
| 12 | 0.083 | 36 | 44 | 54 |

[1] $[\eta]_F$ is the limiting viscosity number measured in formamide.
[2] $[\eta]_{Me_2SO_4}$ is the limiting viscosity number measured in dimethyl-sulphoxide.
[3] $[\eta]_{KOH}$ is the limiting viscosity number measured in 0.15 M aqueous potassium hydroxide solution.

with $B$ being a solvent–solute interaction parameter. Substitution of (4.114) and (4.115) in (4.116) gives the Schaefgen–Flory relation (Flory and Fox 1951):

$$[\eta]^{2/3}/M^{1/3} = K_\theta^{2/3} + C(3/2\pi)^{3/2} B K_\theta^{5/3} (\langle r_0^2 \rangle / M)^{-3/2} (M/[\eta]) \quad (4.117)$$

Thus, if the limiting viscosity number is known as a function of molecular weight over a sufficiently wide range of molecular weights, the parameter $K_\theta$ (and hence $\langle r_0^2 \rangle / M$)—see (4.4) and (4.54)—may be obtained from the intercept of the graph of $[\eta]^{2/3}/M^{1/3}$ as a function of $M/[\eta]$. The data for amylose are shown in this way in figure 4.5.

For each solvent, reasonable linearity is observed over the whole experimentally-accessible molecular weight range. However, the intercept varies with the solvent in a regular manner, decreasing as the solvent power increases. Exactly the same phenomenon is displayed by many polymer–solvent systems (Kurata and Stockmayer 1963), and it is now generally accepted that such observations reflect the deficiencies of (4.117) rather than a solvent-dependence of the unperturbed dimensions.

*Kurata–Stockmayer–Roig Theory.* In the treatment of Kurata, Stockmayer and Roig (1960), (4.114) is used in conjunction with the relation

$$\alpha^3 - \alpha = 1.10 z g(\alpha) \quad (4.118)$$

where

$$g(\alpha) = [4\alpha^2/(3\alpha^2 + 1)] \quad (4.119)$$

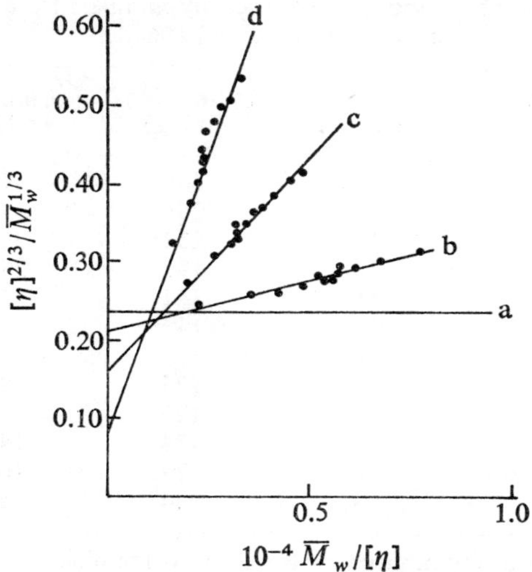


*Figure* 4.5. Graph of $[\eta]^{2/3}/\bar{M}_w^{1/3}$ as a function of $\bar{M}_w/[\eta]$ for amylose in: (a) 0.33 M potassium chloride; (b) formamide; (c) dimethylsulphoxide; (d) 0.15 M potassium hydroxide.

By combining (4.117) and (4.118) and substituting for $z$ we obtain*

$$[\eta]^{2/3}/M^{1/3} = K_\theta^{2/3} + 0.363\Phi_0 B[g(\alpha)M^{2/3}/[\eta]^{1/3}] \qquad (4.120)$$

The parameter $K_\theta$ can then be evaluated from a graph of $[\eta]^{2/3}/M^{1/3}$ against $g(\alpha)M^{2/3}/[\eta]^{1/3}$. The value of $g(\alpha)$ is unknown, therefore an approximate value is obtained for $K_\theta$ from a plot of $[\eta]^{2/3}/M^{1/3}$ as a function of $M^{2/3}/[\eta]^{1/3}$. This then allows $[\eta]_\theta$ to be calculated from (4.108) and subsequently $\alpha$ from (4.114). This derived value of $\alpha$ is inserted in (4.119), enabling $g(\alpha)$ to be calculated, and so allowing (4.120) to be used. The initial value of $g(\alpha)$ to be obtained is usually sufficiently accurate to make unnecessary

*According to the Flory theory $\alpha = ([\eta]/[\eta]_\theta)^{1/3} = (\langle r^2\rangle/\langle r_0^2\rangle)^{1/2}$ i.e. the hydrodynamic radius and the statistical radius increase identically as a result of solvent–solute interaction. For the third power theories this is no longer the case, the hydrodynamic radius increasing less rapidly than does the statistical radius. Strictly, it is then necessary to define *two* expansion coefficients, namely $\alpha_\eta = ([\eta]/[\eta]_\theta)^{1/3}$ and $\alpha = (\langle r^2\rangle/\langle r_0^2\rangle)$, with $\alpha_\eta \leqslant \alpha$. The viscosity 'constant' $\Phi_0$ then becomes a variable, designated $\Phi$, and defined as $\Phi = \Phi_0(\alpha_\eta/\alpha)^3$. The *form* of the relation between $\alpha$ and $z$ is not affected by using $\alpha_\eta$ instead of $\alpha$, but the numerical factors of $z$ are altered. Equation 4.118 is defined in terms of $\alpha_\eta$; if $\alpha$ were used instead it would be necessary to modify this relation to $\alpha^3 - \alpha = 1.33zg(\alpha)$. Similarly, (4.121) is applicable to viscosity results; if it were intended to calculate the unperturbed dimensions using measured values of the radius of gyration, it would be necessary to use a numerical factor of 2.0 rather than 1.55.

any further iterative procedures. Figure 4.6 shows the data for amylose graphed according to (4.120).

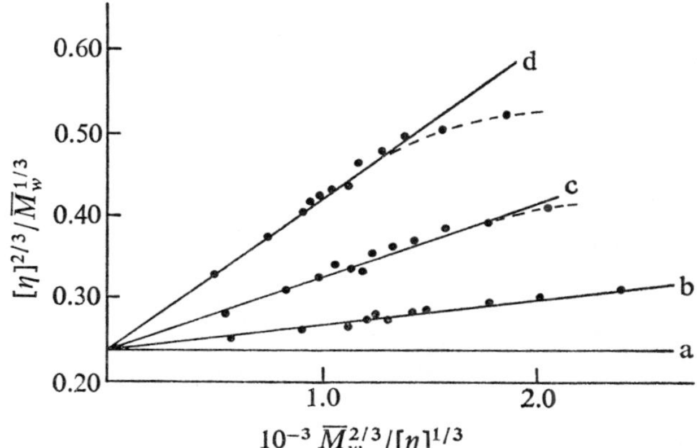


*Figure* 4.6. Graph of $[\eta]^{2/3}/\bar{M}_w^{1/3}$ as a function of $\bar{M}_w^{2/3}/[\eta]^{1/3}$ for amylose in: (a) 0.33 M potassium chloride; (b) formamide; (c) dimethylsulphoxide; (d) 0.15 M potassium hydroxide.

It is found that linearity is not always observed over the entire molecular weight range; the graph bends towards the abscissa at high molecular weights, especially in very good solvents, i.e. at high $z$-values. Various reasons for this curvature will be examined when dealing with the closely-allied relation of Stockmayer and Fixman (1963). For the moment, we note only that the curves through the points may be so constructed as to pass, without strain, through a common intercept on the ordinate axis. It is significant that the results obtained in formamide, in which the value of the exponent of the Mark–Houwink–Sakurada relation is only 0.62 (see equation 4.94) and which is therefore a rather mediocre solvent, do conform to the expected straight line, and that this line has the same intercept as does the non-linear relations derived from measurements in the good solvents, dimethylsulphoxide and aqueous alkali.

*Burchard–Stockmayer–Fixman Treatment.* The remaining theory to be considered is usually known as the Stockmayer–Fixman treatment. Their explicit statement of the equation (Stockmayer and Fixman 1963) occurred some two years after it was first enunciated by Burchard (1961). For accuracy, therefore, we will refer to it as the Burchard–Stockmayer–Fixman equation. It takes the form

$$\alpha^3 = 1 + 1.55z \tag{4.121}$$

which, in conjunction with (4.114) and the definition of $z$ (4.116), gives

$$[\eta]/M^{1/2} = K_\theta + 0.51BM^{1/2} \tag{4.122}$$

Thus a graph of $[\eta]/M^{1/2}$ as a function of $M^{1/2}$ should be linear, and $K_\theta$ may then be evaluated from the intercept on the ordinate axis. Figure 4.7 shows the results graphed according to (4.122) for the various amylose fractions.

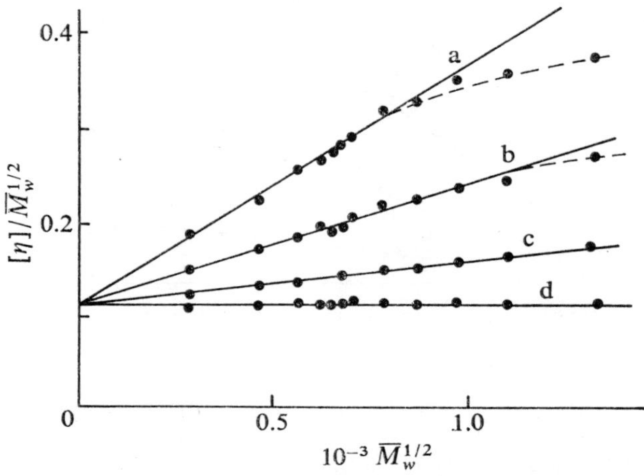


*Figure* 4.7. Graph of $[\eta]/\bar{M}_w^{1/2}$ as a function of $\bar{M}_w^{1/2}$ for amylose in: (a) 0.15 M potassium hydroxide; (b) dimethylsulphoxide; (c) formamide; (d) 0.15 M potassium chloride (Banks and Greenwood 1968d).

Again, as in the Kurata–Stockmayer–Roig treatment, the relation is not linear over a wide range of $z$, but curves towards the abscissa at high values of that parameter. Despite the lack of any apparent similarity between (4.120) and (4.122), the Kurata–Stockmayer–Roig and Burchard–Stockmayer–Fixman theories are, to an excellent approximation, mathematically equivalent over the entire range of $z$. The discussion of the deviations from the relations expressed by (4.120) and (4.122) may therefore be conveniently kept together.

As we showed earlier, the reproducibility in the viscosity–molecular weight relation is not always very good when comparing results from different laboratories. However, if only the results from any one laboratory are considered, a high degree of internal consistency is apparent when using the Burchard–Stockmayer–Fixman relation. For example, we show in figure 4.8 the data of Burchard (1963) graphed according to (4.122). Again, a straight line may be drawn through the points representing low values of $z$, but curvature towards the abscissa is quite apparent at high molecular weights. Using the closely-allied Kurata–Stockmayer–Roig treatment, Cowie (1962) also concluded that a common intercept could be obtained when the results for amylose in dimethylsulphoxide, potassium hydroxide solution, and formamide were graphed in the appropriate manner. Erlander and Purvinas (1968) show a graphical representation of the values of the intercept $K_\theta^{3/2}$

obtained from the analysis of their experimental results according to the Kurata–Stockmayer–Roig theory; within experimental error, the same values are obtained using five different solvents—0.33 M KCl, 4.2 M guanidinium chloride, dimethylsulphoxide, 0.1 M KOH and 0.5 M KOH.

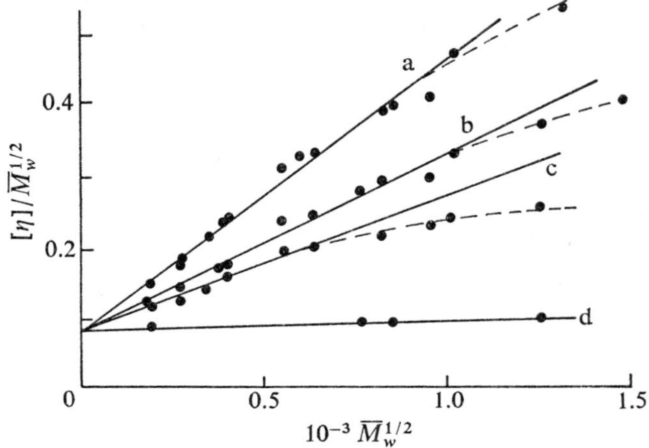


*Figure* 4.8. Graph of $[\eta]/\bar{M}_w^{1/2}$ as a function of $\bar{M}_w^{1/2}$ for amylose in: (a) 0.5 M sodium hydroxide; (b) dimethylsulphoxide; (c) formamide; (d) dimethylsulphoxide–formamide mixture (Burchard 1963).

If we accept that the results for good solvents do agree with (4.122) over the lower range of molecular weights, we must conclude that the common intercept is highly significant; it shows that amylose has the same skeletal structure in both neutral aqueous salt solution and a number of good solvents, including some (4.2 M guanidinium chloride, aqueous alkali, and dimethylsulphoxide) in which helical structures stabilized by intramolecular hydrogen bonding would not be expected to exist.

We have already remarked that the behaviour of 'helical' amylose is decidedly anomalous in being resistant to thermal melting; we must now conclude that it is resistant also to chemical melting. If the validity of the common intercept in figures 4.7 and 4.8 is accepted, the conclusion is inescapable—amylose has the same skeletal structure in water as it does in a good solvent.

Thus if it is argued that amylose has pronounced helical content in neutral aqueous solution, then the polysaccharide must possess that conformation also in good solvent systems. But, since it is recognized that a helical model for amylose in solutions of caustic alkali is not possible, the common intercept shown in figures 4.7 and 4.8 demonstrates that the helical structure is also absent in water.

*Discussion of Deviations from the Burchard–Stockmayer–Fixman Relation.* Our conclusion that the skeletal structure of amylose is independent of the nature of the solvent rests heavily on the emphasis that we give to the fractions of lower molecular weight. We assume that (4.120) and (4.122) are valid at low molecular weight, and that the deviations occur at high molecular weight. The validity of this assertion will now be examined.

The basis of (4.122) is the Yamakawa–Kurata extension of the Kirkwood–Riseman model to non-ideal solvents, which resulted in (4.57) and (4.58). If, in those equations, the functions $XF(X)$ and $p(X)$ have reached their limiting values of 1.26 and 1.55, respectively (i.e. the values expected for an impermeable coil), we may write

$$[\eta]=[\eta]_\theta[1+1.55z-\ldots] \qquad (4.123)$$

which is identical to the Burchard–Stockmayer–Fixman relation, if powers of $z$ higher than the first are omitted. In going from the Kurata–Yamakawa relation to that expressed by (4.121) two assumptions have been introduced, namely, (a) the limit of permeability of the coil has been reached, and (b) the power series by which the excluded volume is represented may be terminated after the linear term for $z$. A deviation may be due, therefore, to one, or both, of these assumptions being in error.

According to Burchard (1961, 1968), the deviation from the linearity predicted by (4.122) is due to the neglect of drainage corrections, i.e. only at the highest values of molecular weight can the drainage parameters $XF(X)$ and $p(X)$ be regarded as having reached their asymptotic limits. If this assertion is indeed valid, our emphasis on the lower molecular weight region is mistaken, the common intercept obtained on graphing the results for various solvents is an artefact, and our conclusion regarding the invariance of the skeletal structure of amylose on changing the solvent is untenable.

For example, if Burchard is correct, it is necessary to introduce a correction for partial free-draining, thus modifying (4.122) to

$$[\eta]/M^{1/2}=K_\theta[1+1.55cM^{1/2}][kM^{1/2}/(1+kM^{1/2})] \qquad (4.124)$$

where

$$c=z/K_\theta M^{1/2} \qquad (4.125)$$

and

$$k=0.52(3/2\pi M_0)^{1/2}(d_h/b_0) \qquad (4.126)$$

To evaluate $k$, Burchard (1968) has suggested the following procedure: equation 4.124 may be re-arranged to give

$$M/[\eta]=(K_\theta k)^{-1}[(1+kM^{1/2})/(1+1.55cM^{1/2})] \qquad (4.127)$$

which in the case of small $M$ may be approximated by

$$M/[\eta]=(K_\theta k)^{-1}[1+(k-1.55c)M^{1/2}] \qquad (4.128)$$

The parameter $(k-1.55c)$ may therefore be obtained from the limiting slope at low molecular weight of a graph of $M/[\eta]$ as a function of $M^{1/2}$. The value of $c$ is obtained by graphing the data according to (4.124), assuming the value of $k$ to be infinite, i.e. $c$ is obtained from the asymptotic slope at high molecular weights. Having thus established the value of $k$, the data may be graphed as $[(1+kM^{1/2})/kM^{1/2}][\eta]/M^{1/2}$ against $M^{1/2}$. These graphs, for amylose in a number of solvents, are shown in figure 4.9.

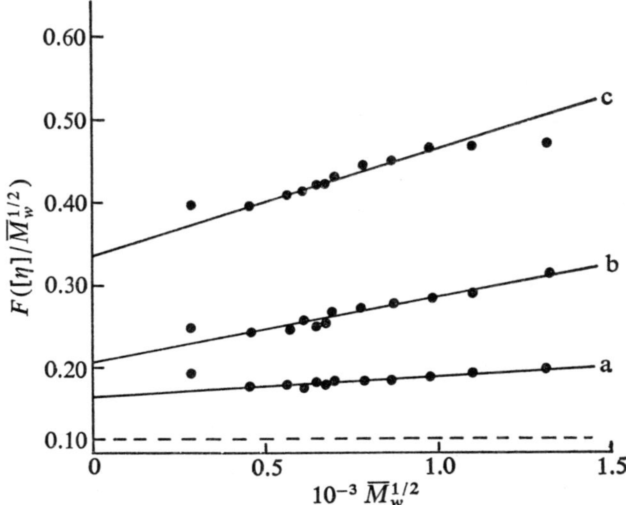


*Figure* 4.9. Graph of $F([\eta]/\bar{M}_w^{1/2})$, where $F=(1+kM^{1/2})/kM^{1/2}$ (see text) as a function $\bar{M}_w^{1/2}$ for amylose in: (a) formamide; (b) dimethylsulphoxide; (c) 0.15 M potassium hydroxide. The broken line represents the relation for amylose in neutral, aqueous solution (Banks and Greenwood 1971b).

In each case, the experimental results approximate to the straight line predicted by (4.124). There is, however, a wide variation in the value of $K_\theta$ obtained by extrapolation:

| | |
|---|---|
| 0.33 M KCl | $K_\theta=0.115$ |
| formamide | $K_\theta=0.166$ |
| dimethylsulphoxide | $K_\theta=0.206$ |
| 0.15 M KOH | $K_\theta=0.332$ |

If these values are to be accepted, the unperturbed dimensions, and therefore the skeletal structure, of amylose must be grossly dependent on the nature of the solvent. Burchard (1961) accepts that such differences are real—he believes the amylose coil to be somewhat inflexible, so that partial free-draining is still significant at degrees of polymerization much higher than that at which it is encountered in the synthetic polymers.

We must accept, therefore, that the same experimental evidence may be interpreted in two ways, which lead to diametrically opposed conclusions. To decide which interpretation is correct reduces to the simple question of whether the Burchard–Stockmayer–Fixman is deficient at high values of the parameter $z$, or at low values, due to a significant contribution from partial free-draining.

For high values of $z$, a specific extrapolation procedure has been suggested for obtaining $K_\theta$. It makes use of the Ptitsyn theory (Ptitsyn and Eisner 1959), in which the excluded volume is given by

$$\alpha^2 = [3.68 + (1 + 9.36z)^{2/3}]/4.68 \qquad (4.129)$$

and $\alpha$ may be related to $[\eta]_\theta$ by the Kurata–Stockmayer–Roig relation

$$\alpha^{5/2} = [\eta]/[\eta]_\theta \qquad (4.130)$$

The mathematical form of (4.129) is such that it cannot be readily manipulated to give the unperturbed dimensions. However, for sufficiently high values of $z$, (4.129) reduces to

$$\alpha^2 = [3.68 + (9.36z)^{2/3}]/4.68 \qquad (4.131)$$

Substitution then gives

$$[\eta]^{4/5}/M^{2/5} = 0.786K_\theta^{4/5} + 0.950K_\theta^{4/5}k^{2/3}M^{1/3} \qquad (4.132)$$

where

$$k = 0.33B[M/\langle r_0^2 \rangle]^{3/2} \qquad (4.133)$$

According to (4.132), a graph of $[\eta]^{4/5}/M^{2/5}$ against $M^{1/3}$ should be linear with an intercept equal to $0.786K_\theta^{4/5}$, provided that $\alpha$ is sufficiently large. Inagaki, Suzuki and Kurata (1966) have suggested that this approximation will be valid for $\alpha \geqslant 1.4$. This criterion is satisfied for all the upper molecular

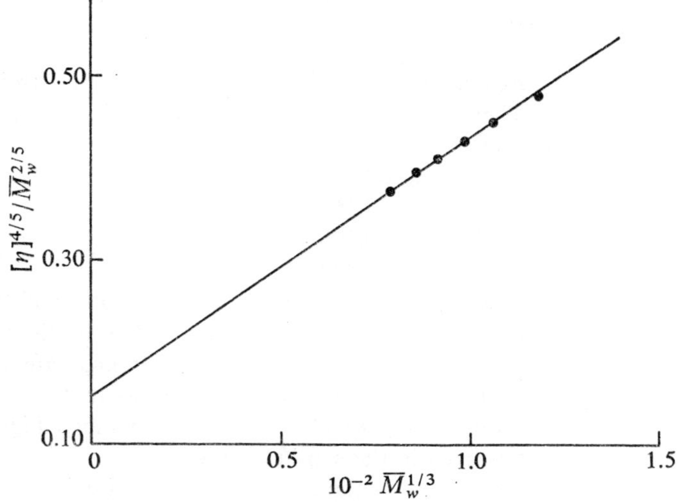


*Figure* 4.10. Graph of $[\eta]^{4/5}/\overline{M}_w^{2/5}$ as a function of $\overline{M}_w^{1/3}$ for amylose fractions of $\overline{M}_w > 0.5 \times 10^6$ in 0.15 M potassium hydroxide (Banks and Greenwood 1971b).

weight fractions of amylose in aqueous alkali, and these results, graphed in accordance with (4.132), are shown in figure 4.10.

The graph is linear, and extrapolation yields an intercept value of 0.150, which gives $K_\theta = 0.125$. The common intercept obtained using the Burchard–Stockmayer–Fixman theory (see figure 4.7) gives $K_\theta = 0.115$. In view of the approximations involved the agreement between these results is satisfactory. Also, the unperturbed dimensions, i.e. the ratio $(\langle r_0^2 \rangle / M)^{1/2}$, enter $K_\theta$ as a cubic power (see equation 4.54) so that any difference between the unperturbed dimensions for amylose in neutral aqueous salt solution and in aqueous alkali cannot be greater than approximately three per cent. This result indicates that the basic skeletal structure of the polysaccharide is the same in both solvents.

It should be pointed out that the combination of the Ptitsyn relation (equation 4.129) with (4.130) has been found to be useful in the case of both synthetic polymers (Ueda and Kajitani 1967) and amylose and its derivatives (Banks and Greenwood 1969b) when high values of $z$ are involved.

The manipulation of the various mathematical relations to yield an expression which, on extrapolating the data from the region in which the Burchard–Stockmayer–Fixman theory breaks down, gives a value of $K_\theta$ that is the same as that obtained experimentally is satisfying, but does not prove that Burchard is incorrect. To do that it is necessary to show that the concept of partial free-draining is not applicable to amylose in the molecular weight region of interest to us, i.e. of degree of polymerization > 400 glucose units. The ways in which this can be done will now be considered.

*Absence of Partial Free-Draining in Amylose.* Kamide and Inamoto (1965), using the Kurata–Stockmayer–Roig expression for the expansion factor in conjunction with the Flory model, showed that

$$-\log K + 1.5 \log\{1 + (4/3)[(a-0.50)^{-1} - 2]^{-1}\} = (a-0.50)\log \bar{M} - \log K_\theta \qquad (4.134)$$

where $K$ and $a$ are the Mark–Houwink–Sakurada parameters, and $\bar{M}$ is the geometric mean molecular weight of the range over which these parameters were determined. If the draining factor is negligible, a graph of the left-hand side of (4.134) as a function of $(a-0.50)$ should be linear, with an intercept of $-\log K_\theta$. Figure 4.11 shows a graph of our data for amylose according to (4.134).

In fact, the results shown in figure 4.11 do conform to a straight line, and the intercept on the ordinate axis has the value associated with neutral aqueous salt solution. These observations strongly suggest that no partial free-draining occurs in any of the amylose fractions used in this investigation, and that the skeletal structure of amylose is similar in all the solvents employed—0.33 M KCl, formamide, dimethylsulphoxide and 0.15 M KOH. Kamide and Inamoto (1965), surveying the published results for amylose, also concluded that the draining effect in amylose is negligibly small.

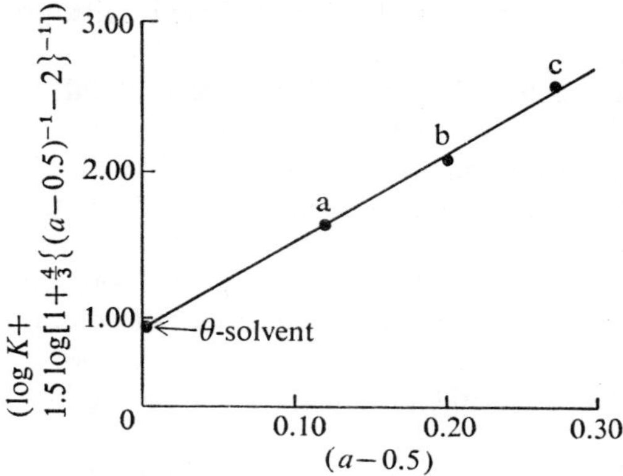


*Figure* 4.11. Kamide and Inamoto theory (equation 4.134). Graph of $(-\log K + 1.5\log[1+\frac{4}{3}\{(a-0.5)^{-1}-2\}^{-1}])$ as a function of $(a-0.50)$ for amylose in: (a) formamide; (b) dimethylsulphoxide; (c) 0.15 M potassium hydroxide (Banks and Greenwood 1968d).

Perhaps the most direct evidence that this conclusion is in fact correct is provided by our ultracentrifugal studies using 0.2 M KOH as the solvent for the amylose fractions (Banks and Greenwood 1969a). The relevant experimental results are shown in table 4.11, and the relation of these hydrodynamic parameters to molecular weight is given by

$$[\eta] = 6.92 \times 10^{-3} \bar{M}_w^{0.78} \tag{4.135}$$

$$S_0 = 2.73 \times 10^{-15} \bar{M}_w^{0.41} \tag{4.136}$$

The hydrodynamic behaviour of these fractions in neutral, aqueous salt solution has already been reported (see section 4.3.3).

*Table* 4.11. The properties of amylose fractions in 0.2 M aqueous potassium hydroxide (Banks and Greenwood 1969a).

| fraction | $\bar{M}_w \times 10^{-6}$ | $[\eta]$ (ml/g) | $S_0 \times 10^{13}$ (s) |
|---|---|---|---|
| 1 | 2.29 | 640 | 11.0 |
| 2 | 1.70 | 500 | 9.6 |
| 3 | 0.80 | 260 | 7.1 |
| 4 | 0.35 | 140 | 5.2 |
| 5 | 0.16 | 80 | 3.75 |

If we summarize Flory's model, we obtain

$$[\eta] = K_\theta M^{1/2} \alpha^3 \tag{4.137}$$

where

$$\alpha^2 = \langle r^2 \rangle / \langle r_0^2 \rangle \tag{4.138}$$

and

$$S_0 = M(1-\bar{V}\rho)/[P\eta_0 N_A \langle r_0^2 \rangle^{1/2} \alpha] \tag{4.139}$$

Comparison of equations 4.15, 4.36 and 4.55 with the generalized expressions given by equations 4.59, 4.60 and 4.61 shows that the exponents in the latter group are related by

$$c = (2-a)/3 \tag{4.140}$$

The experimental value of the exponent $a$ is 0.78 (see equation 4.135), from which $c$ may be calculated to be 0.41, assuming the Flory model to be operative. The experimental figure is also 0.41, in excellent agreement with the predicted value. If, on the other hand, the exponent of 0.78 in (4.135) were due to partial free-draining rather than to the excluded volume effect, application of the Kirkwood–Riseman theory yields a value of $c=0.35$. We believe that this value is so different from the experimental figure as to preclude any pronounced drainage effect.

*Sedimentation Measurements and Unperturbed Dimensions of Amylose.* Sedimentation measurements carried out in good solvents can also be used to obtain the unperturbed dimensions of the polymer coil, a fact first demonstrated by Cowie and Bywater (1965). Stockmayer and Albrecht (1958) related the frictional coefficient at infinite dilution, $f_0$, to the molecular weight by

$$f_0/\eta_0 = K_f M^{1/2} \alpha_f \tag{4.141}$$

where

$$K_f = P_0(\langle r_0^2 \rangle / M)^{1/2} \tag{4.142}$$

and

$$\alpha_f = 1 + 0.609z \tag{4.143}$$

From section 4.2.2, the relation between the sedimentation coefficient and the frictional coefficient is given by

$$S_0 = M(1-\bar{V}\rho)/N_A f_0 \tag{4.144}$$

Substitution of equations 4.15, 4.116, 4.141, 4.142 and 4.143 into 4.144 gives the Cowie–Bywater relation

$$M^{1/2}/S_0 = [\eta_0 N_A/(1-\bar{V}\rho)]K_f + \\ +0.201[\eta_0 N_A/(1-\bar{V}\rho)][\langle r_0^2 \rangle/M]^{-1} P_0 B M^{1/2} \tag{4.145}$$

Thus a graph of $M^{1/2}/S_0$ against $M^{1/2}$ has an intercept equal to $[\eta_0 N_A/(1-\bar{V}\rho)]K_f$. Having evaluated $K_f$, the unperturbed dimensions $(\langle r_0^2 \rangle/M)^{1/2}$ can then be obtained from (4.142).

The viscosity and sedimentation results, analysed according to the Burchard–Stockmayer–Fixman relation and the Cowie–Bywater treatment respectively, are shown in figure 4.12. To aid interpretation, the parameters for amylose in aqueous KCl solution are also included. It will be observed that, for both the viscosity and the sedimentation results, straight lines may be constructed so as to give common intercepts with the values obtained by using neutral aqueous solution.

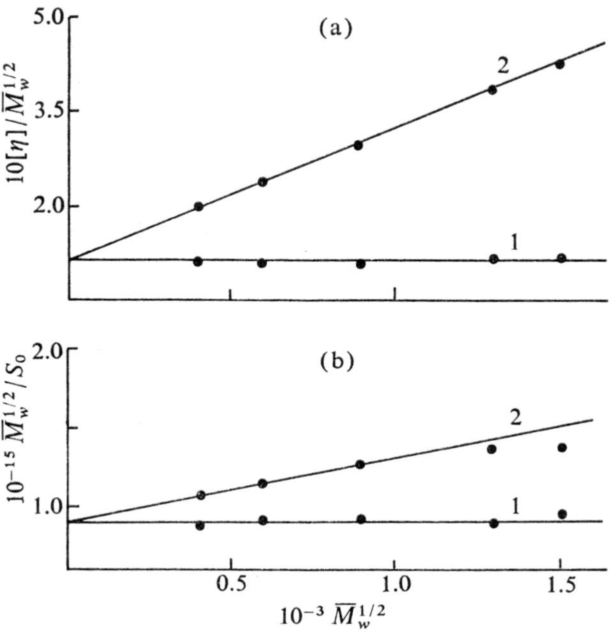


*Figure* 4.12. (a) Graph of $[\eta]/\bar{M}_w^{1/2}$ as a function of $\bar{M}_w^{1/2}$ for amylose in (1) 0.33M potassium chloride solution, and (2) 0.2M potassium hydroxide. (b) Graph of $\bar{M}_w^{1/2}/S_0$ as a function of $\bar{M}_w^{1/2}$ for amylose in (1) 0.33M potassium chloride solution, and (2) 0.2M potassium hydroxide (Banks and Greenwood 1969a).

For the viscosity results the relation is accurately linear over the entire molecular weight range, in contrast with our earlier observations that curvature becomes apparent at high *z*-values. The reason for this seeming anomaly may be found by inverting the normal Burchard–Stockmayer–Fixman procedure, i.e. selecting an arbitrary value for $K_\theta$, giving the interaction parameter $B$ variable values, and constructing the Mark–Houwink–Sakurada relations for a suitable range of molecular weight. In fact, the Burchard–Stockmayer–Fixman relation generates a straight line relation between the logarithm of the limiting viscosity number and that of the molecular weight only at the extreme values of $B$, i.e. when $B=0$ and $B$ is very high. For intermediate values of $B$ continuous curves are generated, but the curvature is so gradual that over a limited range of molecular weight it is indistinguishable from a straight line relation. Any practical determination of the relation between limiting viscosity number and molecular weight is linear when these functions are graphed logarithmically. Hence, the linearity of the resultant Burchard–Stockmayer–Fixman graph of the data is dependent on the molecular weight range involved, and the distribution of experimental points within that range.

The Cowie–Bywater treatment of the sedimentation data shows that in both 0.33M KCl and 0.2M KOH solution a common intercept is obtained (see figure 4.12). Again, the graph curves towards the abscissa at high molecular weights. We suggest that the deviation from linearity results from the approximate nature of the closed expression for $z$, which is in fact a power series from which the higher terms have been omitted. If this assertion is valid, it is legitimate to emphasize the fractions of low molecular weight when extrapolating.

The fact that the results for amylose in the two solvents give a common intercept implies that their unperturbed dimensions differ only by the factor $[\eta_0/(1-\bar{V}\rho)]$. We have shown (Banks and Greenwood 1969a) that the value of $\bar{V}$ is the same in both solvents, and the small differences between $\eta_0$ and $\rho$ are largely self-cancelling, so that the factor differs only by one per cent between the two solvents. Such a difference is negligible, and we conclude that the unperturbed dimensions, and therefore the skeletal structure, are the same in both solvents.

*Validity of Conclusions.* Our conclusion that the skeletal structure of amylose is the same in neutral aqueous salt solution, formamide, dimethylsulphoxide, and caustic alkali solution depends critically upon the validity of the extrapolation procedures represented by equations 4.122, 4.132 and 4.145. All of these relations depend on three assertions, namely that in the unperturbed state (a) the radius of the equivalent hydrodynamic sphere is proportional to a linear dimension of the polymer coil, (b) the ratio of the mean-square radius of gyration (or end-to-end length) of the polymer to the molecular weight is independent of molecular weight, and (c) the chain is of sufficient length for the various approximations inherent in the derivation of the different relations between $\alpha$ and $z$ to be valid.

Assertion (a) implies the absence of partial free-draining. We can devise no test that shows free draining to be significant in the amylose macromolecule: all the data suggest that it is absent. The argument of Burchard (1961, 1968) in favour of partial free-draining rests primarily upon the nonlinearity of the experimental relations graphed according to (4.122). In fact, exactly the same phenomenon has been found by Baumann (1965) for a wide range of synthetic polymers in a variety of solvents. As amylose is not unique in failing to conform to (4.122), then either partial free-draining is very much more commonly encountered than hitherto supposed, or the now widely-recognized deficiencies in that equation arise from its inapplicability at high extensions. The latter explanation would appear to be the more likely.

Assertion (b) demands that the mean-square radius of gyration of amylose in a *theta*-solvent is proportional to the molecular weight. We have indeed established such a proportionality for amylose in 0.33 M KCl, extending to a lower molecular weight of $0.16 \times 10^6$.

It is generally thought that a degree of polymerization of 200–300 units is sufficient to ensure adherence to assertion (c). As the fraction of lowest molecular weight used in our work had a degree of polymerization of 500 glucose residues, and the remainder had degrees of polymerization in excess of 1000 units, there is no readily apparent reason, therefore, why assertion (c) also should not be valid.

Flory (1966) has stressed the inherent weakness of all the extrapolation procedures, namely that one must extrapolate into a region in which every one of the above assertions become invalid. If some experimental measurements have been made in this region and these form the basis of the subsequent extrapolation, large errors could well result. In the present instance, two of the assertions have been shown to be valid for the molecular weight range of the fractions, and we have no reason to believe that the third will be in serious error. We therefore believe that the emphasis we place on the results from lower molecular weights does not lead to the extrapolation procedure being particularly hazardous, as was claimed by Brant and Min (1969). There can be no doubt that the extrapolation procedures are fraught with potential sources of error, as we have demonstrated (Banks and Greenwood 1967c, 1968e, 1971c), but if suitable care is taken to show that the basic assumptions of these procedures are valid, some confidence may be placed in the values of the unperturbed dimensions so obtained.

The extrapolation procedures detailed above by no means comprise an exhaustive list. For a short period in the mid-1960s, a plethora of empirical and semi-empirical methods appeared for obtaining the unperturbed dimensions of the polymer coil from measurements carried out in good solvents. Any one of them is indeed valid for a restricted range of $z$, but none offers any advantage over those detailed here. It is still true to say that there exists no quantitative theory for describing the excluded volume—at best, the theories presently available are valid only for restricted ranges of $z$. As a corollary, we note that none of the results presented here can be used to decide between the third-power law and fifth-power law equations relating $\alpha$ to $z$.

*Summary.* The interpretation of both the viscosity and the sedimentation data leads to the conclusion that amylose has a random conformation in water and in aqueous alkali. Although we could wish that the interpretation of hydrodynamic behaviour was less ambiguous, nevertheless, whilst any individual line of evidence might be suspect, we believe that the entire theoretical treatment of the results obtained in good solvents is inconsistent with the concept of amylose having any tight helical regions stabilized by hydrogen bonding.

**4.3.7.** Interaction with Iodine and Butan-1-ol.

Figure 4.13 shows the changes in viscosity number that occur when iodine and butan-1-ol are added to amylose in neutral aqueous solution to form the appropriate complexes. In both cases, complex formation is associated with a large decrease in the hydrodynamic volume of the macromolecule. For iodine, the minimum viscosity number occurs when the expected stoichiometric amount of reagent necessary to form the complex has been added, i.e. at *ca.* 20 mg of iodine per 100 mg of amylose. The subsequent increase in viscosity number is due to aggregation of the amylose–iodine complex, and after a period of several hours a visible precipitate is formed.

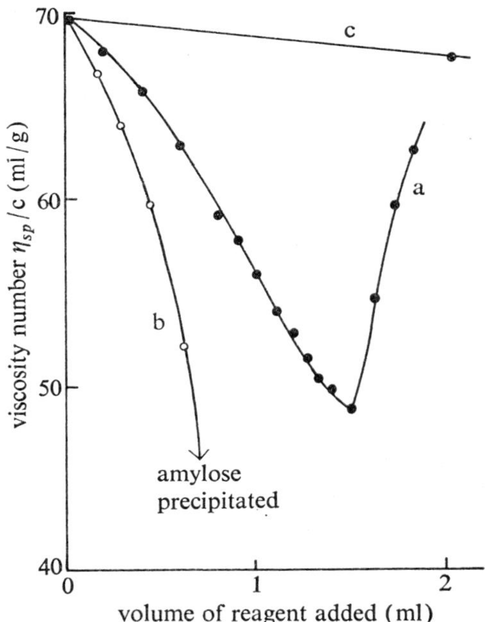


*Figure* 4.13. Changes in the viscosity number $\eta_{sp}/c$ of amylose in aqueous solution (initial concentration$=10^{-3}$ g/ml, initial volume$=$ 10 ml): (a) amylose in 0.015 M potassium iodide+iodine (0.005 M in 0.015 M iodide); (b) amylose in water+butan-1-ol; (c) amylose in water+water (control) (Banks and Greenwood 1971d).

As there is no doubt at all that the amylose–iodine complex is helical in solution—the streaming birefringence experiments of Rundle and Baldwin (1943) are quite unequivocal—the above viscosity study shows that the addition of the complexing agent forces a helical conformation upon amylose. The necessary corollary is that in water, or in neutral aqueous salt solution, the amylose macromolecule possesses no pronounced helical character.

In these experiments, the complexing agent is iodine in the presence of potassium iodide. Iodine alone cannot form a helical complex with amylose;

the blue colour which results when iodine is added to an aqueous solution of amylose is due to hydrolysis of a small amount of iodine to produce free iodide ions, and can be suppressed by addition of mineral acid. Thoma and French (1958) have noted that although amylose will not complex with iodine alone, cycohexa-amylose will, a fact incompatible with amylose having a helical conformation in aqueous solution.

Our experimental observations are radically different from those of Holló and Szejtli (1958c), who reported no change in the viscosity of amylose solutions when iodine was added. We believe that this result was due to a combination of two circumstances—the experiments of Holló and Szejtli (1958c) were performed in such a manner that both the ionic strength of the solution and the concentration of iodine were varied, and, more importantly, they employed the relative viscosity, which contains the contribution of the solvent, and effectively disguises the change in hydrodynamic volume of the amylose. [It is of some interest to trace the quotation of the work of Holló and Szejtli (1958c) back through the literature, and to note the way in which subsequent references to this work gradually assumed that the original observations pertained to the limiting viscosity number!]

*Effect of Heating or Cooling on the Complex.* One of the characteristics of helical biopolymers is that they revert to the random coil model on heating in aqueous solution. It has, of course, been known for a long time that the blue colour of the amylose–iodine complex disappears on heating, and reappears on cooling. Peticolas (1963), in a quantitative study of the relation between the absorbance and temperature of the aqueous amylose–iodine complex, showed a melting phenomenon fairly analogous to that observed in the case of polypeptides and DNA. This work, however, did suggest that the heating and cooling curves follow the same path, in contradistinction to the behaviour exhibited by other biopolymers in which considerable hysteresis is exhibited. It is generally recognized that this hysteresis is a reflection of the nucleation process necessary for the generation of the helical segments. One interpretation, therefore, of the results of Peticolas is that the helical segments in amylose exist at all temperatures, but that the polyiodide complex necessary for the blue colour is destroyed by raising the temperature—an explanation that accounts for the lack of hysteresis. This experimental observation is not in accord with the later work of Weintraub and French (1970), who found marked hysteresis between the heating and cooling cycles.

We have recently carried out the experiment ourselves with the results shown in figure 4.14. Although there are some differences in detail between our results and those of Weintraub and French (1970), they both demonstrate quite pronounced hysteresis. We therefore conclude that the phenomenon shown in figure 4.14 is typical of the helix–coil–helix transitions observed on warming and cooling solutions of, for example, polypeptides.

*Stabilizing Forces in the Complex.* Although the forces stabilizing the helical amylose–iodine complex might be expected to arise principally from hydrogen bonding, this would not appear to be the case. For example, the complex forms quite readily in the presence of 8 M urea, which would not be expected of a structure depending on either hydrogen or hydrophobic bonding.

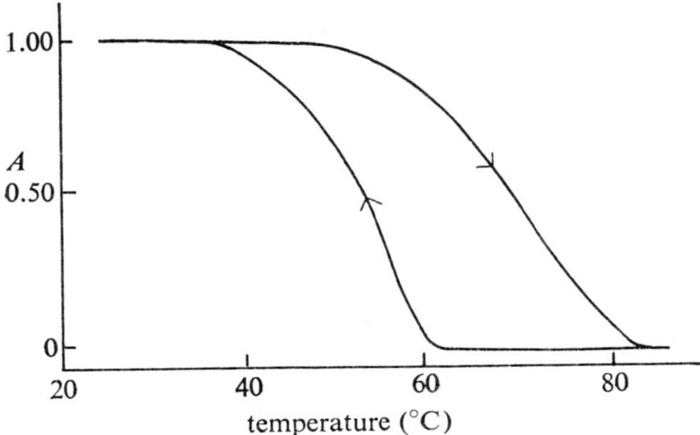


*Figure* 4.14. The relative optical density of the amylose–iodine complex as a function of temperature during the heating and cooling cycle; the value at 25°C is arbitrarily taken as unity (Banks and Greenwood, unpublished work).

Again, in aqueous alkali, the addition of iodine generates the blue amylose–iodine complex (Banks and Greenwood 1972a). In this case the colour is transient, owing to the disproportionation reaction by which free iodine is removed from the system. The reaction of amylose in alkaline solution with butan-1-ol may be studied by measuring the changes in the viscosity number of the polysaccharide as the complexing agent is added. The results of this study are shown in figure 4.15. The viscosity number of amylose in $10^{-2}$M potassium hydroxide solution decreases smoothly and progressively on the addition of butan-1-ol. The sample of amylose used in this work was the same as that employed in obtaining the results shown in figure 4.13. The minimum value of the viscosity number obtained in alkaline solution prior to the appearance of aggregates of the complex was 54 ml/g, in excellent agreement with the corresponding value of 52 m/g of figure 4.13. Thus the viscosity measurements suggest that the same helical form of amylose can be generated by the addition of a complexing agent to both neutral and alkaline aqueous solution.

The important conclusion to be drawn from this work is that a conformational transition *does* take place on adding a complexing agent to an aqueous solution of amylose, at either neutral or alkaline pH; and furthermore, the behaviour of the complex on heating and cooling is typical of that observed in the case of polypeptides. These observations are consistent with our view of amylose in neutral aqueous salt solution as a random coil without any pronounced helical character.

The fact that the conformation of amylose in aqueous solution undergoes a profound change on the addition of iodine has recently been verified (Senior and Hamori 1973; Hamori and Senior 1973), although these workers introduced yet another model for the polysaccharide; an interrupted helix in which the helical segments are much looser than envisaged by Holló and Szetli (1958c). The reason for proposing this model, rather than the more random form envisaged by us, was based on kinetic studies of the amylose–iodine interaction (Thompson and Hamori 1969, 1971) which showed that regions of the polymer already complexed with iodine do not serve as nuclei for further growth when extra iodine is added to the system. However, in the field of polypeptides, it is generally accepted that the stability of a helical region increases with its length: applying this concept to the amylose–iodine interaction would explain the observations of Thompson and Hamori (1969, 1971) without invoking any pre-existent helical structure.

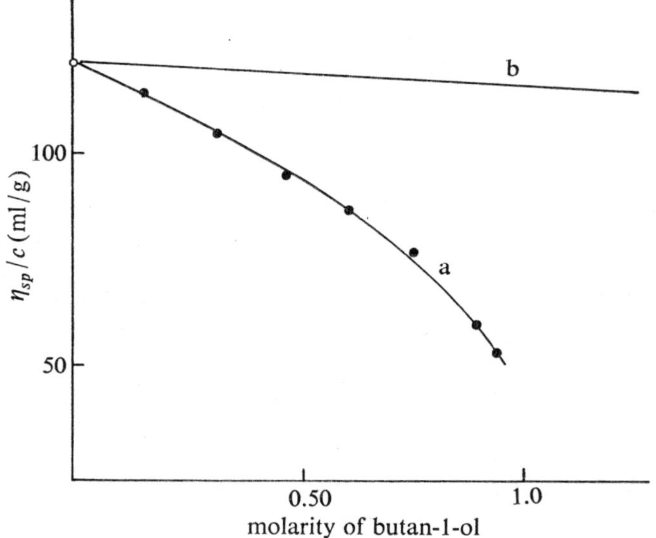


*Figure* 4.15. The viscosity number $\eta_{sp}/c$ of amylose in 0.01 M potassium hydroxide as a function of the addition of successive aliquots of butan-l-ol (curve a) or 0.01 M potassium hydroxide (curve b); the latter experiment constitutes the control (Banks and Greenwood 1972b).

The ease with which the helical form of amylose is generated in alkaline solution is somewhat surprising, for in this solvent the amylose coil is greatly expanded at pH 12 as a result of coulombic repulsion, and yet it still maintains the capacity to form a helix upon the addition of a complexing agent. It could be argued that the ionization of the butan-1-ol suppresses that of the amylose, so that the polysaccharide molecule is uncharged. However, French, Pulley and Whelan (1963) showed that not only polar compounds could act as complexing agents but a wide range of non-polar hydrophobic materials were equally efficient. We have used several of these to complex amylose in alkaline solution, and found them to be as effective as butan-1-ol. In these instances, there is no reason to assume that the ionization of the polysaccharides will be suppressed, but again the typical crystalline aggregates are obtained. We must therefore conclude that some factor so stabilizes the helical form that it can resist the disruptive effects of charge on a proportion of the monomer units.

One such stabilizing influence could be ionic bonding. Doppert and Staverman (1966a) found the pK for the ionization of the hydroxyl groups in amylose to be 12.7. Thus at pH 12, approximately one anhydroglucose unit in every five or six is ionized, i.e. one monomer unit per helical turn. The proposed ionic bonding would then take the form

$$\begin{array}{llllll} O^- & \ldots Na^+ \ldots & O^- & \ldots Na^+ \ldots & O^- & \ldots Na^+ \\ | & & | & & | & \\ R & & R & & R & \end{array}$$

thus tending to stabilize a regular structure. (The groups R are anhydroglucose residues in the same macromolecule, and are adjacent in space when amylose adopts a helical conformation; for example, they could represent the $i$th and $(i-6)$th residues.)

*Solvation Requirements of the Central Helical Cavity.* The major stabilizing influence, however, is in the contribution made by the complexing agents. It is the ability of these materials to satisfy the somewhat complex solvation requirements of the central helical cavity that enables the amylose to adopt this regular conformation. Indeed, we may invert the argument and state the corollary, namely that the solvation requirements of the helical cavity are such that it could not exist in aqueous solution in the absence of a complexing agent.

Although Freudenberg, Schaaf, Dumpert and Ploetz (1939) suggested that the interior of the amylose helix is completely hydrophobic in character, this view was based on the assertion that the monomer units adopted a boat conformation, which we now know to be mistaken. The use of the C1 conformation modifies the model only to the extent that the glycosidic oxygen atoms are now found on the interior of the helix, rather than the exterior. Thus the former surface contains both hydrophobic and hydrophilic character. Model building shows that with amylose in a helical conformation, such as that obtained on the introduction of iodine, the diameter of the central cavity is approximately 6Å. It is our contention that, in developing the model of the helical amylose, too little attention has been paid to the solvation requirements of the internal cavity of the polysaccharide.

The conformation adopted by amylose or any other macromolecule in water must be governed to a considerable extent by the structure of the solvent. Current views of the structure of water (see Blandamer 1970) appear to favour the mixture models, in which there may be water molecules involved in 4, 3, 2, 1, or 0 hydrogen bonds (Nemthy and Scheraga 1967). It is postulated that hydrogen bonding is a co-operative process (Frank and Wen 1957), so that once two molecules associate in this fashion the formation of large hydrogen-bonded clusters, that are in dynamic equilibrium with monomeric water, is favoured. We suggest that water cannot maintain its usual structure within the helical cavity of amylose (Banks and Greenwood 1972a). Therefore, in order to avoid a situation that is energetically unfavourable, the polysaccharide assumes a random conformation. Support for this concept has recently been supplied by Rees (1970) in a study relating the optical rotation of small sugars to their absolute conformation. His calculations show that on dissolving cyclohexa-amylose (the α-Schardinger dextrin) in water the molecule adopts a strained conformation that is relaxed only when a complexing agent occupies the central cavity. In view of the very restricted conformation space available to the six glucose units in this dextrin, the adoption of a strained conformation is somewhat surprising. By analogy, however, it is easy to imagine that the more flexible linear molecule of amylose will adopt a conformation in which the helical cavity is avoided. Only when a complexing agent is present to occupy this cavity can the strain be relaxed, and the amylose takes up its conformation of minimum free energy.

Considerable latitude is allowed in the type of compound that forms a helical complex with amylose. Thus the polysaccharide will accommodate complexing agents of quite different molecular size by forming larger helices having seven or eight glucose residues per turn. A range of either hydrophilic or hydrophobic molecules are equally useful as complexing agents. It would appear that a necessary, although not a sufficient, prerequisite is that the complexing agent have a fairly low solubility in water.

**4.3.8.** Viscosity Number as a Function of pH and Salt Concentration. Figure 4.16(a) shows the variation of the limiting viscosity number of an amylose sample as a function of pH. The limiting viscosity number is independent of pH until pH 11, but after this point the value increases to a maximum at slightly beyond pH 13 before subsequently decreasing.

When a variety of alkalis were studied a set of similar curves are obtained (Banks, Greenwood, Hourston and Procter 1971), as shown in figure 4.16(b). In these experiments, the viscosity number $\eta_{sp}/c$ rather than the limiting viscosity number was employed. This procedure is very much more rapid, but yields comparable results. The three inorganic alkalis behave similarly in that they give an apparent maximum in molecular extension at an alkali concentration of approximately 0.15 M. The curves do, however, differ in detail: those for LiOH and NaOH are virtually identical, but that for KOH shows a lower maximum. The graph for tetramethylammonium hydroxide (TMAH) is the same as that for KOH, within experimental error, up to pH 13, but instead of subsequently decreasing it appears to reach a limiting value.

It is, of course, necessary to prove that alkaline degradation of the amylose does not contribute significantly to the viscometric results presented in figure 4.16. This was accomplished by measuring the limiting viscosity number in 0.15M KOH (found, $[\eta]=455$ ml/g), complexing the polysaccharide with butan-1-ol, determining the limiting viscosity number in 1 M KOH (found, $[\eta]=425$ ml/g), repeating the complexing procedure, and remeasuring the limiting viscosity number in 0.15M KOH (found, $[\eta]=$ 455 ml/g). The identical values found at the beginning and end of this cycle suggest that alkaline degradation of the amylose is not significant within the time-scale of the experiment.

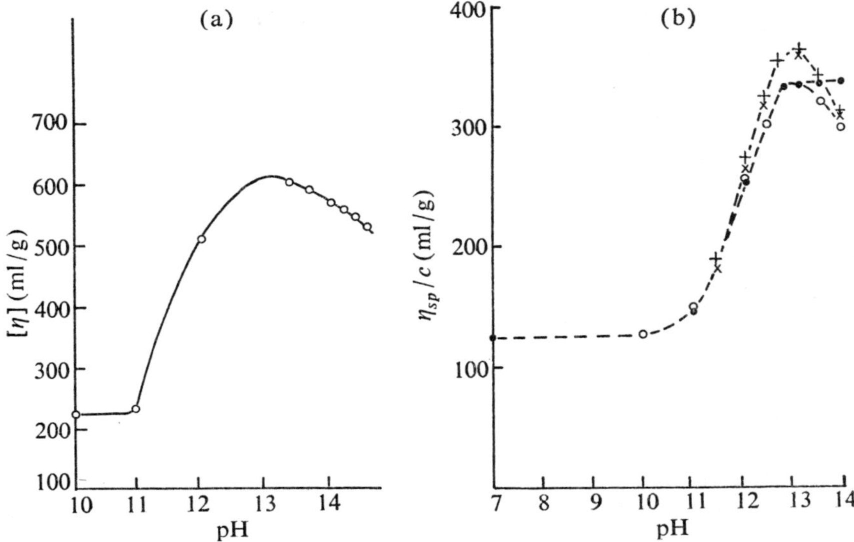


*Figure* 4.16. (a) The limiting viscosity number $[\eta]$ of amylose as a function of pH in KOH. (b) The viscosity number $\eta_{sp}/c$ of amylose as a function of pH in: lithium hydroxide ($\times$); sodium hydroxide (+); potassium hydroxide (open circles); tetramethylammonium hydroxide (dots) (Banks, Greenwood, Hourston and Procter 1971).

The phenomenon of coulombic repulsion (Overbeck and de Jong 1949) explains the form of the curves shown in figure 4.16. Saric and Schofield (1946) showed that at high pH values sugars would be expected to act as weak acids, by virtue of the ionization of their hydroxyl groups. Doppert and Staverman (1966a, b), by means of titration and electrophoresis experiments on amylose solutions and gels, demonstrated the negative charge on the polysaccharide at pH values of 11 and higher. At these high pH values, therefore, the amylose coil possesses polyelectrolyte character, the induced negative charges repel one another, and the coil dimensions, and thus the limiting viscosity number, increase. As the concentration of gegenion increases at higher pH values, a masking effect occurs and the polyelectrolyte structure starts to collapse. For amylose this effect becomes important at an alkali concentration of 0.15M, at which the maximum viscosity is observed.

Whilst the above explanation accounts for the general shape of the graphs of viscosity number as a function of pH for the various alkalis, it does not explain why the effect of KOH is different in detail from that of either LiOH and NaOH. We are dealing with a reaction of the type

$$ROH \rightleftharpoons RO^- + H^+$$

and the $RO^-$ anion will be able to form salts with the cation. However, assuming that the solubility of $RO^-A^+$, where $A^+$ is $Li^+$, $Na^+$, or $K^+$, is in the same order as the corresponding hydroxides, the viscosity would be expected to decrease in the order $K^+ > Na^+ > Li^+$, i.e. the lowest value of the viscosity number at a given concentration of alkali should occur at LiOH, which has the lowest solubility. This is not the case.

As the variation in behaviour cannot be due to the formation of a polyelectrolyte salt, it is highly likely that the explanation is to be found in the changing nature of the solvent with increasing concentration of alkali. When the viscometric behaviour of the alkaline solvents was measured, it was found that at pH values greater than 11–12, both LiOH and NaOH were much more effective in increasing the viscosity of water than was KOH, and both caused an increase of approximately the same magnitude. We therefore tentatively explain the slight differences in the viscosity behaviour of amylose in solutions of the different inorganic alkalis noted in figure 4.16(b) as being due to the structure of the solvent medium.

The behaviour of amylose in TMAH, in which there is a fairly rapid increase in viscosity number between $10^{-4}$M and 0.15M TMAH with a subsequent slow increase, suggests that the masking effect of the cation is not as pronounced in this solvent, i.e. the positive charges are not localized at the site of the induced charge in the polysaccharide to the same extent as with the inorganic alkalis. This effect is probably due to the nature of the cation—the tetramethylammonium ion should be capable of hydrophobic bonding.

It has to be noted that Rao and Foster (1963b) reported that the limiting viscosity number of amylose *decreased* with increase in pH. However, these authors dissolved the amylose in caustic alkali, and brought the solution to the desired pH value by the addition of acid. Hence their solutions contained variable amounts of potassium chloride. For the results shown in figure 4.16, the amylose–butan-1-ol complex was dissolved directly in water, and dilute alkali added to bring the solution to a predetermined pH. In this case, no supporting electrolyte is present. It seems reasonable, therefore, to ascribe the profound differences in the behaviour of amylose as a function of pH recorded by Rao and Foster (1963b) on the one hand, and ourselves on the other hand, to the adventitious presence of salt in the solutions used by the former investigators. To confirm this assumption, we measured the limiting viscosity number as a function of KCl concentration for various molarities of alkali. The results are shown in figure 4.17.

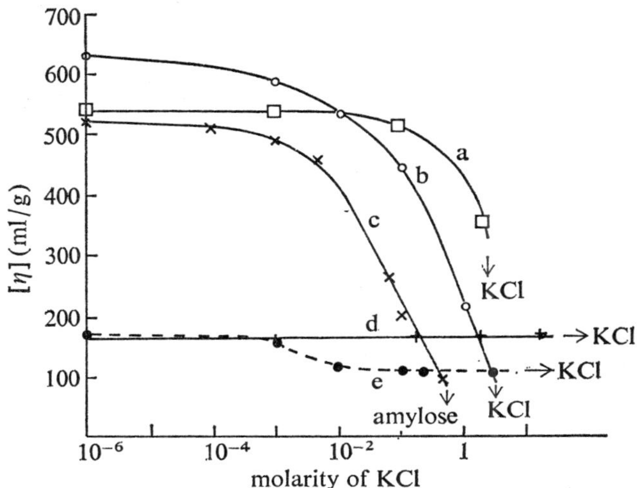


*Figure* 4.17. The limiting viscosity number [$\eta$] of amylose as a function of the molarity of potassium chloride, at several levels of alkali: (a) 1.01 M potassium hydroxide; (b) 0.15 M potassium hydroxide; (c) 0.01 M potassium hydroxide; (d) water; (e) 0.001 M potassium hydroxide (Banks, Greenwood, Hourston and Procter 1971).

In neutral solution, the limiting viscosity number is independent of the concentration of KCl; in alkaline solution, the addition of KCl leads to a decrease in limiting viscosity number. This decrease is, of itself, not surprising—a contraction in the dimensions of a polyelectrolyte normally occurs on the addition of a neutral salt to the solution. What is surprising in the case of amylose, however, is that at certain alkali concentrations the decrease is sufficiently great to bring the limiting viscosity number of the polysaccharide *below* that recorded for the uncharged macromolecule (at pH 7). One would expect the addition of a neutral salt merely to collapse the polyelectrolyte to the dimensions observed in neutral aqueous solution; in solutions which are 0.15M, $10^{-2}$M, and $10^{-3}$M with respect to KOH, the addition of KCl causes the limiting viscosity number to fall below the value recorded for amylose in water or in neutral KCl solutions.

This last phenomenon is dependent on the nature of the salt added to the solution of the polysaccharide, as shown in figure 4.18. To obtain these results concentrated solutions of KCl and KI, and of these salts in 0.01 M KOH, were added successively to amylose in water and amylose in 0.01 M KOH respectively, and the viscosity number $\eta_{sp}/c$ measured.

For amylose in water the addition of KCl solution causes a slight decrease in viscosity number, due to the dilution of the polysaccharide and the fact that $\eta_{sp}/c$ is dependent on concentration, whereas that of KI solution leads to an increase in the viscosity number. In alkali, the addition of either salt causes a rapid decrease in the viscosity number; but with KI the value never falls below that observed for amylose in water, whereas it does with KCl. In view of the profound differences observed with these two salts, the effect of varying the anion and cation was investigated in more detail.

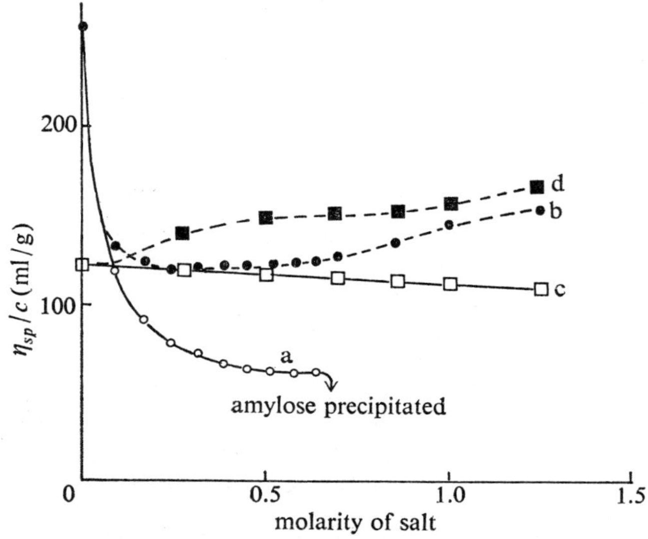


*Figure* 4.18. The viscosity number $\eta_{sp}/c$ of amylose as a function of salt concentration in: (a) 0.01 M potassium hydroxide+potassium chloride; (b) 0.01 M potassium hydroxide+potassium iodide; (c) water+potassium chloride; (d) water+potassium iodide (Banks, Greenwood, Hourston and Procter 1971).

This was done by dissolving the amylose in a series of lithium, sodium and potassium salts (0.25M) and varying the pH at constant salt concentration. In each case, the alkali added had a cation in common with the salt present. The resultant graphs are shown in figure 4.19.

The graphs for the various salts are quite similar in form, and a minimum in the viscosity number of amylose always occurs at pH 12. It should be noted, however, that in neutral solution certain of the salts (LiI, LiBr, NaI and KI) increase the viscosity number of amylose relative to the value observed in water, and that at pH 12 these same salts give a minimum in viscosity number, the numerical value of which is little different from that recorded for the polysaccharide in water. The remainder of the salts do not, at the concentrations employed (0.25M), affect the viscosity number of amylose in neutral solution, but at pH 12 cause it to fall markedly to levels below that observed at pH 7. In all cases, as the alkalinity of the solutions is increased above pH 12 the viscosity number increases, and moreover, at this high pH, the value of the viscosity number is not greatly affected by the nature of the anion.

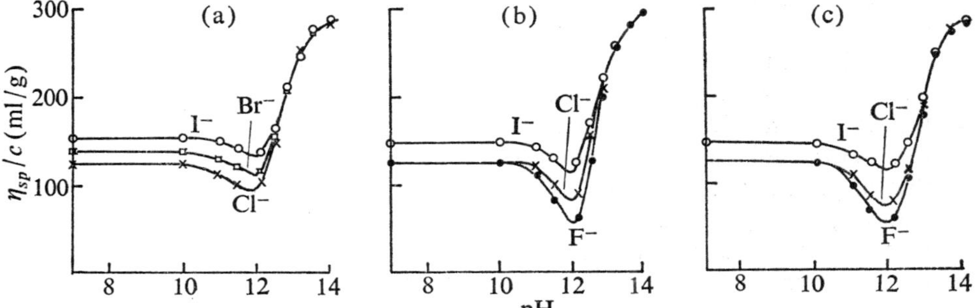


*Figure* 4.19. The viscosity number $\eta_{sp}/c$ of amylose as a function of pH in: (a) LiOH+0.25 M lithium salts; (b) NaOH+0.25 M sodium salts; and (c) KOH+0.25 M potassium salts. The appropriate anions are identified on each curve (Banks, Greenwood, Hourston and Procter 1971).

We noted in section 4.3.7 that the helical conformation is associated with a decrease in the viscosity number to a value much lower than that observed in water. This is exactly the phenomenon observed at pH 12 in the presence of a suitable salt, such as potassium chloride. Furthermore, the addition of potassium chloride to an amylose solution at pH 12 causes the viscosity number to decrease to approximately the same value as is obtained by the addition of complexing agents. This aspect is demonstrated in figure 4.20, in

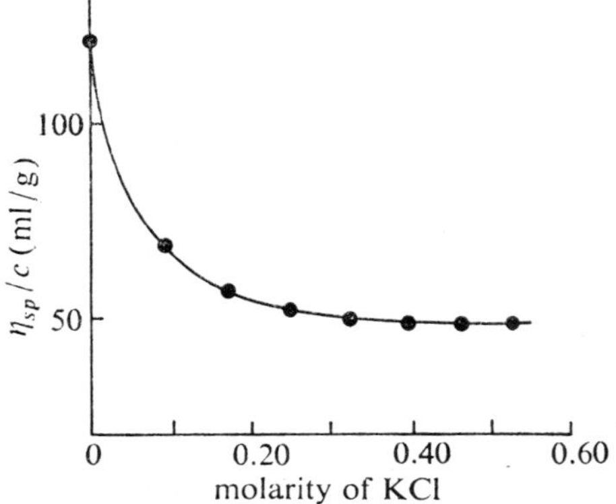


*Figure* 4.20. The viscosity number $\eta_{sp}/c$ of amylose in 0.01M potassium hydroxide as a function of the concentration of potassium chloride (Banks, Greenwood, Hourston and Procter 1971).

which the viscosity number is graphed as a function of the concentration of potassium chloride, at pH 12. The amylose fraction employed in this experiment is also that used to obtain the results shown in figures 4.13 and 4.15; the potassium chloride causes the viscosity number to decrease to the same value (*ca.* 50 ml/g) as does the butan-1-ol and iodine in the case of neutral aqueous solutions, and butan-1-ol in the case of aqueous alkaline solutions of amylose. It is reasonable to conclude that the conformation of amylose at pH 12 in the presence of potassium chloride is, by analogy with the results obtained using complexing agents, basically helical. If we assume that when the viscosity number of amylose falls significantly below the value observed in water the polysaccharide adopts a helical structure, we see from figure 4.19 that at pH 12 in the presence of lithium chloride, sodium fluoride, sodium chloride, potassium fluoride, and potassium chloride, the macromolecule favours the helical state. The numerical values of the minimum viscosity numbers vary somewhat with the salt concerned, indicating that different degrees of helicity must be present.

*Effect of Salts in Neutral Solution.* In neutral solution, the interaction between amylose and water is most probably due to the hydration of the hydroxyl groups of the polysaccharide with free water molecules. Goring (1966) has suggested that polysaccharides may be regarded as structure breakers, because the nature of the repeat unit is such as to exclude it from water clusters. If salts are added to the amylose solution, they will compete with the polysaccharide for the available water. As figure 4.17 shows, amylose must be regarded as having a higher affinity for water than does potassium chloride, because the limiting viscosity number remains constant over the entire range of solubility of the salt. Figures 4.18 and 4.19 show that, at a molarity of 0.25 M, none of the salts investigated is able to lower the viscosity number of amylose at pH 7 relative to the value observed in water. We have previously noted, however, that certain salts are able to increase the value of the viscosity number relative to that in water, i.e. in these specific cases the salt solution is a better solvent medium for amylose than is water. This observation is particularly pertinent in the case of the iodide ion; the viscosity number of the amylose sample in lithium iodide, sodium iodide, and potassium iodide solutions, all at a molarity of 0.25, is approximately 150 ml/g, whereas the corresponding value in water is only 124 ml/g. We suggest that the explanation for this behaviour is to be found in the nature of the iodide ion. Because of its size, this anion possesses the smallest charge per unit surface area of those examined, and hence has the least effect in orienting the water molecules in its immediate vicinity. As a result the hydroxyl groups of the polysaccharide can enter the hydration shell of the iodide ion, so that we are dealing effectively with the solution behaviour of the amylose–iodide complex. Whilst the bromide anion is also capable of increasing the viscosity number of amylose in aqueous solution (see figure 4.19) the effect is less than with the iodide ion. The charge per unit surface area is greater in the bromide ion than in the iodide ion, hence the degree of orientation required of the hydroxyl group, if it is to enter the hydration shell of the former ion, is much greater than in the case of the latter ion. The hydroxyl group is attached to a rather bulky unit (the glucose residue), and therefore will encounter some difficulty in accommodating itself if a pronounced alignment is necessary. As the interaction between the anion and the water molecules becomes more pronounced, the necessary alignment becomes more difficult for the hydroxyl groups of the amylose. Thus one would predict that the small anions (for example, fluoride and chloride ions) would be unable to share their hydration shells with the hydroxyl groups of amylose, and would therefore not swell the polysaccharide macromolecule. This prediction has been verified by experiment (see figure 4.19).

*Effect of Salts in Alkaline Solutions.* The more difficult problem arises in trying to account for the changes in viscometric behaviour that occur as the pH of the various salt solutions is increased, and particularly for the minimum in viscosity number observed at pH 12. Again we may postulate that the type of ionic bonding dealt with in section 4.3.7 may play some part in stabilizing the helical structure. It cannot, however, be the sole contributing influence, because one would then expect similar behaviour irrespective of the nature of the supporting electrolyte—an expectation not realized (see figure 4.18). It seems quite probable that the changing nature of the solvent is the decisive influence in allowing the polysaccharide to adopt a helical conformation. Some support for this hypothesis may be derived from the work of Doppert and Staverman (1966b) who found that the decrease in limiting viscosity number of amylose at pH 12 in the presence of potassium chloride was much less when the solution was 6M with respect to urea. In this case the limiting viscosity number did not decrease to a value lower than that obtained in water, suggesting that no helical structure was formed. The presence of urea is much more likely to radically alter the structure of the aqueous environment than it is to de-stabilize ionic bonding.

That the structure of water should be greatly altered on adding alkali is not really surprising—the flickering clusters must now accommodate the negatively-charged hydroxyl groups. The latter possess a fairly small crystal radius, and hence there is a high degree of orientation of water molecules at their surfaces. This, of course, is equally true of other small anions, such as the fluoride ion, but for the hydroxyl ion there is the additional factor that the negative charge is practically unlocalized within the primary hydration sphere. We then postulate that an aqueous solution of caustic alkali at pH 12 provides an acceptable environment such that amylose would adopt a helical conformation, but for the charge it carries. The introduction of a neutral salt effectively screens the charge, and allows the ordered conformation to be obtained. Obviously, this will happen only if the structure of the solvent is not greatly changed by the addition of the salt. Whether or not there is a change in solvent structure depends upon the hydration behaviour of the anion and cation. In general, the results shown in figure 4.19 are consistent with the hypothesis that the maximum effect at pH 12 is experienced when the salt does not break down the structure of water—the fluoride ion is usually regarded as enhancing solvent structure, and the chloride ion is believed to have only a weakly de-stabilizing influence, whereas the bromide and iodide ions are regarded as pronounced structure-breakers. Thus the results shown in figure 4.19 are in qualitative agreement with the postulate developed above.

The effect of those salts that do not decrease the viscosity number of amylose to values lower than that obtained in water will now be considered; i.e. lithium bromide, lithium iodide, sodium iodide, and potassium iodide. Whilst all these salts give minimal viscosity numbers at pH 12, the absolute values for the amylose sample studied show some variation; e.g. from 134 ml/g for lithium iodide to about 115 ml/g for the other three salts. The initial value of the viscosity number of this amylose in water is 124 ml/g, but carrying out a dilution analogous to that required in order to obtain a solution of pH 12 yields a value of 116 ml/g. Our earlier suggestion that the iodide (or bromide) ion might share its hydration shell with the hydroxyl groups of the polysaccharide will be invalid once the polymer starts to ionize—the ionized hydroxyl groups on the amylose will repel the negatively charged iodide ions. The immediate effect, therefore, of adding lithium iodide is to neutralize the charge on the polymer, causing the viscosity number to decrease. If the salt ROLi rather than $RO^-Li^+$ is formed, the negative charge on amylose is sufficiently reduced for the iodide ion to again act as a 'solvent'. This explanation would account for the minimum value of the viscosity number at pH 12 in the presence of lithium iodide being greater than the corresponding figure for water. In the case of sodium and potassium iodide the net charge on the polymer is sufficiently large to prevent this type of solvation, i.e. $RO^-Na^+$ and $RO^-K^+$ are favoured, and hence the viscosity number falls to the value observed in water. The structure-breaking abilities of the bromide and iodide ions are sufficient to alter the structure of the solvent at pH 12 so that amylose cannot adopt the helical conformation. The above hypothesis is merely an attempt to explain why only lithium iodide from this series of iodides and bromides gives a minimum viscosity number that is greater than the value in water.

In all cases, as the pH is increased beyond 12 the nature of the supporting electrolyte plays progressively less part in determining the value of the viscosity number, and the important factor is the concentration of alkali. Although the viscosity number increases consistently in the region of pH 12–14, it is always less than the corresponding value in the absence of salt. It should be remembered that the curves in figure 4.19 were constructed using a fixed concentration (0.25 M) of neutral electrolyte; increasing this value causes a decrease in the viscosity number. In this region of increasing viscosity both the amylose and the solvent become more polyelectrolyte in character, and these changes are such as to favour the extended conformation rather than the helix.

**4.3.9.** Water-Soluble Derivatives.

The hydrodynamic behaviour of two water-soluble amylose ethers, sodium carboxymethyl amylose and diethylaminoamylose hydrochloride, has been investigated. Unfortunately, the sodium carboxymethyl amylose samples used all differed in their degree of substitution, which makes comparison difficult. We restrict ourselves to the comment that inspection of the values of the Mark–Houwink–Sakurada exponents in table 4.4 would lead to the prediction that the value of this parameter for unsubstituted amylose in aqueous salt solution should be quite high, in contrast to experimental findings. However, the values of the exponents are all consistent with the amylose macromolecule adopting a coiled conformation in solution, even when there is some degree of substitution at carbon atom 6.

The importance of the work on the carboxymethyl and diethylamino derivatives of amylose lies in the fact that it has provided the basis for the comparison between experimentally-derived parameters and those predicted by computer calculation. In particular, the unperturbed dimensions of both derivatives have been obtained as a function of temperature by Goebel and Brant (1970). The relevant values are shown in table 4.12.

*Table* 4.12. The values of the limiting viscosity number $[\eta]$, the expansion factor $\alpha^2$, and the characteristic ratio $C_n$, of sodium carboxymethyl amylose (NaCMA) and of diethylaminoethyl amylose hydrochloride (DEAEA.HCl), as a function of temperature (Goebel and Brant 1970).

| | NaCMA | | | DEAEA-HCl | | |
|---|---|---|---|---|---|---|
| temperature (°C) | $[\eta]$ (ml/g) | $\alpha^2$ | $C_n$ | $[\eta]$ (ml/g) | $\alpha^2$ | $C_n$ |
| 15 | 115.5 | 1.91 | 5.6 | 95.9 | 1.24 | 7.5 |
| 25 | 108.0 | 1.86 | 5.5 | 86.4 | 1.22 | 7.1 |
| 35 | 103.1 | 1.99 | 5.0 | 85.6 | 1.35 | 6.4 |
| 45 | 97.2 | 1.99 | 4.8 | 82.6 | 1.38 | 6.1 |

The characteristic ratio was obtained from measurement of the limiting viscosity number, and the osmotic second virial coefficient, using the relations

$$C_n = ([\eta]M_0/\phi_0 n^{1/2}\alpha^3 b_0^3)^{2/3} \qquad (4.146)$$

and

$$\alpha^2 - 1 = 134A_2M_0^2 n^{1/2}/420N_A(\pi/6)^{3/2} b_0^3 C_n^{3/2} \qquad (4.147)$$

As $C_n$ also appears on the right-hand side of (4.147), an iterative procedure is necessary in order to evaluate the expansion coefficient. The values of the characteristic ratio derived in this fashion are unfortunately heavily dependent on the evaluation of the second virial coefficient, the accurate experimental determination of which is difficult. We are surprised that the values of the expansion coefficient are quite so high as recorded in table 4.12, particularly for the carboxymethyl derivative. As Goebel and Brant (1970) have pointed out, the situation is further confused by the fact that analysis of their viscosity results at 35°C, according to the Burchard–Stockmayer–Fixman relation (equation 4.122), leads to values of 6.9 and 8.0 for the characteristic ratios of the carboxymethyl and the diethylaminoethyl derivative respectively. These values are considerably greater than those recorded in table 4.12. Exactly the same phenomenon was reported in the earlier study of Brant and Min (1969), using carboxymethyl amylose of a higher degree of substitution—the characteristic ratio obtained from (4.146) and (4.147) was 10.0, whereas the Burchard–Stockmayer–Fixman equation yielded a value of 14.0. The suggestion that the fractions were sufficiently monodisperse to enable $\bar{M}_n$ and $\bar{M}_v$ to be equated is tacitly abandoned by Goebel and Brant (1970). The recalculated values are therefore about 20 per cent lower than those given by Brant and Min (1969). For a degree of substitution of 0.30, therefore, we may conclude that $5.5 < C_n < 6.9$, and for a degree of substitution of 0.55, $8 < C_n < 11$, at a temperature of 35°C.

The importance attached to the value of the characteristic ratio in the case of these water-soluble ethers of amylose lies in the fact that Brant and his co-workers have constructed an elegant method of predicting the dimensions of the polysaccharide from the computed conformational energies of the dimer, maltose. Obviously, if the introduction of substituent groups leads to a serious perturbation of the skeletal characteristics of the native amylose, the predictions proceeding from conformational analysis will not be in accord with the experimental results obtained with the derivative. Thus, Goebel and Brant (1970) place more reliance in the characteristic ratios obtained using (4.146) and (4.147) than they do in those derived from the Burchard–Stockmayer–Fixman treatment, mainly, we believe, on the basis of forcing agreement between the characteristic ratio of sodium carboxymethyl amylose and the value obtained using amylose (see section 4.3.3). The temperature-dependence of the characteristic ratio predicted by the computations of Brant and Dimpfl (1970) is in excellent agreement with the results presented in table 4.12 if the glucosidic bond angle is assumed to be 114.0° for the carboxymethyl derivative, and 114.75° for the diethylaminoethyl ether of amylose. Goebel and Brant (1970) have admitted that although the characteristic ratio of carboxymethyl amylose is comparable with that of amylose, the corresponding value for diethylaminoethyl amylose is too high by some 30 per cent. They rationalize this discrepancy by drawing attention to the fact that although one residue in three is substituted with a carboxymethyl unit, and only one in five with a diethylaminoethyl unit, the former is smaller by a factor of about three, so that the perturbation of the skeletal structure relative to amylose is less for the carboxymethyl derivative.

The excellent agreement between the experimental results and the computed parameters led Brant and Dimpfl (1970) to conclude that the correct model had been used for the computations. Their model states specifically that amylose in solution is not helical, unless complexing agents are present. Moreover, it suggests a reason why the regular helix is not the favoured state, namely that although approximately 85 per cent of the conformational space available to the maltose unit would lead to the generation of a left-handed helix, the remainder would lead to a right-handed helix. The fact that both possibilities exist provides the randomizing parameter that enables amylose to conform to the model of the statistical coil.

Whilst it is pleasing to learn of a model so thoroughly in accord with our own observations, we believe it necessary to inject a note of caution. Our principal point of contention is that amylose derivatives rather than the native polysaccharide have been used; to fit the results obtained in this way to a curve generated by consideration of the unsubstituted maltose unit would appear to be a particularly hazardous procedure. According to Brant and Min (1969), the substitution reaction should be confined principally to the primary alcohol group at the $C_6$ position of the anhydroglucose unit, for low degrees of substitution. It is not, however, a general truism that substitution always occurs preferentially at the site of the primary hydroxyl group; we have recently carried out a study of an amylose ether, hydroxyethyl amylose, and confirmed that the most reactive group is the hydroxyl at position $C_2$ (Banks, Greenwood and Muir 1972). Moreover, the rate constant for reaction at this carbon atom is approximately an order of magnitude greater than that for $C_6$. Therefore, the reasoning of Brant and Min (1969), that substitution at $C_6$ would have little effect on the skeletal parameters of the amylose backbone, would not be justified if a considerable proportion of the substituent groups were to be found at $C_2$.

We have already commented on the unusually high values of the expansion coefficient obtained by Goebel and Brant (1970), and to that we must add that their values of $d \ln [\eta]/dT$ for the two derivatives are much higher than have ever been recorded for the unsubstituted amylose (see section 4.3.3) in neutral aqueous solution.

Whilst sophisticated in the mathematical sense, the work does compare experimental results for amylose derivatives with a model calculated for unsubstituted amylose. The reason for not using the native polysaccharide is, we believe, to be found in the nature of the starting material. We have examined similar commercial preparations of amylose and found that whilst they apparently dissolve in hot dilute caustic alkali solution, the material instantly retrogrades when the solution is neutralized. Only by making a derivative can solution be obtained at neutral pH. Brant and Min (1969) also state that a high iodine-binding capacity ensures that their amylose preparation is free from significant amounts of branched material; they evidently equate 'branched material' with amylopectin. These two terms are not synonymous—samples can have a high iodine affinity and still possess a considerable amount of branched material (see section 2.3).

A cautionary note regarding the computational procedures is also necessary. The geometries derived from both cyclohexa-amylose and $\beta$-methyl maltoside give, for a fixed glycosidic bond angle, completely different values of the characteristic ratio; only the use of the former geometry enables the temperature dependence of the characteristic ratio to be predicted.

This observation brings us to our principal criticism of the technique of conformational analysis. It is no exaggeration to say that the polysaccharides provide the most stringent test of this as-yet infant technique. The reason is not hard to find—the geometry of the dimer unit is such that comparatively small changes in the glycosidic bond angle, or in the positions assigned to the groups in the immediate vicinity of the bridge oxygen, generate large changes in the characteristic ratio of the polymer. For example, table 4.13 shows the characteristic ratio of amylose as a function of the glycosidic bond angle, using the different geometries of $\beta$-methyl maltoside and cyclohexa-amylose, the results being interpolated from the graphs given by Brant and Dimpfl (1970). Two features are immediately apparent from inspection of table 4.13, namely the gross dependence of the characteristic ratio on (a) the glycosidic bond angle—the value of the characteristic ratio may change by an order of magnitude for a 6° change in bond angle—and (b) the geometry assigned to the monomer residues—the experimental value for the characteristic ratio of amylose corresponds to a bond angle of 123° using the geometry of the methyl maltoside, but only 114.8° if the calculations are based on the geometry of cyclohexa-amylose.

*Table* 4.13. The computed characteristic ratio $C_\infty$ of amylose as a function of bond angle $\theta$.

| | | 110° | 112° | 114° | 116° | 118° | 120° |
|---|---|---|---|---|---|---|---|
| Brant and Dimpfl (1970) | MM[1] | — | — | — | — | 16 | 7 |
| | CHA[2] | — | — | 7 | 3.5 | 2 | 1 |
| Whittington and Glover (1972) | K[3] | 10.2 | 4.5 | — | 1.1 | — | — |
| | LJ[4] | 17 | 6 | — | 1 | — | — |

[1] MM, monomer geometry taken from the crystal structure of $\beta$-methyl maltoside.
[2] CHA, monomer geometry taken from the crystal structure of the potassium acetate complex of cyclohexa-amylose.
[3] K, calculated using the Kitaigorodsky function for the non-bonded interactions.
[4] LJ, calculated using the Lennard-Jones function for the non-bonded interactions.

As we stressed earlier, there is some diversity of opinion regarding the nature of the mathematical relations by which the potential functions are represented. This too leads to variations in the predicted values of the characteristic ratio. For example, we show in table 4.13 some recent calculations by Whittington and Glover (1972). The characteristic ratio of amylose is given as a function of glycosidic bond angle, the predicted values having been obtained using the same monomer geometry but two different potential functions, due to Kitaigorodsky and to Lennard-Jones respectively. Inspection of table 4.13 is not liable to induce any marked confidence in the technique of conformational analysis as applied to amylose.

Our experimental values for the characteristic ratio of amylose (see section 4.3.3) were found to be in good agreement with those predicted from calculations carried out by Rao, Yathindra and Sundararajan (1969) using a helical model for amylose in neutral aqueous solution. However, the data provide an equally good fit for the random coil model of Brant and Dimpfl (1970). Whittington and Glover (1972) point out that if our experimental figure for the characteristic ratio of amylose is fitted to their mathematical model, it is necessary to assign a value of 111–112° to the glycosidic bond angle, which, they say, '. . . is smaller than that reported for any X-ray determination of the structure of an oligosaccharide of glucose known to the authors'. We suggest that the origin of this discrepancy is simply that their calculations have failed to generate an acceptable model. The experimental determination of the unperturbed dimensions of amylose may be a procedure fraught with error, but the results so obtained are apt to have rather more physical significance than any value presently emanating from a computer!

To the imposing list of adjustable parameters already in use in the conformational analysis of polysaccharides yet another has recently been added. Brant and Goebel (1972) have drawn attention to the fact that the existing computations for cellulose overestimate the characteristic ratio, but underestimate the temperature dependence of this parameter, relative to experimental values. They suggest that all of the glucose units do not exist in the C1 conformation, but rather a considerable fraction are to be found in one of the boat forms—either B1 or B2—and claim that both energetic and entropic factors contribute to the stability of these forms. If this is accepted as a general property of the polysaccharides, the theorist will have at his command sufficient variables to generate a chain having almost any desired characteristic; whether or not the model will then bear any resemblance to the real molecule is debatable.

Perhaps the most realistic attitude towards the conformational analysis of polysaccharides is provided by Whittington and Glover (1972). They suggest that the derivation of the characteristic ratio is subject to so many uncertainties that it is much more useful to draw only broad conclusions on the way the unperturbed dimensions may be influenced by bonding geometry, than to attempt to obtain a close correlation between theory and experiment for particular macromolecules. We would concur with this view.

These criticisms are not meant to imply that the main conclusion of Brant and Dimpfl (1970), namely that the randomizing parameter in the skeletal structure of amylose is the probability that a given bond may be regarded as forming a nascent helix in either the right- or left-hand sense, is wrong. We do, however, believe that the agreement between computer prediction and experiment is by no means as definitive as Brant and his co-workers would have us believe. In fact, the principal reason for rejecting the helical model of amylose remains the detailed hydrodynamic studies that have been carried out over the past decade—studies that have not only failed to uncover any real evidence in favour of the existence of a helical structure for amylose in neutral aqueous solution, but rather are indicative of its absence.

## 4.4. Summary.

The solution behaviour of amylose is that expected of a random coil—of that there is no doubt whatsoever. The debate of the last decade has centred on the amount of helical structure that is associated with the random coil conformation. During this period, five claims have been proposed to establish the basically helical nature of the macromolecule, namely: (a) the large value computed for the length of the statistical segment of the polysaccharide; (b) the anomalous temperature-dependent behaviour of amylose in neutral aqueous solution; (c) the fact that the unperturbed dimensions apparently vary with the solvent used; (d) that no change in hydrodynamic volume is observed as iodine–potassium iodide solution is added to an aqueous solution of amylose; and (e) a helix-to-coil transition, mediated solely by pH, is observed. We have dealt in detail with each of these claims, and, to summarize, have shown that claim (a) tacitly assumes the answer it later generates, claim (b) may be explained without invoking a helical structure, and claims (c), (d) and (e) are wrong. Nothing in the hydrodynamic behaviour of amylose leads one to suppose that there is any marked helical content in aqueous solution in the absence of complexing agents.

The fact that Casu, Reggiani, Gallo and Vigevani (1966) found evidence for hydrogen bond formation between the hydroxyl groups on $C_2$ and $C_3$ for amylose in dimethylsulphoxide solution, using nuclear magnetic resonance, is often quoted as evidence for the existence of a helical structure for the polysaccharide in that solvent. Even accepting that such bonding does exist, it offers no support for the helical model. Our doubts about the conclusions drawn from this work are twofold. First, it is necessary in using the n.m.r. technique to employ a rather concentrated solution of amylose—approximately five per cent, in fact. To extrapolate results obtained at this level to the isolated amylose molecule, with which we deal in hydrodynamic studies, is a somewhat hazardous operation. Our second doubt is related to the fact that Casu *et al.* (1966) found that maltotetraose, maltopentaose and maltohexaose yielded ill-defined n.m.r. patterns from which it was impossible to draw any conclusions regarding hydrogen bonding. This latter result hardly suggests that the conclusions regarding hydrogen bonding in amylose are unambiguous.

Casu, Reggiani, Gallo and Vigevani (1968) reported a study of hydrogen–deuterium exchange of glucose and polyglucoses in dimethylsulphoxide solution, again using the n.m.r. technique. They obtained the equilibrium constants of the exchange reactions

$$2ROH + D_2O \rightleftharpoons 2ROD + H_2O$$

and

$$ROH + CH_3OD \rightleftharpoons ROD + CH_3OH$$

by means of the relations

$$K_W = \frac{[ROD]^2[H_2O]}{[ROH]^2[D_2O]} \qquad (4.148)$$

and

$$K_M = \frac{[ROD][CH_3OH]}{[ROH][CH_3OD]} \qquad (4.149)$$

respectively. The values obtained using (4.148) and (4.149) are shown in table 4.14 for the hydrogen atoms on $O_2$ and $O_3$ of maltose, the Schardinger dextrins and amylose.

*Table* 4.14. The equilibrium constants $K_W$ and $K_M$ of the isotopic exchange reactions for the hydrogen atoms at $O_2$ and $O_3$ for α-1:4-linked glucans (Casu *et al.* 1968).

| | $K_W$ | $K_M$ |
|---|---|---|
| maltose | 0.80 | 0.85 |
| α-Schardinger dextrin | 0.75 | 0.85 |
| β-Schardinger dextrin | 0.65 | 0.75 |
| amylose | 0.85 | 0.90 |

In the absence of a specific stabilizing influence, one would expect the equilibrium constants for the isotopic exchange reaction to be 1.0—only if there is a stabilizing factor would the value be expected to fall materially below this value. Casu *et al.* (1968) therefore interpreted the results shown in table 4.14 as implying that the internal $O_2$ and $O_3$ hydrogen atoms of maltosides are involved in hydrogen bonding. Under the conditions employed by these workers, the value of unity was not obtained for these atoms in the case of glucose, the average figures being $K_W = 0.90$, and $K_M = 0.95$. Using this as a criterion, it may be seen from table 4.14 that the hydrogen bonding in amylose is very weak indeed—less even than in the case of maltose. On the other hand, the values of the equilibrium constants obtained using both cyclohexa- and cyclohepta-amylose indicate quite pronounced hydrogen bonding. The disparity between the values recorded for the Schardinger dextrins on one hand and amylose on the other is, in our opinion, rather good evidence for the lack of any pronounced helical character in the polysaccharide, even under the conditions of concentration employed by Casu *et al.* (1968).

We conclude that all the evidence presented in this chapter shows quite unambiguously that amylose in neutral aqueous solution possesses no helical character. The conformations adopted in various solvent systems can be summarized thus (Banks and Greenwood 1972b):

*Random coil* in water and neutral aqueous potassium chloride.

*Expanded coil* in formamide, dimethylsulphoxide, and aqueous alkali.

*Helical* in neutral, or alkaline, solution in the presence of a complexing agent, and in aqueous solution at pH 12 in the presence of 0.3 M potassium chloride.

# 5. Starch-Degrading Enzymes

## 5.1. Introduction.

Starch-degrading enzymes may be arbitrarily classified as being *exo-acting*, *endo-acting*, or *de-branching*, although this division is by no means rigorous (e.g. *glucoamylase* is an exo-acting enzyme which appears additionally to have the ability to hydrolyse α-1:6-linkages).

The general properties of many of these enzymes have been recently reviewed (Greenwood and Milne 1968a; Thoma, Spradlin and Dygert 1971; Takagi, Toda and Isemura 1971; Lee and Whelan 1971), and here we shall deal in detail only with the action patterns of the most common exo- and endo-acting enzymes, and discuss briefly the various types of de-branching enzymes. In this context, the problem of enzymic reversions is also outlined.

As the common, trivial names of the enzymes being dealt with are so well-established in this field, we shall use these rather than the international nomenclature (for which, see *Enzyme Nomenclature*, International Union of Biochemistry, Elsevier Publishing Company, Amsterdam, 2nd edn., 1972).

## 5.2. Exo-Enzymes.

A number of hydrolytic enzymes degrade amylose and amylopectin by the successive removal of low molecular weight products from the non-reducing chain-end. Prominent amongst these *exo-enzymes* are *beta-amylase*, *phosphorylase*, and *glucoamylase*, which produce maltose, glucose-1-phosphate and glucose respectively. *Beta*-amylase and phosphorylase cannot hydrolyse, or by-pass, the branch linkages present in amylopectin and glycogen, and therefore the action of the enzymes on either substrate results in dextrins of high molecular weight (approximately half that of the starting material) in addition to producing maltose or glucose-1-phosphate. In contrast, glucoamylase has the capacity to hydrolyse both types of linkage, and it quantitatively converts both amylopectin and glycogen to glucose. However, the action of glucoamylase is that of a true exo-enzyme—the introduction of a chromophore group into amylose (Klein, Foreman and Searcy 1969) causes the action of crystalline glucoamylase from *Rhizopus niveus* to stop (Marshall 1970); a low level of periodate oxidation of amylose similarly presents barriers to the progress of the exo-acting enzymes (Smith *et al.* 1970). These methods are excellent for distinguishing between the actions of the exo- and endo-enzymes, or of detecting trace amounts of the latter type in preparations of the former, but they are by no means a recent development—a lightly substituted hydroxyethyl starch was employed over thirty five years ago for detecting *alpha*-amylase activity in the presence of *beta*-amylase (Ziese 1935), and hydroxyethyl amylose was used for the detection of *alpha*-amylase in preparations of phosphorylase (Husemann *et al.* 1958).

### 5.2.1. Action Patterns of Exo-Enzymes on Amylose.

*General Concepts*. The method by which exo-acting enzymes degrade amylose has been the subject of considerable dispute. Basically, two extreme types of action pattern may be envisaged: *single-chain action* and *multi-chain action*. In the former type the enzyme removes successive units from a single amylose molecule until the polysaccharide has been completely degraded, whereas in the latter action pattern the enzyme diffuses away from the substrate molecule after the initial scission and is then capable of attacking any other polysaccharide molecule present.

For a monodisperse substrate, i.e. when all the molecules contain the same number of glucose residues, we can differentiate between these models by following (a) the kinetics of the reactions, and (b) changes in the molecular weight of the residual polysaccharide.

The action patterns of the exo-enzymes are mediated by non-reducing chain-ends. Therefore, the rate at which the low molecular weight products appear will be proportional to the concentration of such chain ends. In the case of single-chain action, one amylose molecule is completely degraded before another is attacked, and hence there is a decrease in the concentration of polymeric chain-ends during the reaction. Consequently, the reaction will kinetically be of the *first order*. On the other hand, during multi-chain attack the concentration of chain ends is constant, and therefore the reaction would be expected to follow *zero order* kinetics.

Similarly, the single-chain mechanism implies that at all stages during the degradation there is a mixture of low molecular weight product and undegraded amylose. The molecular weight of the polysaccharide must therefore be *invariant* throughout the course of the reaction. The multi-chain mechanism leads to the successive shortening of all the chains in the system, hence the molecular weight of the amylose must *decrease* during degradation.

*Early Work on β-Amylolysis.* Kerr (1946) showed that the hydrolysis of amylose by *beta*-amylase was in fact first order, supporting the single-chain hypothesis. Iodine-staining experiments reported by Swanson (1948) and by Swanson and Cori (1948) were in accord with this hypothesis. However, these results were subsequently disputed by Bourne and Whelan (1950) and by Hopkins and Jelinek (1948), and it was suggested that the exo-enzymes operated by a multi-chain action pattern. Cowie *et al.* (1958) showed that the molecular parameters (sedimentation coefficient and limiting viscosity number) of amylose were invariant up to 77 per cent conversion into maltose by *beta-amylase*, and it was concluded that a single-chain mechanism was operative. To add further confusion, Husemann and Pfannemüller (1961) reported that the molecular weight of amylose leached from starch was invariant during hydrolysis by both *beta*-amylase and phosphorylase, whereas amylose prepared enzymically from glucose-1-phosphate exhibited a rapid decrease in molecular weight upon hydrolysis with either *beta*-amylase or phosphorylase. Furthermore, MacGregor (1964) reported that the β-amylolysis of an amylose sub-fraction resulted in an initial marked decrease in molecular weight, which subsequently attained a limiting value that was invariant during further degradation. These three sets of results are shown in figure 5.1.

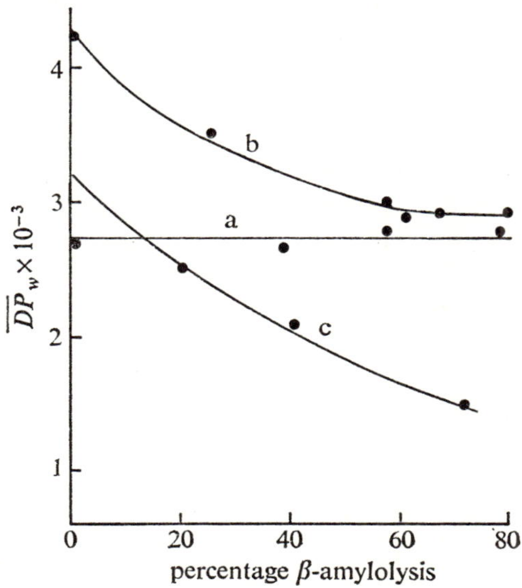


*Figure* 5.1. The change in weight-average degree of polymerization $\overline{DP}_w$ during the degradation of linear amylose with *beta*-amylase for: (a) natural amylose; (b) an amylose sub-fraction; (c) synthetic amylose (Husemann and Pfannemüller 1961; MacGregor 1964).

On subjecting maltoheptaose to β-amylolysis, French, Knapp and Pazur (1950) were unable to detect the intermediate maltopentaose that would be formed were a multi-chain mechanism operative, and hence concluded that the action pattern was single-chain. However, French, Levine, Pazur and Norberg (1950) concurrently reported that the action pattern of *beta*-amylase could be changed by using extremes of either temperature or pH. The concept of using the lower members of the maltodextrin series for determining the action patterns of the exo-enzymes was further developed by Bailey and Whelan (1957), who found definite evidence for an action pattern intermediate between single- and multi-chain. Moreover, using a synthetic amylose (degree of polymerization ~49), Bailey and Whelan (1957) found that the wavelength of maximum absorption of the amylose–iodine complex actually decreased during degradation with the exo-enzymes, signify multi-chain attack in this case also, and these authors presented the schematic representation of the action patterns of exo-enzymes shown in figure 5.2.

Bailey and French (1957), using radioactive synthetic amylose, found that approximately four molecules of maltose were removed per effective encounter with *beta*-amylase. Thus the mode of degradation appears to be intermediate between single- and multi-chain. This intermediate state has been designated *multiple attack*; a term first used by French and Wild (1953). Further studies, reported by French and Youngquist (1963), on the composition of the reaction mixture during the $\beta$-amylolysis of single members of the malto-oligosaccharide series, confirm the validity of the multiple-attack mechanism. These conclusions have received further confirmation from the computer simulation of the action of *beta*-amylase on maltodextrins (Thoma and Spradlin 1970; Thoma, Spradlin and Dygert 1971). This technique involves the numerical solution of the various equations shown in figure 5.2 for the specific case of the $\beta$-amylolysis of maltononaose. The combination of rate constants that generated the product distribution obtained experimentally by French and Youngquist (1963) corresponds to the removal of four maltose units per effective substrate–enzyme encounter under the optimum conditions of pH and temperature.

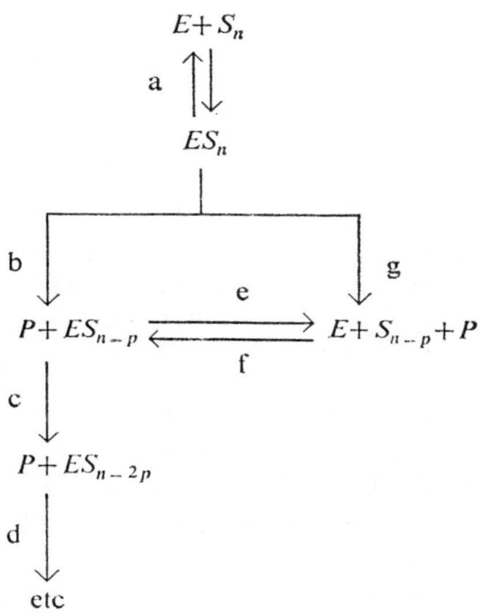


*Figure* 5.2. Schematic representation of the possible model for exo-acting enzymes, where $S_n$ represents a substrate molecule of length $n$ units, $E$ is an enzyme molecule, and $P$ is the degradation product of length $p$ units. *Single-chain* action proceeds by (a), (b), (c), (d), etc. until the substrate is totally degraded; *multi-chain* action results from (a) and (g), or by (a), (b) and (e), followed in (f) by recombination of the enzyme with the same, or with a different, substrate molecule (based on Bailey and Whelan, 1957).

To summarize, the results obtained using maltose oligomers and synthetic amylose molecules of various degrees of polymerization suggest that the action pattern is multi-chain or multiple-attack. On the other hand, the use of natural amylose as substrate leads to a situation in which the molecular weight of the polysaccharide does not change during degradation by the exo-enzymes, a result predicted from a single-chain action pattern. If an amylose sub-fraction is employed, there is the initial decrease in molecular size indicative of multi-chain action followed by a period in which the molecular size is invariant, the latter being indicative of the single-chain mechanism.

*Polymer Theory and the Molecular Weight Distribution.* The debate among carbohydrate enzymologists regarding the action pattern of exo-acting enzymes has its counterpart in synthetic polymer chemistry. Many synthetic polymers can be degraded thermally by a chain-end activated zipping process—the opposite of the mechanism by which they are synthesized. The loss in weight of the polymer during the course of thermal degradation had been found in many instances to follow first order reaction kinetics, and the molecular weight of the residual polymer remained constant during the degradation, both observations apparently signifying single-chain degradation. However, Gordon (1957) drew attention to the importance of the *molecular weight distribution* of the polymer.

If the polymer is regarded as being composed of molecules identical in molecular size, then the distribution function, $f(x)$, i.e. the number of molecules of degree of polymerization $x$, is given by

$$f(x)=1/q \tag{5.1}$$

where $q$ is the reciprocal of the number-average degree of polymerization. Equation 5.1 is merely a convenient mathematical fiction; only in the case of macromolecules constructed using a template mechanism, e.g. proteins, will the molecular weights of all individual members of the population be identical. Although it was accepted that such a distribution function was not realistic, it was thought that the use of the correct distribution function would not be mathematically tractable to solution when applied to degradation kinetics. In fact, as emphasized by Gordon and Shenton (1959) and by Gordon (1960), the use of the true distribution function may considerably simplify the mathematical treatment.

The most widely encountered distribution function is the one obtained for linear polymers formed from monomers in which the functional groups have an equal probability of reacting throughout the lifetime of the reaction. The number-fraction of polymer molecules which contain exactly $x$ units is designated $P_x$, and

$$P_x=(1-p)p^{x-1} \tag{5.2}$$

where $p$ is the probability that a functional group has reacted (and $(1-p)$ is the probability that a functional group has not reacted). For such a system, the number-average degree of polymerization $\overline{DP}_n$ is

$$\overline{DP}_n=1/(1-p) \tag{5.3}$$

From the definition of the parameter $q$ given above, it then follows that

$$q=1-p \tag{5.4}$$

Substituting (5.4) into (5.2) yields

$$P_x=q(1-q)^{x-1} \tag{5.5}$$

In the limiting case of $x-1=x$, i.e. at high degrees of polymerization, (5.5) may be approximated by

$$P_x=q\exp(-qx) \tag{5.6}$$

The distribution function represented by (5.5) is known as the *most probable distribution*; when expressed in the alternative form of (5.6) it is usually referred to as the *exponential distribution*. For all practical purposes (5.5) and (5.6) are identical, so that either terminology may be employed.

It has been found experimentally that linear condensation polymers do possess the distribution of molecular weights predicted by (5.5). The experimental proof is furnished by sub-dividing the polymer into a number of fractions, then using the measurements made on the fractions to reconstitute the original distribution. This manipulation may be exemplified by results obtained with potato amylose.

*The Exponential Distribution of Molecular Weight in Leached Potato Amylose.* The yields and molecular weights of the sub-fractions obtained from leached potato amylose by Banks and Greenwood (1968f) are shown in table 5.1. The parameter $W(M)$ is defined as the cumulative fractional amount to the mid-point of each fraction. The integral and differential mole-

*Table* 5.1. Fractionation data for leached amylose (Banks and Greenwood 1968f).

| fraction | yield (% wt) | $W(M)^{(1)}$ | $\bar{M}_w \times 10^{-6}$ |
|---|---|---|---|
| 1 | 8.5 | 0.042 | 0.084 |
| 2 | 8.5 | 0.127 | 0.214 |
| 3 | 7.5 | 0.207 | 0.321 |
| 4 | 6.8 | 0.279 | 0.382 |
| 5 | 7.5 | 0.350 | 0.427 |
| 6 | 7.0 | 0.423 | 0.461 |
| 7 | 11.0 | 0.513 | 0.489 |
| 8 | 10.1 | 0.618 | 0.612 |
| 9 | 10.2 | 0.720 | 0.753 |
| 10 | 8.4 | 0.813 | 0.944 |
| 11 | 8.2 | 0.893 | 1.22 |
| 12 | 6.3 | 0.968 | 1.74 |
| original | — | — | 0.610 |

[1] Cumulative yield (fractional) to the mid point of each fraction.

cular weight distribution curves for the amylose are shown in figure 5.3; the integral curve is simply a graph of $W(M)$ as a function of molecular weight, and the differential curve is obtained by graphical differentiation of the integral curve.

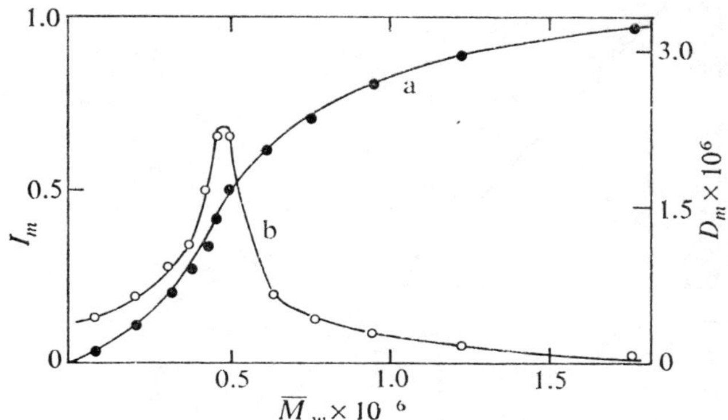


*Figure* 5.3. Curve (a) shows the integral ($I_m = W(\mathrm{M})$) and curve (b) shows the differential ($D_m = dW(\mathrm{M})/dM$) molecular-weight distribution for leached amylose (Banks and Greenwood 1968f).

If amylose leached from potato starch possesses an exponential distribution of molecular weights, the relation

$$dW(M)/dM = (4M/\bar{M}_w^2)\exp(-2M/\bar{M}_w) \tag{5.7}$$

is valid (Gordon 1957), where $M$ is the molecular weight of the fraction and $\bar{M}_w$ is the weight-average molecular weight of the unfractionated starting material. For this type of distribution, the ratio between successive averages is given by

$$\bar{M}_n : \bar{M}_w : \bar{M}_z = 1:2:3 \tag{5.8}$$

If the parameters $q$ and $x$ are substituted into (5.7), we find that

$$dW(M)/dM = q^2 x \exp(-qx) \tag{5.9}$$

Comparison of (5.6) and (5.9) shows that both define the same type of distribution function, but that (5.6) represents the *number* frequency whereas (5.9) represents the *weight* distribution.

The integrated form of (5.7) is given by

$$[1 - W(M)]/[(2M/\bar{M}_w)+1] = \exp(-2M/\bar{M}_w) \tag{5.10}$$

To demonstrate that the data in table 5.1 refer to a material possessing the exponential distribution, it is necessary only to graph $\log[1-W(M)]/[(2M/\bar{M}_w)+1]$ as a function of $2M/\bar{M}_w$, when a linear relation will result if (5.10) is valid. This relation, shown in figure 5.4, demonstrates that amylose leached from potato starch does indeed possess an exponential (or most probable) distribution of molecular size.

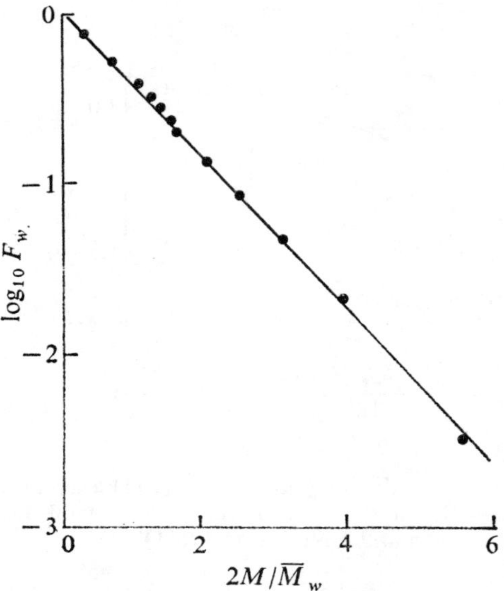


*Figure* 5.4. Exponential distribution: graph of $\log_{10} F_w$ against $2M/\bar{M}_w$ for leached amylose, where $F_w = [1 - W(M)]/[(2\bar{M}/M_w)+1]$ (Banks and Greenwood 1968f).

We suggest that it is very likely that most of the samples of natural amylose used in experiments aimed at elucidating the action patterns of exo-acting enzymes possessed an exponential distribution of molecular weight. Furthermore, the properties of this distribution function, outlined below, explain the conflicting results often obtained with the exo-acting enzymes, particularly *beta*-amylase and phosphorylase.

*Properties of the Exponential Distribution.* If we consider amylose with an exponential distribution of degrees of polymerization, then *single-chain* action will again result in a decrease in chain ends during the course of degradation, and the process would be of the first order kinetically. Also, if the chains are degraded without any preference for molecular size, the weight-average molecular weight (and indeed all the other molecular parameters) will remain constant during the course of the reaction.

Multi-chain action will result in all the polymer molecules being degraded to the same extent, and therefore the smaller ones will be removed whilst remnants of the larger ones still exist. In this case, therefore, there must be a decrease in the concentration of chain ends during the reaction, which will ensure that zero-order kinetics are not observed. Indeed, for a substrate possessing the exponential distribution of molecular weights, degradation by a multi-chain mechanism leads also to a *first order* kinetic scheme. This fact can be shown by considering the simple exponential relation

$$y = \exp(-x) \tag{5.11}$$

shown in figure 5.5. Multi-chain degradation is equivalent to shortening all the chains present by $n$ units, with the consequent disappearance from the system of all chains with $x \leqslant n$. This effect may be obtained by moving the origin in figure 5.5 $n$ units to the right (equivalent to removing $n$ units from each molecule), and normalizing the area under the remainder of the curve to generate a new distribution that is identical to the original, but displaced $n$ units to the right. Effectively, therefore, only the concentration of polymer has altered in the period of time in which $n$ units are removed from each chain. The identity of the curves shows that a constant fraction of chain ends disappear in a given unit of time, confirming that the reaction is kinetically of the first order.

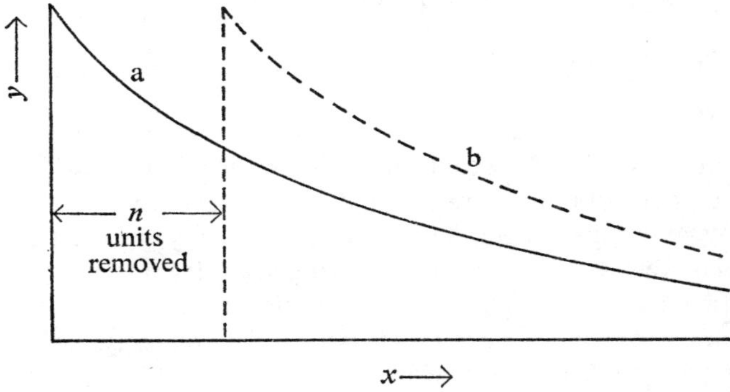


*Figure* 5.5. The generalized exponential curve, $y=\exp(-x)$ is shown in curve (a). Curve (b) shows the effect of cutting off an arbitrary $n$ units from each member of the population in the system. The resultant curve, after normalization, is exactly the same as the original one, i.e. the distribution is invariant to a mathematical process analogous to the multi-chain degradation of a polymer (Banks and Greenwood 1968g).

The above phenomenological analysis suggests that any parameter related to the molecular weight of the amylose will be invariant during multi-chain degradation which is activated at the ends of the molecules. This fact can easily be confirmed by mathematical analysis. The distribution function, $f(x)$, is given in the case of amylose leached from potato starch by

$$f(x) = q \exp(-qx) \tag{5.12}$$

and

$$P_x = q \exp(-qx) \tag{5.6}$$

[Equations 5.6 and 5.12 are identical because the former relation is normalized. By definition

$$P_x = f(x)/\text{total number of molecules present}$$

$$= q \exp(-qx)/q \int_0^{\infty} \exp(-qx)\,dx$$

$$= q \exp(-qx)/q(1/q)$$

$$= q \exp(-qx).]$$

If the origin is moved $n$ units to the right as shown in figure 5.5

$$f'(x) = q \exp[-q(x+n)]$$

and

$$P'_x = \{q \exp[-q(x+n)]\}/q \int_n^{\infty} \exp(-qx)\,dx$$

$$= q \exp(-qx)$$

$$= P_x$$

For a substrate possessing the exponential distribution of molecular weights, we have the unique solution therefore that the number-fraction of $x$-mer is unchanged by multi-chain degradation, and hence any parameter derived from molecular weight will be invariant during such degradation.

This treatment can be criticized because it refers to a continuous distribution function, in which all degrees of polymerization between zero and infinity are represented. In the case of a real polymer this situation is not liable to be strictly valid. For example, in the case of the amylose referred to above, molecules with a degree of polymerization of less than 200 anhydroglucose units might not be complexed with butan-1-ol, and so would be lost. Also, the leaching process is liable to place an upper limit on the molecular weight of the amylose obtained in this manner, so that the mathematical representation of the distribution is no longer strictly correct at the extremes of high and low molecular weight. In fact, these missing fractions do not influence the above conclusions, because only very small amounts of material are involved.

For example, let us suppose that in a given amylose sample there are present no chains having a degree of polymerization greater than $10\overline{DP}_n$, or less than $\overline{DP}_n/10$, that is $\overline{DP}_n/10 \leqslant DP \leqslant 10\overline{DP}_n$. The number-fraction $P_m$ missing at the high molecular weight 'tail' of the distribution is then given by

$$P_m = q \int_{10/q}^{\infty} \exp(-qx)\,dx$$

The corresponding weight-fraction $W_m$ missing is

$$W_m = q^2 \int_{10/q}^{\infty} x \exp(-qx)\,dx$$
$$= 11 \exp(-10)$$
$$= 0.05 \text{ per cent}$$

Similarly, the weight-fraction $W'_m$ missing at the low molecular weight end of the distribution is given by

$$W'_m = q^2 \int_{0}^{1/10q} x \exp(-qx)\,dx$$
$$= 1 - 1.1 \exp(-0.1)$$
$$= 0.47 \text{ per cent}$$

Thus the missing fractions constitute only approximately 0.5 per cent of the theoretical weight-fraction distribution, and the small changes occasioned to the molecular parameters by their absence are well within the experimental error involved in molecular weight determinations.

The above analysis demonstrates that if the starting material has an exponential distribution of molecular weights, as do most samples of natural amylose, then measurements of neither the kinetic nor the molecular parameters during chain-end activated degradation can possibly distinguish between single- and multi-chain action. Failure to appreciate this fact led to the controversy regarding the action patterns of phosphorylase and *beta*-amylase.

*Analysis of Results in Figure 5.1.* Figure 5.1 showed that degradation by *beta*-amylase of (a), *leached amylose*, results in constancy in molecular weight; of (b), an *amylose fraction*, results in an initial decrease in molecular size which then attains a constant, limiting value; and of (c), *synthetic amylose*, results in a continual decrease in molecular weight.

*Curve (a).* As shown above, the constancy of the molecular parameter in (a) reflects only the fact that the initial material had an exponential distribution of molecular weights, and *no conclusion* can be drawn regarding the enzyme action pattern.

*Curve (b).* The fact that the molecular weight of the substrate decreases initially and then remains constant during $\beta$-amylolysis is related to the molecular weight distribution of the starting material, a point with which we will now deal in some detail.

When a polymer is fractionated the fractions still retain a distribution of molecular sizes, the distribution being less broad than that of the parent sample. In the case of linear polymers, the distribution of molecular size within a single fraction is usually given in terms of the Schulz (1939) exponential function

$$f(x) = [y^h/(h-1)!]\,x^{h-1} \exp(-xy) \tag{5.13}$$

where $h$ is a breadth parameter defined by

$$y=h/\overline{DP}_n=(h+1)/\overline{DP}_w=(h+2)/\overline{DP}_z \tag{5.14}$$

and $\overline{DP}_w$ and $\overline{DP}_z$ are the weight-average and $Z$-average degrees of polymerization respectively. In the particular case when $h=1$, (5.13) reduces to

$$f(x)=y \exp(-xy) \tag{5.15}$$

From (5.14), when $h=1$

$$y=q \tag{5.16}$$

Substituting (5.16) into (5.15) yields

$$f(x)=q \exp(-qx) \tag{5.12}$$

Thus when $h=1$, the Schulz distribution function is equivalent to the most probable, or exponential, distribution.

In fact, (5.13) is derived from (5.12) by the mathematical process of *convolution*, a process best described as follows. Consider a system in which the distribution of molecular weights is defined by the exponential function. If from this system, pairs of molecules are withdrawn in a random manner, linked together end-to-end and deposited in another system, the process being repeated until the molecules originally present are exhausted, a new distribution will have been generated. This is the *convoluted exponential distribution function* defined by

$$f(x)=[(q/2)^2/1!]x \exp(-qx/2) \tag{5.17}$$

where $q$ is again the reciprocal number-average degree of polymerization of the original most probable distribution. More generally, the chains may be withdrawn from the original exponential distribution in groups of $\lambda$ and linked through their ends to give a $\lambda$-fold convolution, for which the relation

$$f(x)=\{[q/(\lambda+1)]^{\lambda+1}/\lambda!\}x^{\lambda} \exp[-qx/(\lambda+1)] \tag{5.18}$$

is obtained. The parameter $\lambda$ is characteristic of the breadth of the distribution, being defined by

$$\overline{DP}_w/\overline{DP}_n=1+[1/(\lambda+1)] \tag{5.19}$$

From (5.14)

$$\overline{DP}_w/\overline{DP}_n=(h+1)/h \tag{5.20}$$

Substituting (5.20) in (5.19) gives

$$h=\lambda+1 \tag{5.21}$$

and, by definition (see equation 5.14)

$$y=hq \tag{5.22}$$

Substituting (5.21) and (5.22) into (5.18) results in the generation of (5.12). Thus we obtain the valuable result that the Schulz exponential relation, successfully employed in describing the distribution function of polymer fractions, is related by the process of convolution to the most probable distribution.

The importance of this conclusion becomes apparent when the above argument is reversed, and we consider what will happen when a polymer fraction is degraded. If the degradation is random, as, for example, in the initial stages of the hydrolysis of amylose by *alpha*-amylase, the process is the exact converse of convolution. Therefore, the random degradation of a fraction must rapidly generate the exponential distribution function. What is not quite so obvious is the fact that chain-end activated *multiple attack* will have the same consequence, although it will not be attained quite so rapidly.

The term *multiple attack* describes the situation by which several units are removed from the polysaccharide at a single encounter of enzyme and substrate. Gordon (1957) has studied the analogous 'zipping effect' in synthetic polymers, and has shown that the ease with which the initial exponential distribution is generated depends on the breadth of the initial distribution, and also on the ratio of the mean zip-length to the mean chain-length. As this ratio tends to zero, the action pattern of degradation becomes multi-chain; as it approaches infinity, the action pattern becomes single-chain. For intermediate finite values of the ratio, the situation corresponds to multiple attack in the enzymic case.

Gordon, from his mathematical treatment, proved the existence of a general dictum, namely that 'zipping undoes convolution'. If the exponential distribution is self-convoluted $\lambda$-times (i.e. its chains are joined endwise in groups of $(\lambda+1)$ original chains to make chains of mean length $(\lambda+1)$ times greater), and if these chains are submitted to chain-end activated degradation, then the distribution will revert asymptotically to the original unconvoluted exponential distribution. Accordingly, the mean chain-length approaches the limiting value $L$ given by

$$L=1/(1+\lambda) \tag{5.23}$$

Whether or not this asymptotic limit is attained depends on several factors, the principal one being the value of $\lambda$: as $\lambda$ increases, and the distribution becomes narrower, the probability of the asymptotic limit being attained during the course of chain-end degradation becomes less.

These observations explain why the molecular weight initially decreases on $\beta$-amylolysis of a fractionated amylose sample, and then reaches a constant limiting value. The amylose sample, obtained by a fractional precipitation process, has a molecular weight distribution in which the ratio of the principal averages, i.e. the quotient $\overline{DP}_w/\overline{DP}_n$, is 1.5. Inserting this value in (5.20) yields $h=2$. The distribution of molecular sizes within the fraction will therefore be approximated by (5.13), with $h=2$. From the identity existing between the Schulz relation (equation 5.13) and the mathematical process of convolution, we see that the value of $h=2$ is equivalent to $\lambda=1$ (see equation 5.21), i.e. the fraction of amylose used has a distribution of molecular weights equivalent to that obtained from a one-fold convolution of the original exponential (most probable) distribution. Thus, from (5.23), the distribution of molecular sizes within the fraction should revert to the general exponential form when the number-average degree of polymerization has decreased by a factor of two. Figure 5.1 shows the molecular size in terms of the weight-average degree of polymerization, but the corresponding values of the number-averages are easily calculated from: (a) the starting fraction, $\overline{DP}_n=\overline{DP}_w/1.5$; (b) the final exponential distribution, $\overline{DP}_n=\overline{DP}_w/2$. For the amylose fraction, the initial value of $\overline{DP}_w$ is 4200 glucose residues, so that the corresponding value of $\overline{DP}_n$ is 2800 units; the asymptotic limit reached on $\beta$-amylolysis corresponds to a value of $\overline{DP}_w=2900$ units, which, assuming that an exponential distribution has now been generated, corresponds to $\overline{DP}_n=1450$. We have already noted that the exponential distribution is theoretically predicted to occur when the initial value of $\overline{DP}_n$ has decreased by a factor of two, i.e. when $\overline{DP}_n=2800/2=1400$ residues. The experimental value calculated from $\overline{DP}_w$ is 1450 residues. The excellent agreement between the predicted value and that calculated from the experimental results demonstrates quite unambiguously the correctness of Gordon's mathematical treatment of chain-end activated degradation.

*Curve* (*c*). In this curve the molecular size of the amylose decreases over the entire course of enzymic degradation. The synthetic amylose employed has a rather narrow distribution of molecular weights, defined by

$$f(x)=[(1/q)^x \exp(-1/q)]/x! \tag{5.24}$$

This *Poissonian distribution function* is obtained from the process of 'living polymerization', i.e. all the chains are initiated simultaneously and grow without any termination step (other than lack of monomer). In the presence of a suitable primer (a maltosaccharide containing more than three glucose residues) and excess glucose-1-phosphate, phosphorylase catalyses this type of polymerization to yield an amylose with a narrow distribution of molecular sizes. (The careful work of Whelan and Bailey (1954) demonstrated that a multi-chain mechanism was operative during the synthesis of amylose by this procedure, and thus that the product would have a Poissonian distribution function. A much wider distribution is obtained if maltotriose is used as the primer, or if the material retrogrades from solution during synthesis.) The breadth of the various types of distribution discussed in this book may be classified using the *inhomogeneity factor*, *U*, defined as

$$U=(\overline{DP}_w/\overline{DP}_n)-1 \tag{5.25}$$

Calculated values of this parameter for the different distribution functions are shown in table 5.2.

The inhomogeneity factor for amylose (with $\overline{DP}_n = 3200$) with a Poissonian distribution is ~0.05; to obtain a similar value by the process of convoluting the exponential distribution would require, approximately, a twenty-fold convolution. From (5.23), therefore, one would predict that the exponential distribution would be regenerated only when the number-average degree of polymerization has decreased by a factor of twenty, i.e. from the initial value of 3200 units to a final one of 160 units. From the shape of curve (c) in figure 5.1, it follows that this value can be achieved only when less than one per cent of the starting material remains, which is in a range inconvenient for experimental observation.

*Table* 5.2. The inhomogeneity factor *U* for various types of distribution.

| type of distribution | distribution function $f(x)$ | inhomogeneity factor $U$ |
|---|---|---|
| monodisperse | $1/q$ | 0 |
| Poissonian | $(1/q)^x \exp(-1/q)/x!$ | $\sim 0.05^{(1)}$ |
| $\lambda$-fold convolution of the exponential | $\{q/(\lambda+1)\}^{\lambda+1} x^{\lambda} \exp\{-qx/(\lambda+1)\}/\lambda!$ | $1/(\lambda+1)$ |
| exponential | $q \exp(-qx)$ | 1 |

[1] The numerical value of the inhomogeneity factor for the Poissonian distribution decreases with increasing degree of polymerization; the quoted value corresponds to a polymer with a degree of polymerization of 3200 residues.

*Conclusions.* The apparent anomaly inherent in figure 5.1, in that exo-enzymes can apparently degrade by either a single- or a multi-chain action pattern, is explained on the basis of the difference in the distribution functions for the various substrates. The most important point to emerge from this treatment is that if an amylose (or any other polymer) having the exponential distribution of molecular weights is subjected to chain-end attack, then there is no way of distinguishing between single- and multi-chain action patterns (French 1961). It is also apparent from figure 5.1 that, using a substrate having a distribution of molecular weights that enables a meaningful conclusion to be drawn, *beta*-amylase does not function by a single-chain, but either by a multi-chain or a multiple attack mechanism. The same is true also for phosphorylase. And there is little doubt that the use of the proper substrates will show it to be equally true of other exo-acting, starch-degrading enzymes such as glucoamylase.

It is of interest to note that later work by Husemann and Pfannemüller (1961, 1965), using mixtures of fractions (each having a Poissonian distribution) so that overall the ratio of $\bar{M}_w/\bar{M}_n$ was two, confirmed that the single-chain mechanism was not operative. However, this work suggested that the short chains in the mixture were degraded preferentially by both *beta*-amylase and phosphorylase. The earlier work by Thoma and Koshland (1960) had shown that cyclohexa-amylose, cyclohepta-amylose, and the internal segments of starch, functioned as competitive inhibitors for *beta*-amylase. The exceedingly detailed studies carried out by Husemann and Pfannemüller (1961, 1963, 1965), and by Pfannemüller (1968), confirmed that the Michaelis parameters $V_{max}$ and $K_m$ are dependent on chain-length. Nevertheless, it is difficult to explain why such inhibition should lead to the preferential removal of the smaller molecules in a mixture of the two. The situation is further confused by the fact that Dygert (1971) reported that $K_m$ was dependent on the chain-length of the substrate, but that $V_{max}$ was not. It is not clear whether the source of this discrepancy lies in the use of enzymes from different sources (Dygert used sweet potato *beta*-amylase, whereas the German workers employed the enzyme extracted from barley), or is more fundamental in origin.

**5.2.2.** Action Patterns of the Exo-Enzymes on Amylopectin and Glycogen. Substrates such as amylopectin and glycogen, in which end groups comprise approximately 5 and 10 per cent respectively of the total molecule, are rapidly attacked by the exo-enzymes.

*General Concepts.* Various action patterns may be envisaged, e.g. all the available units may be removed from one molecule before another is attacked; this is *single-molecule* attack. This type of action pattern is inherently unlikely, because of the vast number of chain ends within any one molecule of amylopectin or glycogen. The alternative *multi-molecular* attack is intuitively favoured, where at any stage in the degradation all molecules would be expected to be hydrolysed to the same extent. However, in the case of these multiply-branched polysaccharide structures the enzyme, having attacked one chain, might degrade that chain completely before diffusing away to attack either another chain within the same molecule, or in another substrate molecule (a *single-chain* mechanism); or it might remove only a single unit per effective encounter (*multi-chain* action). In addition, the possibility of the intermediate state of *multiple attack* also exists.

*Molecular Degradation.* Single- and multi-molecule mechanisms can be distinguished, in principle, by isolating the high molecular weight residue at various stages of degradation, measuring the number-average molecular weights $\bar{M}_n$, and comparing these with the appropriate theoretical values. The difficulty arises in experimentation, because the large molecular weights of amylopectin and glycogen make the accurate determination of $\bar{M}_n$ virtually impossible. Unfortunately, the determination of the weight-average molecular weight, which is ideally suited for such materials, does not differentiate between the two mechanisms.

In the case of glycogen, however, it has been possible to show that a multi-molecule mechanism occurs on $\beta$-amylolysis by measuring the distribution of molecular weights during degradation using ultracentrifugation (Bryce, Cowie, Greenwood and Jones 1958). The virtually ideal solution behaviour of glycogen makes such a study comparatively simple; the highly pronounced non-ideality of dilute solutions of amylopectin makes the technique too inaccurate to be applied to this polymer.

Figure 5.6 shows the molecular weight distribution curves for (a) the original glycogen, (b) a material isolated after 13 per cent conversion into maltose, (c) the glycogen β-amylolysis limit dextrin (41 per cent conversion into maltose), and (d) the curve calculated for the limit dextrin. The very close similarity between the experimental and calculated distribution for the limit dextrin shows that the basic assumption underlying the calculations, namely that the degree of β-amylolysis is independent of the molecular weight, is most probably valid, and emphasizes the accuracy of the experimental technique.

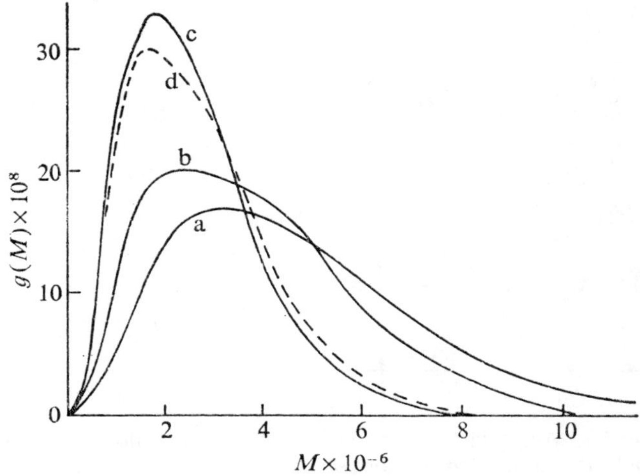


*Figure* 5.6. Molecular weight distributions for glycogen and its β-amylolysis degradation products: (a) original glycogen; (b) intermediate β-amylolysis dextrin (13% conversion of original into maltose); (c) β-amylolysis limit dextrin (41% conversion of original into maltose); (d) curve calculated for the β-amylolysis limit dextrin, assuming that all of the original glycogen molecules are degraded to the same extent (41% conversion into maltose) (Bryce, Cowie, Greenwood and Jones 1958).

In the case of the intermediate dextrin, theoretical curves for the distribution function were calculated assuming (i) that 32 per cent of the molecules over the entire molecular weight range were converted to the limit of 41 per cent maltose (i.e. the percentage of molecules necessary to account for the observed limit of 13 per cent) and the remainder were unchanged, and (ii) that all the molecules were degraded to the same extent. The curves obtained using both assumptions are shown in figure 5.7.

The curve (b), calculated on the basis of the multi-molecule mechanism, is seen to exhibit a reasonable correlation with the experimental results (curve (a)). The assumption of a single-molecule attack, (curve (c)), although giving the same maximum in the distribution function as obtained experimentally, predicts much less material in the molecular weight range of $(3–6) \times 10^6$ but more in the high molecular weight region of the distribution than is actually present. There can be no doubt that curve (b) provides the better fit to the experimental results, indicating that the degradation of glycogen by *beta*-amylase proceeds by a multi-molecule mechanism.

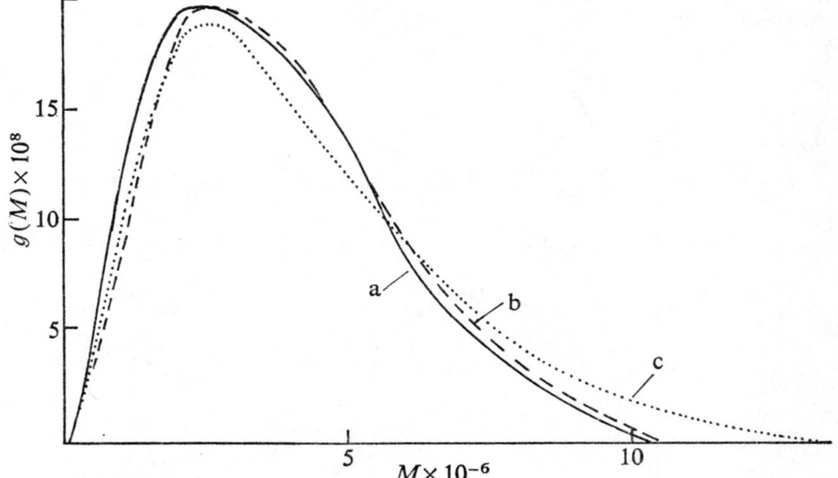


*Figure* 5.7. Molecular weight distributions for the intermediate $\beta$-amylolysis degradation product of glycogen: (a) experimental value for the intermediate dextrin; (b) calculated on the assumption that all the molecules are degraded to the same extent; (c) calculated on the assumption that 32% of the molecules have been converted to the $\beta$-amylolysis limit dextrin, and the remainder have not been attacked (Bryce, Cowie, Greenwood and Jones 1958).

The analogous phosphorolysis of glycogen was reported by Larner, Ray and Crandall (1956) to show evidence that the largest molecules were degraded preferentially. However, Cori (1958) demonstrated that there was no molecular weight specificity—the changes in the distribution curves during phosphorolysis showed all molecular weight species to be degraded to the same extent, suggesting multi-molecule attack. Geddes (1965) has also studied the $\beta$-amylolytic degradation of glycogens, isolated by quite different techniques and therefore differing greatly in the distribution of molecular weights, and found that a multi-molecule mechanism was operative in all cases.

Experimental difficulties prevent a similar study being carried out on amylopectin. All that may be said is that there is no doubt that the extent of $\beta$-amylolysis is again independent of molecular weight. This is shown in table 5.3 by the fact that the weight-average molecular weights during degradation, $\bar{M}_{w(t)}$, are in excellent agreement with the values calculated from the initial value of this parameter, $\bar{M}_{w(0)}$, using the relation

$$\bar{M}_{w(t)} = (1 - w_t)\, \bar{M}_{w(0)} \qquad (5.26)$$

where $w_t$ is the weight-fraction of polysaccharide hydrolysed by the enzyme.

*Table* 5.3. Molecular weight of amylopectin as a function of degradation with *beta*-amylase; a comparison of calculated and experimental values (Banks and Greenwood, unpublished work).

| $\bar{M}_w \times 10^{-6}$ experimental | $[\beta]^{(1)}$ | $\bar{M}_w \times 10^{-6}$ calculated$^{(2)}$ |
|---|---|---|
| 350 | 0 | 350 |
| 300 | 0.08 | 322 |
| 275 | 0.22 | 273 |
| 225 | 0.37 | 221 |
| 150 | 0.56 | 154 |

[1] $[\beta]$ is the weight fraction of amylopectin converted to maltose by *beta*-amylase.
[2] Calculated using equation 5.26.

*Chain Degradation.* It is much more difficult to decide whether the enzyme action pattern conforms to a single- or a multi-chain mechanism. However, some information on this aspect may be deduced from a study of the kinetics of the degradation of glycogen and amylopectin by *beta*-amylase shown in figure 5.8. In the case of glycogen, first order kinetics are followed for only some 10 per cent of the total reaction, and thereafter the rate continually decreases. For amylopectin, the graph is linear over some 80 per cent of the reaction. Thus, whilst it is not possible in the case of amylopectin to rule out multiple attack, the kinetic evidence shows that it is more difficult to remove maltose units near a branch point, and therefore it is exceedingly unlikely that a chain is completely degraded in a single enzyme–substrate encounter. Similarly, the kinetic evidence indicates that the removal of maltose units becomes increasingly difficult as $\beta$-amylolysis proceeds, suggesting multi-chain attack.

For glycogen, at least, there is little doubt that during degradation by *beta*-amylase only single maltose units can be removed at each enzyme–substrate encounter because of the increasing difficulty in removing successive units. It follows from this observation that all molecules must be degraded to the same extent at intermediate stages of the reaction; i.e. the single observation that it becomes much more difficult for the enzyme to remove successive units, almost certainly because of steric factors, ensures not only a multi-chain but also a multi-molecule mechanism.

In the case of amylopectin, little evidence regarding the action pattern has yet been forthcoming. Lee (1971) suggests that the $\beta$-amylolysis of both glycogen and amylopectin proceeds in two stages, a rapid initial production of maltose, followed by a much slower release of the product. He offers convincing evidence that the second slow stage is the hydrolysis of *A*-chain residues that have been reduced to four glucose units in the rapid primary stage.

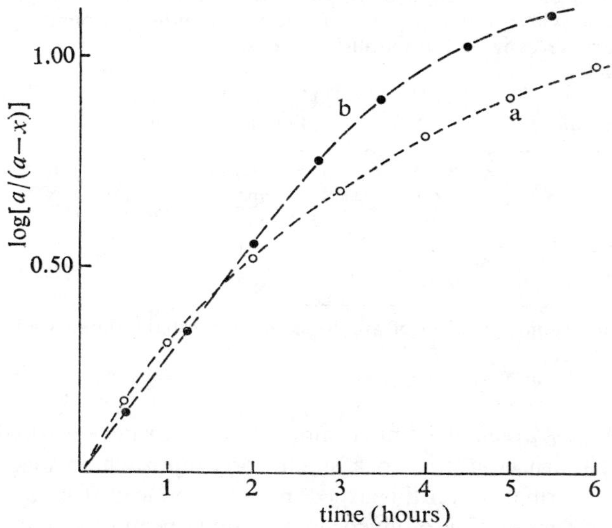


*Figure* 5.8. The kinetics of the degradation of amylopectin and glycogen by *beta*-amylase. The data are graphed according to a first-order rate law for the production of maltose, the value of *a* being the limiting amount of maltose produced on prolonged incubation of the substrates (a) glycogen, and (b) amylopectin, with *beta*-amylase (Banks 1960).

### 5.2.3. Other Exo-Enzymes.

In addition to the well-known exo-enzymes detailed above, two others have been reported recently.

Robyt and Ackerman (1971) isolated an exo-amylase from *Pseudomonas stutzeri* which liberates maltotetraose from amylose, amylopectin, and glycogen (leaving limit dextrins of high molecular weight in the case of the last two polysaccharides), and which specifically hydrolyses the fourth bond from the non-reducing ends of maltohexaose and maltoheptaose. When exposed to large concentrations of the enzyme for prolonged periods, maltotetraose is hydrolysed to give either glucose plus maltotriose, or two molecules of maltose.

Kainuma *et al.* (1972) isolated from *Aerobacter aerogenes* an amylase liberating maltohexaose. Again, the action of this enzyme on amylopectin resulted in the formation of only the product maltohexaose and a residual limit dextrin of high molecular weight, with no intermediate products. The use of radioactive end-labelled members of the maltosaccharide series showed a specific hydrolysis of the sixth bond from the non-reducing ends. It was also reported that the enzyme preparation slowly hydrolysed maltohexaose (to maltotetraose and maltose), but whether this activity is associated with that producing maltohexaose, or is merely a trace contaminant, has not been established.

## 5.3. Endo-Enzymes.

The term *endo-enzymes* applies almost exclusively to the *alpha-amylases*, which are so-called because the product of hydrolysis assumes the α-configuration at the anomeric carbon atom. The action pattern of the exo-enzymes on amylose produces a large amount of small sugars without any marked change in the iodine stain of the polysaccharide; the endo-enzymes, on the other hand, produce a rapid decrease in iodine stain but only a comparatively small increase in reducing power.

It should be noted that one endo-enzyme isolated from *Bacillus polymyxa* has been reported to produce β-maltose in high yield (Robyt and French 1964). Furthermore, Shibaoka and co-workers (1971) have demonstrated that the hydrolysis of both *p*-nitrophenyl-α-maltoside and *p*-tert.-butylphenyl-α-maltoside by *Bacillus subtilis alpha*-amylase can result in the liberation of β-maltose.

The fact that the anomeric form of the product can alter according to the nature of the aglycone is difficult to reconcile with current concepts of enzyme specificity, but if the effect is found to be general it should provide a valuable insight into the catalytic mechanism of the amylases.

### 5.3.1. Action Patterns of Endo-Enzymes on Linear Amylose.

*General Concepts.* The ultimate products from the α-amylolysis of amylose, using any crystalline *alpha*-amylase, are glucose and maltose. However, the ways by which this common end are achieved vary with the source of the enzyme, and in one case at least, with the digest conditions: Pazur, French and Knapp (1950) reported that the iodine-stain–reducing-power relation for the digestion of amylose by porcine pancreatic *alpha*-amylase changed appreciably on altering the pH from 7.0 to 10.3. At the same time, Myrbäck (1950) suggested that the action patterns of plant and mammalian *alpha*-amylases were quite different, in particular pointing out that malted barley *alpha*-amylase required a substrate containing at least six to eight residues in order to exhibit maximum activity, whereas the requirement was less for mammalian enzymes. An earlier theory (Meyer and Bernfeld 1941; Meyer and Gonon 1951) postulated that *alpha*-amylase is unable to attack the linkage at the end of the substrate molecule, but that all other bonds are equally susceptible to hydrolysis. Bird and Hopkins (1954), using paper chromatography to follow the degradation of amylose and amylopectin by *alpha*-amylases from malted barley, *B. subtilis*, and human saliva, showed that at an intermediate stage in the reaction the cereal amylase yielded mainly a mixture of oligosaccharides containing six to eight glucose residues and maltose; the bacterial enzyme gave the same oligosaccharides but maltotriose rather than maltose; and salivary *alpha*-amylase produced predominantly maltose and maltotriose. These authors postulated that the five bonds at the non-reducing chain-end of the substrate are fairly resistant to hydrolysis by the cereal and *B. subtilis alpha*-amylases, but that the enzymes have higher affinities for the bonds at the reducing chain-end, the plant enzyme having the ability to hydrolyse the penultimate bond whereas the bacterial enzyme has not (the ultimate bond is resistant to both amylases), and that all bonds other than those specified are equally susceptible to hydrolysis. Similarly, they suggested that the salivary amylase could not easily attack the first two linkages at the non-reducing chain-end, or the linkage involving the reducing end-group.

*Table* 5.4. A comparison of the degradation products of amylose present when the digest no longer gives a colour reaction with iodine, using *alpha*-amylases from broad bean and human saliva (Greenwood and Milne 1968a,b).

| enzyme source | wt. % of sugar[1] G1 | G2 | G3 | G4 | G5 | G6 | >G6 |
|---|---|---|---|---|---|---|---|
| broad bean | 2 | 8 | 10 | 6 | 6 | 14 | 54 |
| human saliva | 1 | 28 | 30 | 24 | 1 | 1 | 15 |

[1] G1 is glucose, G2 is maltose, etc.

The use of maltosaccharides (Pazur 1953; Pazur and Budovich 1956; Greenwood, MacGregor and Milne 1965d) has given strong support to the theory of Bird and Hopkins (1954), that the action pattern of *alpha*-amylases on small sugars is very similar, but that this similarity is not reflected in their action on amylose (see table 5.4). In the case of the plant *alpha*-amylase there is a considerable build-up of oligosaccharides containing six or more units, an observation easily reconciled with the report that plant amylases hydrolyse maltohexaose approximately six times slower than they do starch (Svanborg and Myrbäck 1953). In the case of the mammalian *alpha*-amylase, oligomers larger than the tetrasaccharide are attacked at the same rate as starch, hence the main products of this enzyme in the later stage of degradation are maltose, maltotriose, and maltotetraose, with only some material larger than the hexaose present (Greenwood and Milne 1968b,c). From these and similar studies it became clear that all bonds in the substrate are not hydrolysed with the same ease, as shown in figure 5.9.

The most detailed study of the action of *alpha*-amylase on oligosaccharides

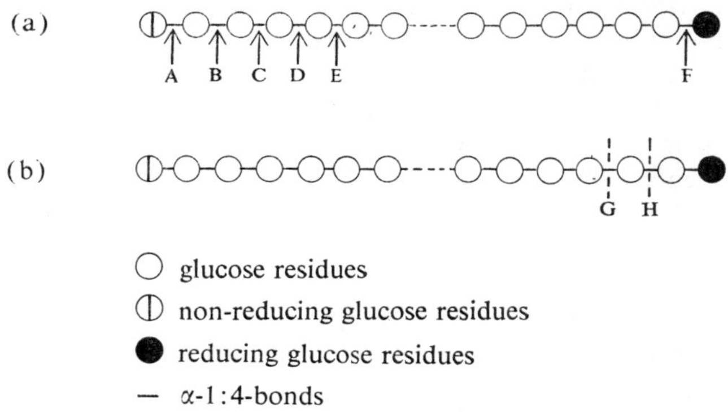

*Figure* 5.9. Bonds in a linear substrate that are hydrolysed non-randomly by *alpha*-amylases. (Based on Greenwood and Milne 1968c.)
(a) *Bonds resistant to α-amylolysis:* A and B are resistant to attack by cereal, bean, pig pancreatic, and *B. subtilis alpha*-amylase; C, D and E are resistant to attack by cereal, bean and *B. subtilis alpha*-amylases; F is resistant to attack by cereal, pig pancreatic and *B. subtilis alpha*-amylases.
(b) *Bonds preferentially hydrolysed by alpha-amylase:* G is labile to bean and *B. subtilis alpha*-amylases; H is labile to cereal, bean, and pig pancreatic *alpha*-amylases.

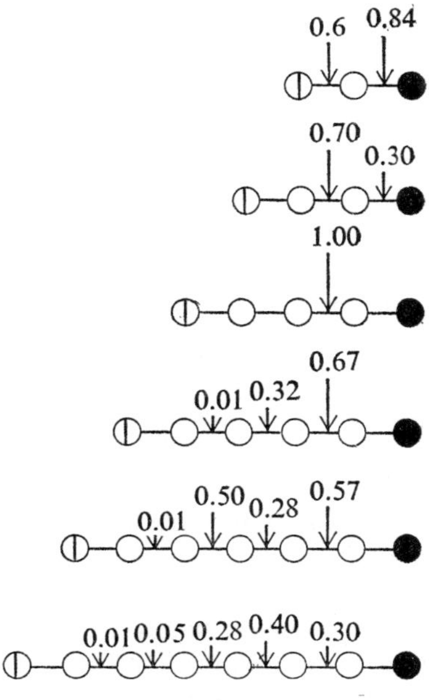

*Figure* 5.10. The frequency of bond hydrolysis during the degradation of various oligosaccharides, labelled at the reducing-end, by pig pancreatic *alpha*-amylase (Robyt and French 1970).

has been carried out by Robyt and French (1970) using porcine pancreatic enzyme and radioactive substrate oligomers. These elegant studies have enabled a deduction to be made of the frequency distribution of bond cleavage for the oligosaccharides containing from four to eight glucose residues. Their results are shown in figure 5.10, and of the oligomers tested only maltopentaose was hydrolysed by a single-chain mechanism. It must be emphasized that the frequency distributions are valid for comparison only within a given substrate, and that no attempt was made to quantify the absolute rates of degradation for the various species beyond the observation that the degradation of maltotriose was very much slower, and that of maltotetraose slower, than the rates observed with the higher oligomers.

*Binding Site Models for Alpha-Amylase.* The non-random nature of the hydrolysis of oligosaccharides has led to the development of sub-site models for the different *alpha-amylases*, i.e. the enzyme is assumed to possess a series of binding sub-sites that are distributed asymmetrically about the point at which catalysis occurs, and which are complementary to the monomer units of the polyglucose substrate. The first suggestion of the number of binding sites necessary was made by Myrbäck and Johansson (1945), who postulated that malt barley *alpha*-amylase must contain binding sites for eight glucose units. Similarly, Robyt and French (1963) suggested that the *B. subtilis* enzyme contained nine sub-sites. Greenwood, MacGregor and Milne (1965d), from a study of the breakdown products of oligomers containing four to ten glucose residues, suggested that the higher plant *alpha*-amylases contain eight substrate binding sites. The manner in which the various substrate species can accommodate themselves in these sub-sites is demonstrated in figure 5.11, in which (a) shows binding modes that are favourable and (b) shows those that are unfavourable.

The hypothetical schemes shown in figure 5.11 postulate that the active centre of the enzyme (the portion containing the pH-dependent nucleophilic and electrophilic groups which cause catalytic scission of the glycosidic bond) contains eight sub-sites, with six lying to one side and two to the other of the point of scission. A further necessary postulate is that the sites designated N and R have, relative to the other sites, an increased affinity for non-reducing and reducing terminal units respectively; sites $X_1$ and $X_2$ show a decreased affinity for non-reducing chain-ends, and $X_3$ for reducing chain-ends. This scheme, by arbitrarily balancing the binding contributions of the various sites, can satisfactorily explain the distribution of degradation products from the oligosaccharide series. As a chain of at least eight glucose residues is required for optimum binding, the above scheme also explains the decreased rate of hydrolysis observed for the smaller oligosaccharides.

More recent work (Thoma, Brothers and Spradlin 1970; Thoma, Rao, Brothers, Spradlin and Li 1971) has elaborated considerably on the earlier studies of *B. subtilis** *alpha*-amylase. The energetics of sub-site–monomer

*There is some confusion regarding the source of the enzyme. Thus the

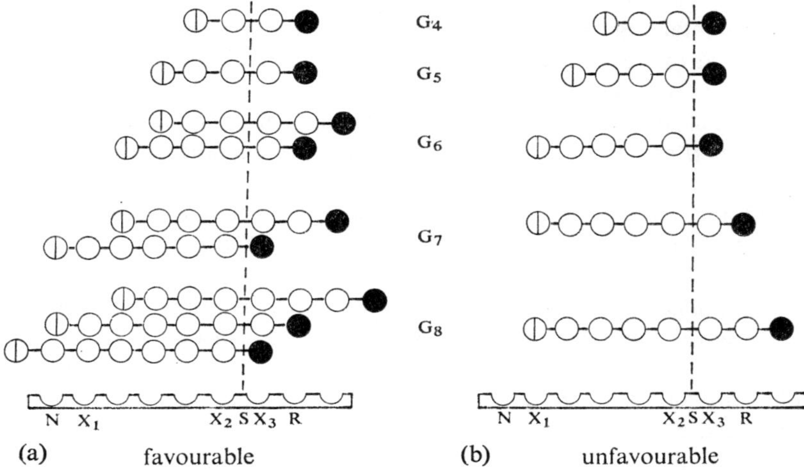


*Figure* 5.11. Interaction of various maltodextrins with the hypothetical active site of cereal *alpha*-amylase: (a) favourable complex formation; (b) unfavourable complex formation.
N=site that preferentially binds a non-reducing chain-end; $X_1$ and $X_2$=sites that are specifically unfavourable for binding non-reducing end-groups; S=point at which hydrolysis occurs; $X_3$=site that is specifically unfavourable for binding a reducing end-group; R=site that preferentially binds a reducing chain-end.
(Based on Greenwood and Milne 1968c.)

residue interaction on the enzyme have been established, and the resulting energy contour used to compute the population distribution of productive and non-productive positional isomers for any set of the enzyme–substrate complexes.† This information makes possible the calculation of the theoretical dependence of the Michaelis parameters on chain length. Very poor agreement between the calculated and experimental values was noted unless the model was modified by introducing the concept of substrate induced strain (Thoma 1968; Wakim, Robinson and Thoma 1969). With this concept,

bacterium from which the enzyme is isolated, previously thought to be *B. subtilis*, is most probably, according to Welker and Campbell (1967), *B. amyloliquefaciens*. Immunochemical studies show that the *alpha*-amylases from these two sources are not related. They also differ in the pH of optimum activity (pH 6.3 for *B. subtilis alpha*-amylase, and pH 5.9 for *B. amyloliquefaciens alpha*-amylase), and in the amount of enzyme produced per ml of culture medium, the *B. amyloliquefaciens* producing some 80 times as much. However, until the manufacturers correct the labelling of their product, it will probably avoid confusion if the original nomenclature is used.

† This work parallels that carried out on lysozyme, which has been reviewed by Chipman and Sharon (1969).

however, agreement between the computed and experimental data for substrates containing one to twelve glucose residues was excellent. Using the postulate that the residue undergoing hydrolysis is probably highly strained, Thoma and co-workers suggested that the longer chains should be subjected to increased strain and therefore were more easily hydrolysed, in agreement with the experimental observation. This work represents the first occasion on which the Michaelis constants for an enzyme substrate reaction have been explained theoretically, and, as such, must provide a milestone in the development of our understanding of the mechanism of enzyme action.

Robyt and French (1970) concluded that the active centre of porcine pancreatic *alpha*-amylase is capable of binding a maximum of five glucose units. Moreover, they draw attention to the asymmetric placing of the catalytic site relative to the binding sub-sites. Although quite profound differences in the number of substrate-binding sub-sites in the various *alpha*-amylases exist, the asymmetric placement of the catalytic site is such that the majority of binding sub-sites are always found at the non-reducing end of the oligosaccharide. Robyt and French (1970) pointed out that the side of the active site having the greater affinity for the chain fragment will, immediately after a hydrolytic scission, tend to hold that fragment, whilst the one on the other side of the catalytic centre will diffuse away. The remaining fragment may then rearrange about the active site, so giving an opportunity for a second hydrolytic event. Thus the direction (polarity) of any multiple attack will be towards the non-reducing end-group in the case of *alpha*-amylases. (In contrast, the polarity is towards the reducing end for *beta*-amylase; French and Youngquist 1963.)

*Enzyme Action Pattern.* The discussion so far has dealt only with the variations in action pattern noted at a fairly late stage in the α-amylolysis of amylose. The initial number-average degree of polymerization of amylose is approximately 2000 glucose residues, so that the fact that a few bonds (six to eight) are resistant to hydrolysis by the enzyme implies that the situation will not be observably different from that expected for a purely random action. Inherent in this statement is the postulate that only a single bond is hydrolysed (randomly) at a given enzyme–substrate encounter. The observation by Pazur, French and Knapp (1950), that the iodine-stain–reducing-power ratio for the digestion of amylose by porcine pancreatic *alpha*-amylase changed dramatically on going from pH 7.0 to pH 10.5, gave rise to a model challenging that basic postulate. This change has since been repeatedly verified, and gives rise to relations of the form shown in figure 5.12. For a given decrease in intensity of iodine stain, the use of a neutral or slightly acid pH gives a much greater reducing power than does the adverse alkaline pH. Also, it was noted that at the achroic point (that is, the stage in the degradation at which the digest does not give a visible colour with iodine) the distribution of molecular species at pH 7.0 was quite different to that obtained at pH 10.3. Pazur *et al.* (1950) suggested that the enzyme might be capable of attacking the same

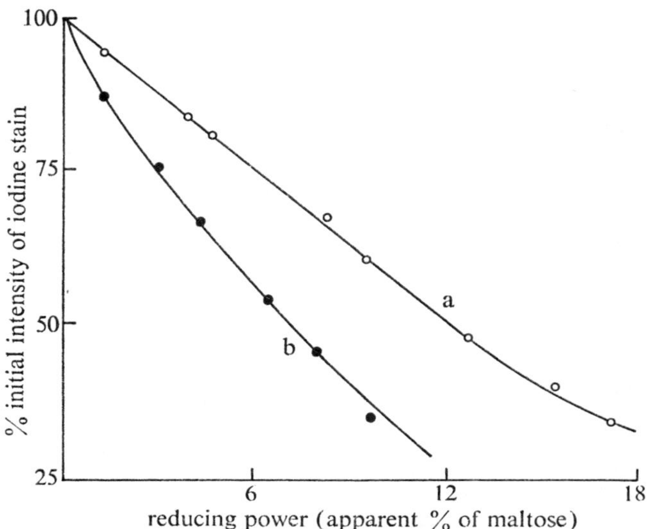


*Figure* 5.12. Intensity of iodine stain as a function of reducing power for the degradation of amylose by porcine pancreatic *alpha*-amylase: (a) pH 7.0; (b) pH 10.6 (Banks, Greenwood and Khan 1971b).

molecule more than once at a single enzyme–substrate encounter. If the number of attacks were a function of pH, the form of figure 5.12 could be explained. Thus, just as in the case of the exo-enzymes, we must consider the various types of action pattern shown in figure 5.13, i.e. *single-chain*, *multiple attack*, and *multi-chain* (Abdullah, French and Robyt 1966; Robyt and French 1967).

In the case of *alpha*-amylase, however, the term *single-chain* implies that after hydrolysing a bond randomly, one of the newly-produced chains (that having its reducing end-group just exposed) is completely degraded by the successive removal of small molecules such as maltose and maltotriose. The polarity of this repetitive attack, i.e. the fact that the enzyme moves from the reducing end towards the non-reducing end of the molecule, has been established only for porcine pancreatic *alpha*-amylase (Robyt and French 1971). For a polymer of sufficient size so that end-effects may be neglected, *multi-chain attack* implies that one bond is hydrolysed (randomly) in a single enzyme–substrate encounter, and *multiple attack* implies that in addition to the single random hydrolytic scission a number of non-random attacks take place with the formation of small sugars. The type of action envisaged is presented schematically in figure 5.14. This model, suggested by Robyt and French (1963), assumes that the fragment having the newly exposed reducing end-groups is retained on the enzyme surface slightly longer than the other fragment, and that the former may then re-align itself with respect to the binding sub-sites.

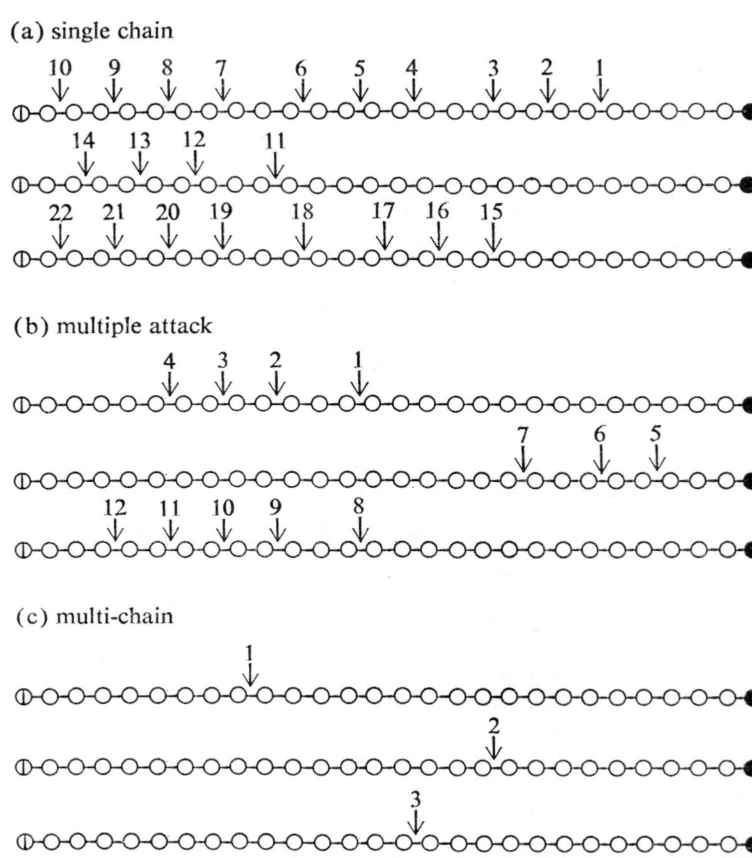


*Figure* 5.13. Various types of action-pattern possible with *alpha*-amylase. In each of the three cases, the numbered arrows refer to the sequence of events experienced by a single enzyme molecule. It has been arbitrarily assumed that the main product of repetitive attack is maltose, with maltotriose as a minor product. The polarity of multiple attack is shown according to the results of Robyt and French (1971) for porcine pancreatic *alpha*-amylase.

The iodine-stain–reducing-power relations obtained for a number of *alpha*-amylases are shown in figure 5.15. The point at which the intensity of the iodine stain falls to 50 per cent of its original value varies from 8 to 18 per cent apparent conversion into maltose. One explanation for this variation is that the degree of multiple attack depends on the type of *alpha*-amylase. However, a second viable explanation is the concept of *preferred attack*. According to this model, the internal bonds of the amylose macromolecule are hydrolysed in a random fashion, but as the product becomes smaller the specific end-effects associated with each *alpha*-amylase become dominant

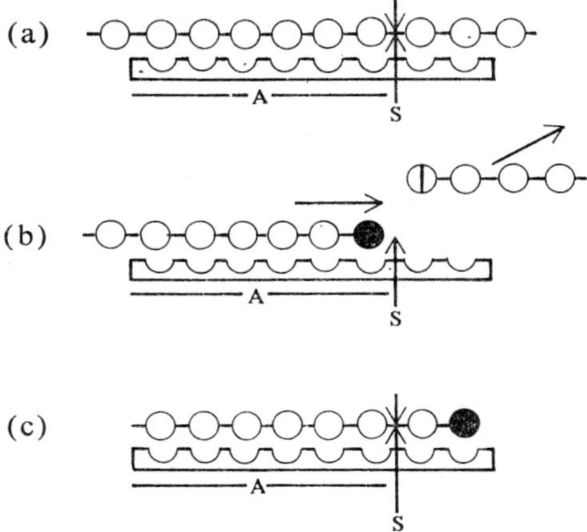


*Figure* 5.14. The sequence of events envisaged in multiple attack: (a) an internal segment of amylose is bound to the active site of the enzyme, and hydrolytic scission occurs at point S; (b) after scission, the fragment bearing the newly-exposed non-reducing chain-end diffuses away, but the remaining fragment is more strongly bound to portion A of the enzyme, and hence remains longer on this surface; (c) the substrate rearranges itself on the active site, and a second hydrolytic event occurs. It has been assumed, arbitrarily, that maltose is the product of this repetitive attack (based on Greenwood and Milne 1968c).

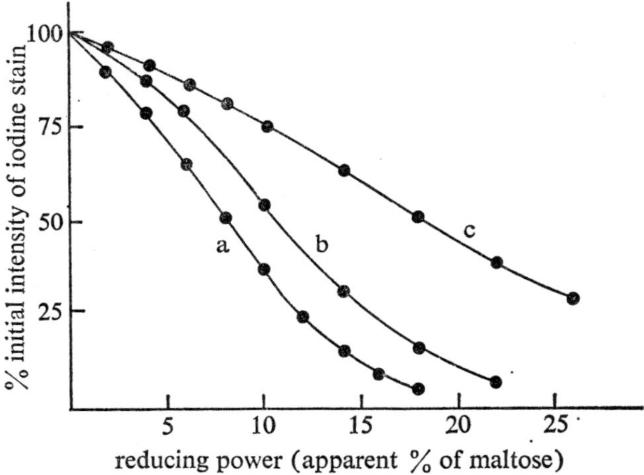


*Figure* 5.15. Intensity of iodine stain as a function of reducing power (apparent % maltose) for: (a) *B. subtilis alpha*-amylase; (b) human salivary *alpha*-amylase; (c) porcine pancreatic *alpha*-amylase (drawn from the tabulated data of Kung, Hanrahan and Caldwell 1953).

(as first suggested by Bird and Hopkins 1954), so that the distribution of chain lengths is a function of the source of the amylase. The evidence in favour of the two possible models—multiple attack and preferred attack—will now be considered.

*Initial Stages of the α-Amylolysis of Amylose.* Viscosity may be used to follow the enzyme reaction at very low amounts of hydrolysis, for the rate is then inversely proportional to the degree of polymerization for both zero and first order de-polymerization processes. The linear form of the relation between $(\overline{DP}_\eta)^{-1}$* and reaction time found for a number of *alpha*-amylases was interpreted as signifying random (i.e. multi-chain) hydrolysis by Greenwood and co-workers (Greenwood, MacGregor and Milne 1965c, d; Milne 1966). However, as Robyt and French (1967) pointed out, the use of the viscometric technique effectively ignores small molecules produced as a result of multiple attack, and the degradation should ideally be followed by two techniques enabling both the number-average $(\overline{DP}_n)$ and weight-average $(\overline{DP}_w)$ degrees of polymerization to be measured in conjunction. If the former average decreased more rapidly than the latter, then multiple attack would be indicated. Banks, Greenwood and Khan (1970c) reported that the value of the ratio $\overline{DP}_w/\overline{DP}_n$ was approximately two (as predicted for an exponential distribution) throughout the early stage of the action of *alpha*-amylases from porcine pancreas, human saliva, malted rye, and *B. subtilis* on amylose. These results suggest that multiple attack does not occur.

For experimental convenience, however, the α-amylolysis was carried out in the presence of 40 per cent aqueous glycerol, and later work (Banks, Greenwood and Khan 1971b) showed that glycerol greatly alters the form of the iodine-stain–reducing-power relation for porcine pancreatic *alpha*-amylase (other *alpha*-amylases gave the same curves, within experimental error, in the presence, or absence, of 40 per cent glycerol). In fact, the change in action pattern is virtually the same as that occasioned by using an adverse alkaline pH. The ratio $R = \overline{DP}_w/\overline{DP}_n$ was redetermined, therefore, in the absence of glycerol for the porcine and *B. subtilis* enzymes. Table 5.5 shows that whilst the value of $R$ is constant at approximately two for amylose digested with *B. subtilis alpha*-amylase, it increases throughout the course of the reaction for the porcine pancreatic enzyme. This result indicates very clearly that, at a stage when the amylose is sufficiently large for end-effects to be neglected, there is a rapid build-up of small sugars when the polysaccharide is hydrolysed by the pancreatic enzyme at pH 4.8. Quantitative analysis of the sugars present in this digest, and in one incubated at pH 10.5, gave the results recorded in table 5.6. The large build-up of small sugars occurring when amylose is hydrolysed by the porcine enzyme at pH 4.8, in contrast to the situation at pH 10.5, indicates that porcine pancreatic *alpha*-amylase functions at pH 4.8 by a most pronounced multiple-attack mechanism, whereas the use

* $\overline{DP}_\eta$ is the viscosity average molecular weight; it is much more closely allied to the weight-average than to the number-average.

of an adverse pH, or the presence of glycerol, causes alteration to an essentially multi-chain action pattern. In contrast, the *alpha*-amylases from human saliva, *B. subtilis*, and malted rye, function by multi-chain mechanisms over a wide range of conditions.

*Table* 5.5. Action of *B. subtilis* and porcine pancreas *alpha*-amylases on amylose in absence of glycerol: value of $R = \overline{DP}_w / \overline{DP}_n$ (Banks, Greenwood and Khan 1970c).

| *B subtilis* | | | | porcine pancreas | | | |
|---|---|---|---|---|---|---|---|
| sample | $\overline{DP}_w$ | $\overline{DP}_n$ | $R$ | sample | $\overline{DP}_w$ | $\overline{DP}_n$ | $R$ |
| 1 | 1820 | 810 | 2.2 | 1 | 2110 | 850 | 2.5 |
| 2 | 1540 | 685 | 2.2 | 2 | 1760 | 565 | 3.1 |
| 3 | 1280 | 595 | 2.2 | 3 | 1540 | 575 | 3.0 |
| 4 | 1110 | 505 | 2.2 | 4 | 1430 | 410 | 3.5 |
| 5 | 940 | 450 | 2.1 | 5 | 1230 | 300 | 4.1 |

*Table* 5.6. Action pattern of porcine pancreatic *alpha*-amylase: yields of maltodextrins produced from amylose as a function of pH and decrease in absorbance (Banks, Greenwood and Khan 1970c).

| sample | pH of digest | decrease in absorbance (%) | moles of maltodextrin[1] per ml ($\times 10^9$) G1 | G2 | G3 | G4 | G5 | G6 |
|---|---|---|---|---|---|---|---|---|
| 1 | 4.8 | 27 | 2 | 290 | 127 | 72 | 38 | 29 |
| 2 | 4.8 | 50 | 4 | 620 | 268 | 142 | 86 | 70 |
| 3 | 4.8 | 71 | 6 | 910 | 300 | 222 | 135 | 66 |
| 1a | 10.5 | 28 | 4 | 15 | 16 | 8 | 9 | 6 |
| 2a | 10.5 | 47 | 4 | 47 | 30 | 15 | 14 | 11 |
| 3a | 10.5 | 68 | 6 | 75 | 48 | 33 | 25 | 21 |

[1] G1 is D-glucose, G2 is maltose, etc.

The above conclusions do not agree with the earlier studies of Robyt and French (1967). The experimentation of these authors involved withdrawing samples as a function of time, measuring the total reducing power $R_t$, precipitating the polymeric material by adding a standard amount of alcohol, and measuring the concentration and reducing power $RV_p$ of this fraction. The degree of repetitive attack was then calculated from $r = RV_t / RV_p$, where $r$ is defined as the number of bonds hydrolysed during a single effective substrate–enzyme encounter. Values of $r > 1$ signify multiple attack. The $r$ values obtained by Robyt and French (1967) were: 1M sulphuric acid (60°), 1.9; porcine pancreatic *alpha*-amylase (pH 6.9), 7.0; porcine pancreatic *alpha*-amylase (pH 10.5), 1.7; human salivary *alpha*-amylase (pH 6.9), 3.0; *Aspergillus oryzae alpha*-amylase (pH 5.5), 2.9. The degree of multiple attack found for acid hydrolysis was said to represent the fact that the terminal bonds in the polymer are more susceptible to acid hydrolysis than any other. The measurements indicated multiple attack for all the enzymes studied—even in the case of porcine pancreatic alpha-amylase at pH 10.5. The conclusions are based on the assumption that the formation of oligosaccharides as a result of primary action is insignificant; an assumption which it is claimed breaks down only when the degree of polymerization of the polymeric substrate, i.e. the material precipitated by the addition of ethanol, is less than 30–40 units.

This assumption may easily be verified by calculating the amount of material having $DP \leqslant x$ formed during the random degradation of the exponential distribution. In practice, ethanol will not cause a sharp separation, and quite a range of degrees of polymerization will be found in both the precipitate and supernatant. Our own (unpublished) results suggest that the lowest fraction present in the precipitate has $\overline{DP}_n \sim 23$, a value in reasonable agreement with the figure of 20 quoted by Robyt and French (1967). Calculations of weight percentage $w_m$ of material left in solution, on the basis that all chainsw ith $DP > 15$ or 20 units are precipitated, were made using (5.12) for various arbitrary degrees of degradation, which enabled the parameter $r$ to be evaluated. The results shown in table 5.7 indicate that a limited degree of multiple attack apparently occurs, despite assuming a multi-chain mechanism in the calculation. Thus the method of Robyt and French can lead to the conclusion that a limited degree of multiple attack is operative when the hydrolysis is random, due to multi-chain action contributing appreciably to the number of chains with $DP < 20$ glucose residues.

It is rather more difficult to explain the discrepancy between the results of Robyt and French (1967) and of Banks, Greenwood and Khan (1970c) for the proposed action pattern of human salivary *alpha*-amylase. Certainly, the build-up of non-precipitable material as a result of random degradation is not sufficiently great to explain the differences. As has been emphasized by Thoma, Spradlin and Dygert (1971), the technique of Banks *et al.* for measuring $\overline{DP}_n$ depends on the release during multiple attack of a strict equivalence of even- and odd-numbered members of the maltosaccharide series: an enzyme that released only maltose, for example, as the product of multiple attack would appear by the technique of Banks *et al.* to function by a multi-chain mechanism, because even-numbered members of the series are not counted (Banks and Greenwood 1968b). However, when digests were analysed for small sugars at early stages in the $\alpha$-amylolysis of amylose, the results (shown in table 5.8) indicated that the preference for producing even-numbered members of the oligosaccharide series was not particularly marked (Banks, Greenwood and Khan 1970c). It is thus unlikely that the conclusion, that the action-pattern of salivary *alpha*-amylase does not involve multiple attack, is due to an experimental artefact.

Recently Thoma, Rao, Brothers, Spradlin and Li (1971) have concluded that the *alpha*-amylase from *B. amyloliquefaciens* adopted a multi-chain

*Table* 5.7. Calculations showing that the conditions of Robyt and French (1967) would appear to give some degree of multiple attack for a completely random degradation (Banks and Greenwood, unpublished work).

(a) Chains of $DP \leqslant 20$ are not removed (see text)

| $\overline{DP}_n$ (1) | $w_m$ (2) | $\overline{DP}_m$ (3) | $\overline{DP}_p$ (4) | $RV_t$ (5) | $RV_p$ (6) | $r$ (7) |
|---|---|---|---|---|---|---|
| 200 | 0.48 | 10 | 220 | 10 | 9 | 1.1 |
| 100 | 1.76 | 10 | 119 | 20 | 16.5 | 1.2 |
| 80 | 2.65 | 10 | 99 | 25 | 20 | 1.25 |
| 60 | 4.15 | 10 | 77 | 33 | 25 | 1.3 |
| 40 | 9.03 | 10 | 57 | 50 | 32 | 1.6 |
| 20 | 26.42 | 10 | 31 | 100 | 48 | 2.1 |

(b) Chains of $DP \leqslant 15$ are not removed (see text)

| $\overline{DP}_n$ | $w_m$ | $\overline{DP}_m$ | $\overline{DP}_p$ | $RV_t$ | $RV_p$ | $r$ |
|---|---|---|---|---|---|---|
| 150 | 0.48 | 8 | 164 | 13 | 12 | 1.1 |
| 75 | 1.76 | 8 | 88 | 27 | 22 | 1.2 |
| 60 | 2.65 | 8 | 73 | 33 | 26 | 1.3 |
| 45 | 4.15 | 8 | 56 | 44 | 34 | 1.3 |
| 30 | 9.03 | 8 | 41 | 67 | 44 | 1.5 |
| 15 | 26.42 | 8 | 22 | 133 | 67 | 2.0 |

[1] Number-average degree of polymerization of digest.

[2] Weight percentage of amylose having a degree of polymerization less than, or equal to, the stated limit, and which is not precipitated by 66.7% ethanol.

[3] Calculated number-average degree of polymerization of that fraction not precipitated by 66.7% ethanol.

[4] Calculated number-average degree of polymerization of the polymeric fraction, that is, the fraction precipitated by 66.7% ethanol. Calculated from

$$\frac{w_m}{\overline{DP}_m} + \frac{100 - w_m}{\overline{DP}_p} = \frac{100}{\overline{DP}_n}$$

[5] Reducing value of digest, calculated as μg maltose/ml, assuming an initial polysaccharide concentration of 1 mg/l. Calculated from $RV_t = 2(1000/\overline{DP}_n)$.

[6] Reducing value of polymeric fraction, calculated as μg maltose/ml from $RV_p = 2(1000 - 10w_m)/\overline{DP}_p$.

[7] The ratio $r$ is a measure of multiple attack, being defined by $r = RV_t/RV_p$.

action pattern over a wide range of pH values, a conclusion in excellent agreement with our own work on the bacterial enzyme.

*Table* 5.8. Action pattern of human salivary *alpha*-amylase: yields of maltodextrins produced from amylose as function of decrease in absorbance at pH 4.8 (Banks, Greenwood and Khan 1970c).

| sample | decrease in absorbance (%) | moles of maltodextrin[1] per ml ($\times 10^9$) | | | | | |
|---|---|---|---|---|---|---|---|
| | | $G_1$ | $G_2$ | $G_3$ | $G_4$ | $G_5$ | $G_6$ |
| 1 | 26 | 3 | 32 | 30 | 24 | 10 | 15 |
| 2 | 45 | 3 | 62 | 68 | 39 | 16 | 30 |
| 3 | 64 | 6 | 146 | 195 | 130 | 52 | 70 |

[1] $G_1$ is glucose, $G_2$ is maltose, etc.

*Later Stages of the α-Amylolysis of Amylose: Theoretical Concepts.* The action pattern of the *alpha*-amylases may be examined also by studying the composition of the digest at a fairly late stage in the degradation, and comparing the actual oligosaccharide composition with that predicted by various theories based on different assumptions. A comparison of the experimental and predicted oligosaccharide yields are shown in tables 5.9–11 for the *alpha*-amylases from *B. subtilis*, pig pancreas, and malted wheat. The basis of each of the theories used in predicting the oligosaccharide yield is detailed below.

(a) *Random degradation.* Kuhn (1930) has shown that the weight fraction $W_i$ of an oligomer of length $i$ units, formed by the random degradation of a polymer of degree of polymerization $x$, is in the limiting case of $i \ll x$ given by

$$W_i = is^2(1-s)^{i-1} \tag{5.27}$$

where $s$ is the degree of scission, i.e. the number of bonds broken per original polymer molecule.

Equation 5.27 may be used to calculate the yield of a specific oligosaccharide at any stage during the degradative process, if the value of $s$ is known. Although this parameter is not known exactly, it may be approximated by choosing a value such that agreement is forced between experiment and theory for the yields of the larger oligosaccharides, i.e. those containing more than six glucose residues in the case of the bacterial and cereal enzymes, and more than four residues for the pancreatic *alpha*-amylase.

(*b*) *Random degradation, with a lower limit to the oligosaccharide size that may be degraded.* Statistical considerations show that the weight fraction $W_i$ of *i*-mer produced from the random scission of polymer of degree of polymerization $x$ is

$$W_i = (i/x)[2s(1-s)^{i-1} + (x-i-1)s^2(1-s)^{i-1}] \tag{5.28}$$

Differentiating (5.28) with respect to $s$ gives

$$dW_i/ds=(i/x)[2(1-s)^{i-1}-2s(i-1)(1-s)^{i-2}+2s(x-i-1)(1-s)^{i-1} \\ -(x-i-1)(i-1)s^2(1-s)^{i-2}] \qquad (5.29)$$

In (5.29), positive terms represent the formation of $i$-mer by the degradation of larger oligomers, and negative terms the disappearance of the $i$-mer as it is degraded to smaller fragments. Painter (1963) has shown that if the $i$-mer is resistant to further degradation, but that all higher oligomers may be degraded, the final yield of $i$-mer is given by the integral of the positive terms in (5.29), that is

$$W_i=(i/w)\int_0^1 [2(i-s)^{i-1}+2s(x-i-1)(i-s)^{i-1}]ds \\ =2/(i+1) \qquad (5.30)$$

Equation 5.30 shows that the yield of $i$-mer is independent of the size and size distribution of original polymer, and hence represents the weight fraction of totally degraded polymer present as $i$-mer.

Similarly, if all the oligomers smaller than the $i$-mer are resistant to hydrolysis, the weight-fraction of the $(i-j)$-mer is

$$W_{i-j}=2/(i-j+1)\prod_{k=0}^{j-1}(i-k-1)/(i-k+1)$$

which simplifies to

$$W_{i-j}=2(i-j)/i(i+1) \quad \text{when } j<1 \qquad (5.31)$$

Equation 5.31 is valid only when the polymer is *completely* degraded to oligomers of size $i$ and smaller; for intermediate stages of the degradation

$$W_{i-j}=[2(i-j)/i(i+1)]x$$

where $x$ is the fraction of the total polymer that has been degraded to oligomers of size $i$ and smaller.

(c) *Preferential attack.* Certain bonds in the substrate are especially susceptible to hydrolysis.

(d) *Multiple attack.* More than one bond is hydrolysed in a single effective substrate–enzyme encounter.

*Oligosaccharide Yield from B. subtilis alpha-Amylase.* Table 5.9 shows the maltodextrin yield obtained at two different stages of α-amylolysis, and compares them with the amounts predicted by the various theories. It appears that the products of the degradation of amylose at either of the two stages examined cannot be explained by random attack. This observation in no way affects our earlier conclusion that in the initial stages random degradation is operative, because both the samples are taken at fairly advanced stages of hydrolysis when little polymer remains, and the dominant influence is the effect of chain-ends.

The result of assuming that sugars containing five and six glucose residues (and the lower members of the series) are resistant to degradation whilst there are larger molecules still present, are shown in lines $R_5$ and $R_6$ respectively: neither postulate provides a very good fit for the experimental results.

*Table* 5.9. Experimental and theoretical yields of maltodextrins from amylose; *B. subtilis alpha*-amylase action (Greenwood and Milne 1968c).

| theory[1] | % by weight of | | | | | | |
|---|---|---|---|---|---|---|---|
| | $G_1$ | $G_2$ | $G_3$ | $G_4$ | $G_5$ | $G_6$ | $>G_6$ |
| *E* | 2 | 10 | 13 | 6 | 10 | 21 | 38 |
| *R* | 7.6 | 11.0 | 11.9 | 11.6 | 10.5 | 9.1 | 38.3 |
| $R_5$ | 2.7 | 5.5 | 8.2 | 10.9 | 13.7 | — 59 — | |
| $R_6$ | 3.0 | 5.9 | 8.9 | 11.8 | 14.8 | 17.8 | 38 |
| *S*(*a*) | 1 | 4 | 20 | 5 | 5 | 27 | 38 |
| *S*(*b*) | 1 | 5 | 21 | 5 | 5 | 25 | 38 |
| *E* | 2 | 12 | 24 | 8 | 14 | 30 | 10 |
| *R* | 20.3 | 22.4 | 18.4 | 13.5 | 9.3 | 6.2 | 9.9 |
| $R_5$ | 4.0 | 8.0 | 12.0 | 16.0 | 20.0 | — 40.0 — | |
| $R_6$ | 4.3 | 8.6 | 12.9 | 17.1 | 21.4 | 25.7 | 10.0 |
| *S*(*a*) | 4 | 7 | 20 | 5 | 5 | 48 | 10 |
| *S*(*b*) | 4 | 7 | 21 | 5 | 5 | 48 | 10 |
| *S*(*c*) | 3 | 9 | 20 | 5 | 11 | 41 | 10 |

[1] *E*, experimental
*R*, random degradation
$R_5$, random degradation with $G_5$ resistant
$R_6$, random degradation with $G_6$ resistant
*S*, theory concerning specificity of enzyme: for theories (*a*), (*b*) and (*c*), see text.

For preferential attack, it is assumed that the third bond from the reducing end of the substrate is particularly labile, so yielding large amounts of maltotriose. The calculations *S*(*a*) in table 5.9 assume (Greenwood and Milne 1968c) that this bond is four times more easily hydrolysed than any other. Comparing the calculated and experimental values shows that this theory predicts too much maltotriose for the first stage, and not enough for the second. Also, less maltohexaose is found than is predicted, which of course is to be expected in view of the fact that this product is still a substrate for the enzyme and will be slowly degraded during the reaction, even if larger oligosaccharides are present. Additionally, the calculations suppose that maltoheptaose is degraded quantitatively to glucose and maltohexaose, whereas it is known that a second reaction producing maltose and maltopentaose occurs. Assuming, in fact, that the degradation of maltoheptaose may proceed by either mechanism with equal probability, improves the fit between theory and experiment by predicting a larger yield of maltose—see calculations $S(b)$ of table 5.9.

For the model of multiple attack, Robyt and French (1963) suggested that when scission produces a dextrin of length greater than maltooctaose attached to part $A$ of the enzyme (see figure 5.14), it rearranges on the active site of the enzyme and subsequently produces a molecule of *maltotriose*. Calculations $S(c)$ based on this assumption give the oligosaccharide yields shown in table 5.9: the agreement between theory and experiment is certainly no worse than that obtained using the model of preferential attack.

Thus the analysis of the products at a fairly late stage in the degradation of amylose by the *alpha*-amylase from *B. subtilis* cannot distinguish between the models of multiple and preferential attack. However, the enzymic technique of measuring $\overline{DP}_n$ employed by Banks, Greenwood and Khan (1970c) would be especially sensitive to the presence of maltotriose, and therefore $\overline{DP}_n$ would decrease much more rapidly than $\overline{DP}_w$ if the multiple attack envisaged by Robyt and French (1963) did occur. Also, Thoma, Rao, Brothers, Spradlin and Li (1971), from studies of the form of the iodine-stain–reducing-power relation as a function of pH, concluded that the action mechanism did not involve repetitive attack. The balance of evidence would appear to favour a multi-chain action pattern for *B. subtilis alpha*-amylase.

*Oligosaccharide Yield from Porcine Pancreatic alpha-Amylase*. The mammalian enzyme hydrolyses polymeric and oligomeric substrates with equal facility, as shown by the fact that only 12 per cent of the sugars present at the achroic limit contain more than six glucose units, whereas the corresponding value for the cereal *alpha*-amylases is in excess of 60 per cent. The results in table 5.10 show that neither random attack nor the theories predicting a random process, but with either maltotriose or maltotetraose resistant to further degradation, are likely.

The model of preferential attack in this case assumes that the linkage second from the reducing end of the substrate molecule is three times less difficult to hydrolyse than any other bond, and the value for the degree of scission has been chosen so as to force agreement between the experimental and theoretical figure for the yield of maltotetraose. At both stages of the hydrolysis, this process predicts less maltose and more maltotriose than is actually obtained (see table 5.10, $S(a)$).

For the multiple-attack model, the calculations assume that each scission giving a fragment containing more than six glucose residues bound to portion $A$ of the enzyme (see figure 5.14) will be followed by two more non-random hydrolytic events, each producing maltose; if the fragment retained at $A$ contains four to six residues, then one molecule of maltose will be released by a second scission. Again, agreement has been forced between the experimental and theoretical yields for maltotetraose. It is obvious from $S(b)$ in table 5.10 that the multiple-attack model accurately predicts the amounts of maltose and maltotriose formed during hydrolysis.

*Table* 5.10. Experimental and theoretical yields of maltodextrins from amylose: hog pancreas *alpha*-amylase action (Greenwood and Milne 1968c).

| theory[1] | % by weight of | | | | | | |
|---|---|---|---|---|---|---|---|
| | $G_1$ | $G_2$ | $G_3$ | $G_4$ | $G_5$ | $G_6$ | $>G_6$ |
| *E* | 1 | 42 | 25 | 15 | 3 | 2 | 12 |
| *R* | 19.4 | 21.8 | 18.3 | 13.7 | 9.6 | 6.4 | 10.8 |
| $R_3$ | 11.3 | 22.7 | 34.0 | 32 | | | |
| $R_4$ | 8.3 | 16.6 | 24.9 | 33.2 | 17 | | |
| *S*(*a*) | 0 | 38 | 30 | 15 | 17 | | |
| *S*(*b*) | 0 | 43 | 24 | 15 | 17 | | |
| *E* | 1 | 44 | 31 | 21 | 1 | 1 | 1 |
| *R* | 15.4 | 18.7 | 17.1 | 13.9 | 10.6 | 7.7 | 16.6 |
| $R_3$ | 12.7 | 25.3 | 38.0 | 24.0 | | | |
| $R_4$ | 9.7 | 19.4 | 29.1 | 38.8 | 3.0 | | |
| *S*(*a*) | 0 | 40 | 36 | 21 | 3 | | |
| *S*(*b*) | 0 | 45 | 30 | 21 | 3 | | |

[1] *E*, experimental
*R*, random degradation
$R_3$, random degradation with $G_3$ resistant
$R_4$, random degradation with $G_4$ resistant
*S*, theory concerning specificity of enzyme: for theories (*a*) and (*b*), see text.

It is now generally accepted that porcine pancreatic *alpha*-amylase does indeed function by a pronounced multiple-attack mechanism at pH values near the optimum for this enzyme. However, the use of adverse pH (Robyt and French 1967), high concentrations of glycerol (Banks, Greenwood and Khan 1970c), erythritol and methyl-α-D-glucopyranoside (Banks, Greenwood and Khan 1971*c*) cause the enzyme to adopt a multi-chain mechanism. These results indicate that the binding between enzyme and substrate can be altered in subtle ways. The use of glycerol (40 per cent v/v) in the digest also demonstrates that multiple attack is a property of the enzyme rather than of the substrate. The diffusion of the amylose away from the enzyme after a successful encounter must be comparatively slow in such a highly viscous medium, and should therefore provide ideal conditions for this kind of hydrolysis. But contrary to any such expectation, glycerol suppresses such attack in the case of the pancreatic enzyme, and does nothing to encourage it in the other *alpha*-amylases tested (those from *B. subtilis*, human saliva, and malted rye).

Another possible explanation of the action pattern of porcine pancreatic *alpha*-amylase has been advanced (Banks, Greenwood and Khan 1970a), namely that the various component isoenzymes of this system (Rowe,

Wakim and Thoma 1968) could differ slightly in their action pattern. More recent work (Cozzone, Pasero and Marchis-Mouren 1970; Banks, Greenwood and Khan 1971c) suggests that these isoenzymes are liable to be artefacts: indeed the various samples produced from the pancreatic enzyme by passage through an ion-exchange column possess identical action patterns (Banks, Greenwood and Khan 1971*c*). Nevertheless, the possibility of isoenzyme-specific action patterns should be considered when planning experiments.

*Oligosaccharide Yield from Malted Wheat alpha-Amylase*. The experimental and calculated values for the yields of oligosaccharides are shown in table 5.11. It is again apparent that the process of random attack, with or without resistant residues, could not give rise to the results obtained experimentally.

*Table* 5.11. Experimental and theoretical yields of maltodextrins from amylose: germinated wheat *alpha*-amylase action (Greenwood and Milne 1968c).

| theory[1] | % by weight of | | | | | | |
|---|---|---|---|---|---|---|---|
| | $G_1$ | $G_2$ | $G_3$ | $G_4$ | $G_5$ | $G_6$ | $>G_6$ |
| *E* | 1 | 9 | 4 | 5 | 5 | 16 | 60 |
| *R* | 4.0 | 6.4 | 7.7 | 8.2 | 8.2 | 7.9 | 57.6 |
| $R_6$ | 1.9 | 3.8 | 5.7 | 7.6 | 9.5 | 11.5 | 60.0 |
| *S*(*a*) | 1 | 7 | 5 | 6 | 6 | 16 | 59 |
| *S*(*b*) | 1 | 14 | 3 | 3 | 4 | 15 | 60 |
| *E* | 2 | 15 | 5 | 8 | 8 | 35 | 27 |
| *R* | 11.1 | 14.8 | 14.8 | 13.2 | 11.0 | 8.8 | 26.3 |
| $R_6$ | 3.5 | 6.9 | 10.4 | 13.9 | 17.4 | 20.9 | 27.0 |
| *S*(*a*) | 2 | 14 | 7 | 7 | 6 | 37 | 27 |
| *S*(*b*) | 2 | 20 | 6 | 5 | 5 | 35 | 27 |

[1] *E*, experimental
*R*, random degradation
$R_6$, random degradation with $G_6$ resistant
*S*, theory concerning specificity of enzyme: for theories (*a*) and (*b*), see text.

The model of preferred attack in this instance assumes that the cereal *alpha*-amylase does not hydrolyse the first five linkages at the non-reducing end of the substrate molecule, that it has a low affinity for the scission of the ultimate bond at the reducing end of the molecule, but possesses a high affinity for hydrolysing the penultimate bond at the reducing end. The calculations are based on a rate of hydrolysis for the ultimate bond that is half that of a bond removed from either end, whereas the penultimate bond is assumed to be hydrolysed twice as easily as any other bond in the system.

Table 5.11 shows that there is good agreement between the experimental results and the values calculated on the basis of preferential attack (compare $E$ and $S(a)$).

For malted wheat *alpha*-amylase, the multiple-attack calculations $S(b)$ in table 5.11 are based on a model which assumes that each hydrolytic event produces a fragment containing more than seven glucose units bound to portion $A$ of the enzyme, and this event is followed by the rearrangement of the substrate so that, in the following scission, a molecule of maltose is produced. The yields of oligosaccharides calculated on this basis ($S(b)$) consistently predict the presence of more maltose than is actually produced. In this case, therefore, it is most unlikely that the model of multiple attack is applicable.

*Conclusions.* For the various *alpha*-amylases examined in detail—with the exception of the one isolated from the pig pancreas—the evidence, although by no means overwhelming, indicates that preferential hydrolysis is occurring. Obviously, much more detailed work is necessary before the action patterns of these *alpha*-amylases are fully understood. In contrast, the *alpha*-amylase from pig pancreas is apparently unique in possessing a pronounced capability for multiple attack. It would therefore be of some interest to investigate whether the properties of pancreatic *alpha*-amylases from other sources are similar.

The distinction between *alpha*- and *beta*-amylases is becoming less marked as more information is forthcoming. Thus the endo-acting enzyme from *B. polymyxa* liberates mainly $\beta$-maltose, and has a high degree of repetitive attack (Robyt and French 1964). An enzyme system has been found in bovine serum (Banks, Mazumder and Spooner 1971) that also possesses the ability to engage in a most pronounced form of multiple attack, but in this case the configuration at the anomeric carbon atom of the product is unknown. Both enzymes discussed above, however, have the ability to by-pass the $\alpha$-1:6-branch points of amylopectin and glycogen, indicating that they are endo-enzymes. Robyt and French (1970) have speculated that the polarity of the multiple attack may be related to the configuration of the anomeric carbon atom of the product. This hypothesis would be confirmed if future work indicates that the multiple attack of the endo-amylase from *B. polymyxa* is directed in the same sense as is that of *beta*-amylase.

**5.3.2.** Action Patterns of Endo-Enzymes on Amylopectin and Glycogen. The study of the action patterns of the endo-enzymes on amylopectin and glycogen presents a rather more intractable problem than does the analogous study using amylose, because, as we have emphasized earlier, the structures of the branched polysaccharides are not yet defined. Furthermore, not only are the branch points themselves resistant to hydrolysis by the *alpha*-amylases but they also confer some degree of resistance on neighbouring $\alpha$-1:4-linkages. The smallest residual branched dextrins obtained by the $\alpha$-amylolysis of amylopectin are shown in figure 5.16: the size and type are typical of the enzyme, e.g. only a cereal enzyme will yield a branched trisaccharide, $6^2$-α-glucosylmaltose (the nomenclature adopted for the branched oligosaccharides is that recommended by Whelan 1960b). In addition, the degradation products from the α-amylolysis of amylopectin and glycogen include other large branched structures (Roberts and Whelan 1960; Heller and Schramm 1964; Brammer, Rougvie and French 1972) as well as linear oligosaccharides, the latter being ultimately converted to glucose and maltose.

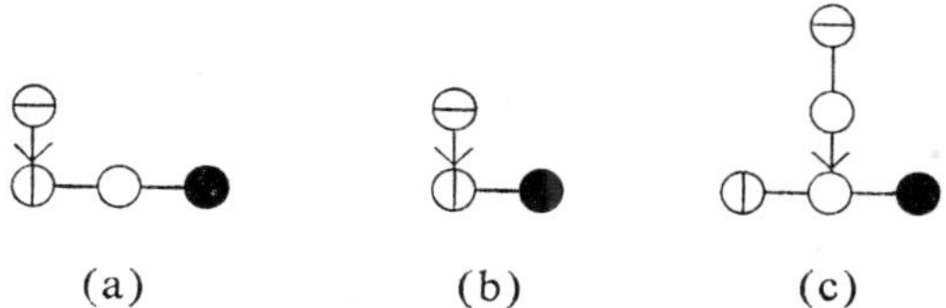


*Figure* 5.16. The smallest resistant dextrins obtained on the α-amylolysis of amylopectin: (a) $6^3$-α-glucosyl maltotriose—obtained using *alpha*-amylases from pig pancreas, human saliva, and *Aspergillus oryzae* (Whelan 1960a; Nordin and French 1958); (b) $6^2$-α-glucosyl maltose—obtained using malted barley *alpha*-amylase (Whelan 1960a); (c) $6^2$-α-maltosyl maltotriose—obtained using *B. subtilis alpha*-amylase (Hughes, Smith and Whelan 1963; French, Smith and Whelan 1972). (Nomenclature of the oligosaccharides is based on the recommendations of Whelan 1960b.)

A detailed study by Robyt and French (1963), using the enzyme from *B. subtilis*, enabled comparative values to be obtained for the oligosaccharides present at equivalent stages in digests of amylose and amylopectin. The results shown in table 5.12 indicate that the absolute yields of individual oligosaccharides are quite different in the two cases, although the same

*Table* 5.12. The oligosaccharides released on the α-amylolysis of amylose and amylopectin by the enzyme from *B. subtilis* (Robyt and French 1963).

| sugar[1] | amylose | | amylopectin | |
|---|---|---|---|---|
| | 60 min (wt. %) | 180 min (wt. %) | 60 min (wt. %) | 180 min (wt. %) |
| $G_1$ | 2 | 5 | 1 | 3 |
| $G_2$ | 10 | 12 | 6 | 8 |
| $G_3$ | 13 | 22 | 8 | 11 |
| $G_4$ | 6 | 11 | 1 | 3 |
| $G_5$ | 10 | 15 | 5 | 7 |
| $G_6$ | 21 | 30 | 14 | 27 |
| $G_7$ | 15 | 5 | 10 | 9 |
| $>G_7$ | 23 | 0 | 55 | 32 |

[1] All the sugars up to $G_7$ are linear.

general pattern is followed. In particular, the enzyme has a pronounced tendency to produce large amounts of maltotriose and maltohexaose during the course of both reactions. As we have noted previously, the external chain-length of amylopectin is 10–15 glucose units, and some 60 per cent of the residues are to be found in such chains. Thus on a purely statistical basis, which neglects steric effects, approximately half of the effective enzyme–substrate encounters in the initial stages of the degradation should result in the formation of linear oligomers. If this statistical fact is considered in conjunction with the preferential attack model for *B. subtilis alpha*-amylase, it is obvious that the larger linear maltodextrins must arise from the external chains. In fact, by comparing the degradation of the $\beta$-amylolysis limit dextrin of amylopectin with that of the native polysaccharide, Robyt and French (1963) concluded that the maltose and maltotriose arose from attacks on segments between branch points, and that the higher members of the series came from degradation of the external chains. Similar experiments using glycogen and glycogen $\beta$-amylolysis limit dextrin also showed that linear oligomers such as maltohexaose and maltoheptaose must come exclusively from the peripheral chains.

Robyt and French (1963) measured the initial rates for the production of reducing sugars by *B. subtilis alpha*-amylase using glycogen, amylopectin and amylose as substrates; their results are summarized in table 5.13, in which the rates of attack on glycogen and amylopectin are quoted relative to that on amylose. The values show comparatively little variation, bearing in mind the quite profound differences in the structures of the three polysaccharides.

*Table* 5.13. The relative rates of initial production of reducing-power obtained using *B. subtilis alpha*-amylase on amylose, amylopectin, glycogen (Robyt and French 1963).

| substrate | $\frac{\text{rate(amylose)}}{\text{rate(substrate)}}$ |
|---|---|
| amylose | 1.0 |
| amylopectin | 2.0 |
| glycogen | 2.7 |

Geddes (1968) has compared the initial relative rates for the hydrolysis of the various polysaccharides and their $\beta$-amylolysis limit dextrins using the technique of light scattering, and employing a number of plant *alpha*-amylases. The results of this work (table 5.14) show that the rates of attack on the various polysaccharides vary considerably. The order obtained is the same as that noted by Robyt and French (1963).

It would be particularly useful if both reducing-power and light-scattering measurements were carried out in conjunction, so evaluating the number- and weight-average molecular weights, respectively, during the early stages

*Table* 5.14. Experimental rates of enzyme attack on various substrates, relative to amylose (data from Geddes 1968).

| substrate | *alpha*-amylase | | |
|---|---|---|---|
| | soya bean | malted barley | barley |
| amylose | 1.00 | 1.00 | 1.00 |
| amylopectin | 0.58 | 0.68 | 0.67 |
| amylopectin, $\beta$-limit dextrin | 0.46 | 0.41 | 0.40 |
| glycogen | 0.32 | 0.23[(1)] | 0.04 |
| glycogen, $\beta$-limit dextrin | 0.10 | 0.09[(1)] | 0.02 |

[1] Activity of enzyme increased sevenfold, therefore it is necessary to divide these values by seven in order to obtain figures comparable with the others quoted.

of the reaction. Knowledge of these parameters would enable at least a qualitative insight to be gained of the relative ease of hydrolysis of internal and external chains in amylopectin.

## 5.4. De-Branching Enzymes.

The study of de-branching enzymes is complicated by lack of knowledge of their exact specificities. Here, we briefly outline the characteristics of *R*-enzyme, pullulanase, the isoamylases, and amylo-1:6-glucosidase.

### 5.4.1. *R*-Enzyme.

A debranching enzyme, *R*-enzyme, was isolated from broad beans by Hobson, Whelan and Peat (1951), and was reported to hydrolyse (incompletely) amylopectin, amylopectin $\beta$-amylolysis limit dextrin, and various branched $\alpha$-amylolysis limit dextrins, but to have *no* action on glycogen or its $\beta$-amylolysis limit dextrin. Later work by MacWilliam and Harris (1959), using preparations from both broad beans and malted barley, suggested that the originally defined *R*-enzyme was composed of two quite distinct activities: one of these was specifically adapted for the de-branching of macromolecular substrates, and was re-defined as *R*-enzyme, whilst the other hydrolysed the branch points in $\alpha$-amylolysis limit dextrins, and so was classed as a *limit dextrinase*. Manners and Sparra (1966) and Manners and Rowe (1969) confirmed these results for malted barley, but subsequently (Manners, Marshall and Yellowlees 1970) modified their view of the specificity of the limit dextrinase by suggesting that it had the ability to de-branch amylopectin $\beta$-amylolysis limit dextrin—a finding in direct contradiction to that of MacWilliam and Harris (1959). Drummond, Smith and Whelan (1970) proposed that these conflicting results could be rationalized by assuming that the limit dextrinase activity was in fact merely a more dilute solution of the *R*-enzyme activity. These authors showed that the de-branching activity of a preparation from potato could be diluted so as to selectively lose its ability to hydrolyse amylopectin, whilst retaining its activity towards the $\alpha$- and $\beta$-amylolysis limit dextrins. Lee, Marshall and Whelan (1971) subsequently showed that a similar selective loss of activity could be observed if the de-branching enzyme from sweet corn were diluted. Manners (1971) vigorously opposes the above explanation.

**5.4.2.** Pullulanase.

Perhaps the most widely studied of the de-branching enzymes is that elaborated by *Aerobacter aerogenes*, discovered by Bender and Wallenfels (1961) and named *pullulanase*. The nomenclature arose because of the activity of the enzyme towards pullulan, a linear polysaccharide composed of maltotriose units linked by $\alpha$-1:6-bonds. Pullulanase hydrolyses this substrate by an endo-mechanism (Drummond, Smith, Whelan and Tai 1969). The first detailed study of the enzyme's specificity (Abdullah, Catley, Lee, Robyt, Wallenfels and Whelan 1966) showed that its minimum substrate was $6^2$-$\alpha$-maltosylmaltose: the enzyme was incapable of removing a single glucose residue joined by an $\alpha$-1:6-bond to a chain of residues linked by $\alpha$-1:4-bonds, or conversely of hydrolysing an $\alpha$-1:6-bond joining a chain of $\alpha$-1:4-linked residues to a single glucose unit. In the case of macromolecular substrates, it was established that the enzyme hydrolysed amylopectin and its $\beta$-amylolysis limit dextrin (both so that $>90$ per cent maltose was released on subsequent $\beta$-amylolysis), and glycogen and its $\beta$-amylolysis limit dextrin, although the action on glycogen was not very pronounced. Subsequent reports have tended to suggest that the observed limited degradation of glycogen was an artefact (see Lee and Whelan 1971) and that *A. aerogenes* pullulanase has no effect on native glycogen. On the other hand, Walker (1968) isolated a pullulanase from *Streptococcus mitis* that did de-branch glycogen to a limited extent. This enzyme possessed slightly different specificities to that from *A. aerogenes*, as demonstrated in table 5.15. The *A. aerogenes* enzyme shows a much greater specificity for the hydrolysis of 6-$\alpha$-maltosylmaltosaccharides than does its counterpart from *S. mitis*, but as the length of the *A* side-chain increases *S. mitis* pullulanase becomes more effective in de-branching. However, the difference in specificity decreases again, because the intensity of iodine stain of amylopectin follows the same course with both enzymes. The low specificity of *S. mitis* pullulanase for the removal of maltosyl side-chains ensures that the initial rate of hydrolysis of the $\beta$-amylolysis dextrins is slower than when the *A. aerogenes* enzyme is used. These differences between the pullulanases from the two sources are comparatively trivial, and there is no doubt that the basic enzyme activity is identical.

In the presence of *beta*-amylase and either of the above pullulanases, both amylopectin and glycogen are quantitatively degraded to maltose and glucose (the latter sugar arising from those chains containing an odd number of glucose units). This forms the basis of the enzymic determination of chain-length using *beta*-amylase and pullulanase (Lee and Whelan 1966; Adkins,

Banks and Greenwood 1966; Banks and Greenwood 1970).

In contrast to pullulanase, *R*-enzyme (as originally defined by Hobson, Whelan and Peat), even acting in concert with *beta*-amylase, has no effect on glycogen. Despite this marked difference between *R*-enzyme and *A. aerogenes* pullulanase, Lee and Whelan (1971) now classify both enzymes as pullulanases, because they are so similar in their other activities (plant *R*-enzyme was shown to hydrolyse pullulan). But until the present controversy surrounding *R*-enzyme is resolved, it will probably cause less confusion if the original distinct terminologies are retained.

*Table* 5.15. Relative rates of hydrolysis of various substrates using pullulanase from *Streptococcus mitis* and from *Aerobacter aerogenes* (Walker 1968).

| substrate | relative rate of hydrolysis by pullulanase from: | |
|---|---|---|
| | *S. mitis* | *A. aerogenes* |
| $6^2$-α-maltosylmaltose | 3 | 26 |
| $6^3$-α-maltosylmaltotriose | 6 | 61 |
| $6^3$-α-maltosylmaltotetraose | 18 | 188 |
| $6^2$-α-maltotriosylmaltose | 44 | — |
| $6^2$-α-maltotetraosylmaltose | 79 | — |
| $6^2$-α-maltopentaosylmaltose | 73 | — |
| $6^2$-α-maltohexaosylmaltose | 71 | — |
| $6^2$-α-maltoheptaosylmaltose | 61 | — |
| Pullulan | 78 | 111 |
| $6^3$-α-maltotriosylmaltotriose | 100 | 100 |
| $6^3$-α-maltotetraosylmaltotriose | 162 | 163 |
| $6^3$-α-maltotriosylmaltotetraose | 220 | 123 |

### 5.4.3. Isoamylases.

The *isoamylases* form a third group of de-branching enzymes, and again there is some degree of confusion regarding their precise activity—a state not greatly aided by Kobayashi (1969) attempting to re-classify *R*-enzyme and *A. aerogenes* pullulanase as plant isoamylase and bacterial isoamylase respectively. The nomenclature was first used by Maruo and Kobayashi (1951) to describe an activity originally reported (Nishimura 1930, 1931) as an 'amylose synthease', so called because on incubation with solutions of glutinous rice starch (composed solely of amylopectin) it caused the iodine stain of the substrate to change from red to blue-purple. Maruo and Kobayashi (1951) re-investigated the enzyme system, and decided that it was a de-branching enzyme. However, since the enzyme apparently causes a large increase in the chain length, the substrate having a chain length of 18 and the product one of 100 glucose residues (Maruo and Kobayashi 1951; Kobayashi 1969), it is, as pointed out by Lee and Whelan (1971), impossible for there to be only a de-branching activity present.

Manners and co-workers (Manners and Maung 1955; Gunja, Manners and Maung 1961) examined *yeast isoamylase*, and noted its ability to de-branch various systems containing α-1:6-linkages. A later investigation (Lee, Nielsen and Fischer 1967; Lee, Carter, Nielsen and Fischer 1970) showed that all of the reducing power liberated by the yeast preparation could be accounted for by glucose. This result is not explicable by a direct de-branching of the polysaccharide; rather it is characteristic of the joint actions of a transferase and a de-branching enzyme, the former transferring short chains and so exposing a single glucose residue, joined by an α-1:6-link to the *B* chain, to the action of the de-branching entity (a situation analogous to that existing for amylo-1:6-glucosidase; see below). Subsequently, Bathgate and Manners (1968) showed that yeast extracts possessed at least two types of polysaccharide de-branching activity. One of these (isoamylase) degrades the outer regions of glycogen, releasing a series of oligosaccharides in the range maltopentaose to maltononaose with a concomitant comparatively small increase (*ca.* 50 per cent) in the intensity of the iodine stain, whilst the second activity is in all respects similar to that of amylo-1:6-glucosidase. These authors pointed out that yeast also contains a limit dextrinase, an enzyme specific for the de-branching of the *alpha*-amylase limit dextrins. It is quite obvious, however, that the definition of isoamylase employed by Kobayashi (1969) covers both the specificity for removing chains directly by means of hydrolysis of the α-1:6-linkage, and by means of the amylo-1:6-glucosidase activity.

The term isoamylase has also been applied to the de-branching enzymes isolated from *Pseudomonas SB-15* by Japanese workers (Harada, Yokobayashi and Misaki 1968; Yokobayashi, Misaki and Harada 1969, 1970) and from *Cytophaga sp.* by Whelan and his co-workers (Gunja-Smith, Marshall, Smith and Whelan 1970). These preparations can hydrolyse all the α-1:6-linkages in both amylopectin and glycogen. This enzyme provides the simplest method yet devised for measuring chain length, whereby the polysaccharides are totally de-branched and the reducing power measured (Gunja-Smith, Marshall and Smith 1971). Enzymes from both sources are not capable of removing maltosyl *A*-chains, and hence the β-amylolysis limit dextrins of both glycogen and amylopectin are incompletely degraded.

**5.4.4.** Amylo-1:6-Glucosidase.

The term *amylo-1:6-glucosidase* was used to describe an enzyme system found in rabbit muscle (Cori and Larner 1951) which de-branched glycogen phosphorylase limit dextrin with the production of glucose. The simultaneous action of phosphorylase and amylo-1:6-glucosidase resulted in the complete conversion of glycogen to glucose-1-phosphate and glucose (estimation of the latter sugar enabled the chain length to be determined). To reconcile these observations with the known effect of *beta*-amylase on the phosphorylase limit dextrin, it was suggested that the latter possessed the structure shown in figure 5.17(a).

A later study of this type of limit dextrin (Walker and Whelan 1960), in which *R*-enzyme was allowed to act on the analogous amylopectin phosphorylase limit dextrin, enabled the correct structure (figure 5.17(b)) to be established. From this model, it became apparent that the de-branching enzyme possesses two activities: (a) acting as a *transferase*, it moves maltotriosyl and maltosyl residues from a donor chain to other linear or branched acceptor molecules of a suitable structure; and (b) although the enzyme cannot transfer a single glucose unit, it can hydrolyse a single glucose unit joined by an α-1:6-bond to a chain of α-1:4-bonded glucose residues in its capacity as a *hydrolase*. The existence of this dual specificity has been repeatedly demonstrated for the rabbit muscle preparation (Brown and Illingworth 1962; Brown, Illingworth and Cori 1963; Abdullah and Whelan 1963), for the yeast enzyme (Lee, Nielsen and Fischer 1967; Lee, Carter, Nielsen and Fischer 1970), and for a preparation from rabbit liver (Gordon, Brown and Brown 1972). Moreover, both the transferase and hydrolase activity reside in a single protein molecule (Gordon, Brown and Brown 1972), hence the enzyme is called *transferase-glucosidase* or, more properly, oligo-α-1:4-glucan-4-glycosyl-transferase-amylo-1:6-glucosidase. There can be little doubt that this system is the most thoroughly studied, and understood, of the de-branching enzymes.

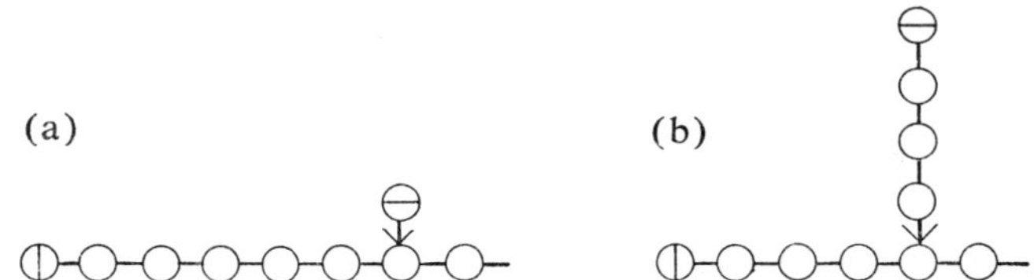


*Figure* 5.17. Structure of the phosphorylase limit dextrin proposed by (a) Cori and Larner (1951), and (b) Walker and Whelan (1960).

**5.4.5.** Summary.

The above brief survey demonstrates that some caution may be necessary when interpreting the work of different authors, because they employ what is at first sight common terminology to define quite distinct activities (for example, compare the résumés of specificity given by Kobayashi 1969, table 2; Manners 1971; and Lee and Whelan 1971, table 9).

## 5.5. Enzymic Reversions and Glycosidic Bond Synthesis.

Enzyme-mediated reactions involving carbohydrates follow the general relation

$$\text{D-OR} + \text{H-OA} \rightleftharpoons \text{D-OA} + \text{H-OR},$$

where DOR represents the donor substrate and HOA the acceptor substrate; although D must be a carbohydrate residue, groups R and A need not necessarily be so. This reaction is reversible, as indeed is demanded by the law of microscopic reversibility, but with the exception of phosphorylase—

which can function in a degradative or synthetic capacity depending on the ratio of free phosphate to glucose-1-phosphate—we have considered only the degradative action of various enzymes. There is now, however, considerable evidence indicating that *all* the enzymes discussed above can, under appropriate conditions, act synthetically.

Two quite distinct types of synthetic action may occur, namely *transfer* and *condensation*, as illustrated in figure 5.18. In the transfer reaction a bond is broken, but instead of the acceptor being a water molecule, as it is in the reactions we have previously described, the donor group is transferred to another carbohydrate moiety; i.e. there is an exchange of one glycosidic bond for another so that the total number of such bonds in the system is invariant. On the other hand, in condensation a new bond is formed in the system. If such syntheses occurred to any great extent there might be an effect on the distribution of degradation products in the case of *alpha*-amylase, so invalidating some of the conclusions based on the quantitative yields of the oligosaccharides. It is therefore pertinent to consider the relative importance of synthetic reactions during the α-amylolysis of amylose.

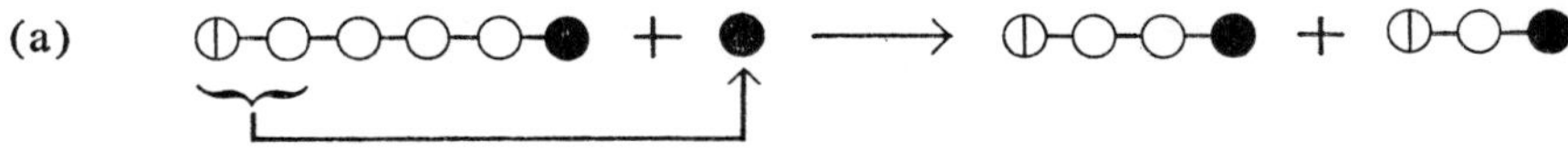


(b)

*Figure* 5.18. Illustration of bond synthesis by means of (a) transfer, and (b) condensation reactions. In (a), a maltosyl unit is transferred to glucose, so synthesizing maltotriose; in (b) two molecules of maltose are condensed to form one molecule of maltotetraose.

French and Abdullah (1966a) suggested that malted barley *alpha-amylase* had a specific transfer activity involving the donation of a maltosyl unit; Pazur and Okada (1966) independently reached a similar conclusion for *B. subtilis alpha*-amylase. Greenwood, Milne and Ross (1968) studied, therefore, the transfer activities of various *alpha*-amylases by allowing the amylolysis of maltohexaose or amylose to proceed in the presence of radioactive glucose or maltose. Glucose was found to be an ineffective acceptor, but the results for maltose are summarized in tables 5.16 and 5.17. It should be noted that to obtain these results relatively high concentrations of *alpha*-amylase were employed; enzymes used at the concentrations necessary to obtain the results shown in tables 5.9–11 showed no evidence of any transfer of radioactivity.

The distributions of radioactivity on incubating radioactive maltose with maltohexaose are shown in table 5.16. Recovery of this marked sugar after incubation with *alpha*-amylase and maltohexaose varies from 82–99 per cent of the original. The lowest value is found on using an extract of malted wheat, and is probably due to the presence of a maltose-splitting impurity because the yield of radioactive glucose declines markedly after purification of the enzyme. However, even on using a thrice-recrystallized enzyme (from pig pancreas) there is decided evidence of a synthetic reaction. It is not possible to decide whether synthesis occurs only by transfer, because the results in table 5.16 indicate that some degree of condensation may have taken place; the highest yield of radioactive product with the *B. subtilis alpha*-amylase was

*Table* 5.16. Percentage of total radioactivity found in maltodextrins after incubation of *alpha*-amylase with $G_6$ and $G_2$* (Greenwood, Milne and Ross 1968).

| time of incubation (hours) | source of *alpha*-amylase | maltodextrins[1] | | | | | | | |
|---|---|---|---|---|---|---|---|---|---|
| | | $G_1$ | $G_2$ | $G_3$ | $G_4$ | $G_5$ | $G_6$ | $G_7$ | $>G_7$ |
| 9 | *B. subtilis* | 3.0 | 92.5 | 1.0 | 0.5 | 1.8 | 1.0 | 0.1 | 0.1 |
| | hog pancreas | 2.0 | 96.1 | 1.8 | 0.1 | 0 | 0 | 0 | 0 |
| | saliva | 0.6 | 99.1 | 0.3 | 0 | 0 | 0 | 0 | 0 |
| | malted barley | 0 | 96.9 | 0.2 | 1.0 | 0.4 | 1.3 | 0.2 | 0 |
| | malted wheat | 0 | 98.3 | 0.1 | 0.4 | 0.2 | 1.0 | 0 | 0 |
| | malted wheat extract | 15.9 | 82.3 | 1.3 | 0.3 | 0.1 | 0 | 0 | 0.1 |
| 24 | *B. subtilis* | 3.4 | 89.4 | 1.6 | 1.3 | 3.7 | 0.2 | 0 | 0.4 |
| | malted barley | 0 | 95.3 | 0.5 | 2.2 | 0.5 | 1.3 | 0.1 | 0.1 |
| | malted wheat | 0 | 97.5 | 0.3 | 0.5 | 0.2 | 1.2 | 0.2 | 0.1 |

[1] $G_1$ is glucose, $G_2$ is maltose, $G_3$ is maltotriose, etc.

*Table* 5.17. Percentage of total radioactivity found in maltodextrins after incubation of *alpha*-amylase with amylose and $G_2$* (Greenwood, Milne and Ross 1968).

| time of incubation (hours) | source of *alpha*-amylase | maltodextrin | | | | | | | |
|---|---|---|---|---|---|---|---|---|---|
| | | $G_1$ | $G_2$ | $G_3$ | $G_4$ | $G_5$ | $G_6$ | $G_7$ | $>G_7$ |
| 9[1] | *B. subtilis* | 0.7 | 88.6 | 5.7 | 1.7 | 0.7 | 1.0 | 1.4 | 0.2 |
| | hog pancreas | 0.5 | 95.5 | 3.8 | 0.1 | 0 | 0 | 0 | 0 |
| | saliva | 0.5 | 98.2 | 1.1 | 0.1 | 0 | 0 | 0 | 0.1 |
| | malted barley | 0.5 | 95.8 | 1.2 | 0.6 | 0.5 | 0.6 | 0.6 | 0.2 |
| | malted wheat | 0.3 | 95.5 | 1.3 | 0.6 | 0.4 | 0.4 | 0.6 | 0.9 |
| 24 | *B. subtilis* | 1.7 | 86.9 | 6.3 | 1.8 | 1.0 | 1.6 | 0.6 | 0.1 |
| | hog pancreas | 0.7 | 95.5 | 3.7 | 0.1 | 0 | 0 | 0 | 0 |
| | saliva | 0.9 | 98.2 | 0.8 | 0 | 0 | 0 | 0 | 0 |
| | malted barley | 0.7 | 94.3 | 1.2 | 1.0 | 0.7 | 1.3 | 0.7 | 0.1 |
| | malted wheat | 0.5 | 95.6 | 1.2 | 0.7 | 0.5 | 0.6 | 0.7 | 0.2 |

[1] Samples removed at the achroic point.

maltopentaose, whereas the purified cereal amylases yield relatively high amounts of radioactivity in maltohexaose. The preferred mode of degradation of maltooctaose by *B. subtilis alpha*-amylase yields maltopentaose, whereas the action of the cereal enzyme yields maltohexaose preferentially. Thus the results shown in table 5.16 could be due in some measure to the reaction.

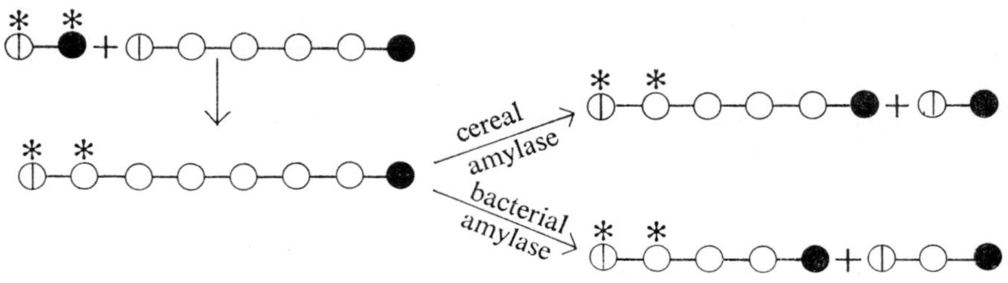

cereal amylase

bacterial amylase

Table 5.17 shows the distributions of activity obtained on incubating amylose with radioactive maltose in the presence of various *alpha*-amylases. The only enzyme preparation to show pronounced transfer activity is that from *B. subtilis*. We would again stress, however, that the experimental conditions employed in obtaining the tabulated data were such as to emphasize synthetic reactions, and there is no reason to believe that these reactions have significantly affected the oligosaccharide yields on which the model of preferential attack by *alpha*-amylase is largely based.

Condensation reactions brought about by starch-degrading enzymes, particularly the amylases, have repeatedly been noted, but relatively little consideration appears to have been given to them. For example it has been reported that *beta*-amylase catalysed the synthesis of maltotetraose from maltose (Abdullah and French 1966); that maltose and isomaltose were formed from glucose under the influence of glucoamylase (Watanabe, Kawamura, Tanno and Matsuda 1968); and that porcine pancreatic *alpha*-amylase mediated the condensation of maltotriose (Robyt and French 1970).

Although the product-stereospecificity has long been appreciated in carbohydrate enzymology, little concern has until recently been shown for the related topic, namely the configuration necessary to produce the optimum condensation reaction. It remained to Hehre, Okada and Genghof (1969) to demonstrate unambiguously that the reversal of hydrolysis by *alpha*-amylase, *beta*-amylase, and glucoamylase depended upon the donor substrate having the correct anomeric form. Thus: glucoamylase condenses $\beta$-glucose but not $\alpha$-glucose; *alpha*-amylase condenses $\alpha$-maltose but not $\beta$-maltose; and *beta*-amylase condenses $\beta$-maltose but not $\alpha$-maltose. Moreover, it was shown in all cases that the mechanism was one of glycosyl transfer—the same mechanism as is operative during hydrolysis by these enzymes.

It has also been demonstrated that the de-branching enzymes can mediate condensation reactions under appropriate conditions. Abdullah and French

(1966) demonstrated that pullulanase could condense maltose and maltotriose, and showed (French and Abdullah 1966b) the hexasaccharide obtained from condensation of maltotriose to be a mixture of $6^3$-, $6^2$-, and $6^1$-$\alpha$-maltotriosylmaltotriose. At a substrate level of five per cent, Walker (1968) found that the condensation of maltotriose by pullulanase from *S. mitis* was comparable to that obtained on using the *Aerobacter aerogenes* enzyme, but that on doubling the substrate concentration the streptococcal enzyme produced some five to ten times as much reversion product as did the *Aerobacter* enzyme. Similarly, amylo-1:6-glucosidase can also introduce single glucose units into glycogen (Larner and Schliselfeld 1956), the converse of its normal de-branching activity.

This very brief outline of some synthetic reactions of 'starch-degrading' enzymes stresses that some care must be taken in choosing experimental conditions, so that unwanted side-reactions do not obscure the action pattern of the enzyme under investigation.

# 6. The Structure and Biosynthesis of the Starch Granule

## 6.1. Introduction.

All the properties of the starch granule are related to a combination of two factors, namely: (a) its *chemical constitution*, depending on the presence of the two major discrete components, amylose and amylopectin; and (b) its *physical constitution*, involving the organization of these polymers to form a unique structural entity. Necessarily, the structure of the starch granule will also reflect the biosynthetic pathway. In this chapter, therefore, we consider in conjunction the closely-related topics of structure and biosynthesis.

## 6.2. The Starch Granule.

Starch is a solid material with a density of approximately 1.5 g/ml. The granules themselves vary considerably in shape (from spheres to rods) and size (from a diameter of 2 μm in the pollen starch of amylomaize to 175 μm for canna starch), so much so that the source of a starch can be identified merely from its microscopic appearance. However, to characterize the starch granule fully, it is necessary to determine (a) the ratio of amylose to amylopectin, (b) the fine structures of both components, and (c) the way in which the individual molecules are arranged with respect to each other. We have dealt in earlier chapters with the first two of these points, and now wish to examine the third—the arrangement of molecules within the granule.

### 6.2.1. The Crystalline Nature of Starch.

It has been accepted that starch is a semi-crystalline material since the classic work of Katz and his collaborators (Katz and van Itallie 1930; Katz and Derksen 1933; Katz 1937). Using X-ray diffraction techniques these authors distinguished three types of crystalline structure in intact starch granules, giving diffraction patterns which were designated as *A*-, *B*-, and *C*-patterns. The structural type depends on the botanical source of the starch: the *A*-pattern is given by most starches of cereal origin; the *B*-pattern by potato and amylomaize starches (and by retrograded starch); and the *C*-pattern by smooth pea and various bean starches. Diffractometer traces showing each type of pattern are given in figure 6.1.

The *C*-pattern is in fact intermediate in form between the *A*- and *B*-types. Hizukuri (1969) has shown that, by altering the proportions of maize and potato starch in a mixture of the two, it is possible to obtain a gradual change in the diffraction pattern from an *A*-type to a *B*-type, through the *C*-type. Starches from soya bean seedlings grown at various temperatures exhibit different X-ray patterns: at 30°C rather more than 95 per cent of the crystalline material is in the *A*-form, by 22°C there are equal amounts of the *A*- and *B*-forms, and at 13.5°C a *B*-pattern is obtained (Hizukuri, Fujii and Nikuni 1961, 1965). A similar effect was obtained on growing the roots of sweet potato at a series of constant temperatures, whereas the crystalline patterns of the starches of potato and rice did not show any change when the

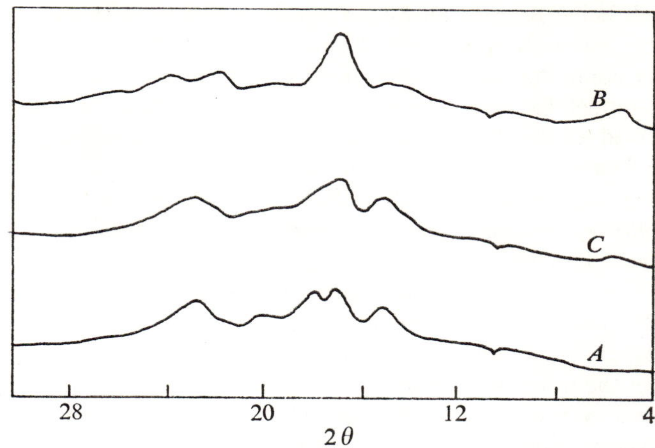


*Figure* 6.1. Typical diffractometer patterns for starches giving *A*-, *B*-, and *C*-type X-ray spectra.

plants were grown over a range of temperatures.

Thirty years ago Sair and Fetzer (1944) demonstrated that a heat-moisture treatment could convert the *B*-pattern of potato starch to an *A*-type, with concomitant changes in the physical properties of the granules. Controlled drying of a heated starch gel can produce any of the crystalline forms, depending on the temperature of drying, and Hizukuri (1969) showed that a dextrin with a number-average degree of polymerization of only 12–15 glucose units could yield any of the three patterns, depending on its concentration and the temperature at which it was crystallized.

The above evidence emphasizes the interrelations that exist between the various crystalline forms found in starch.

Whilst X-ray examination of granular starches enables them to be consigned to the appropriate crystal pattern, they have not allowed identification of the actual structures involved. For this purpose, it is necessary to use either fibres or films of amylose, the orientation enabling the repeat unit along the fibre axis to be evaluated, so facilitating estimation of the unit-cell size and identification of the parent crystal system. On this basis, Rundle, Daasch and French (1944) established for the *B*-pattern that the fibre axis was 10.6 Å, which they claimed was the length of two glucose units, with the lateral dimensions being 16.0 Å and 9.2 Å, the unit cell being orthorhombic.

No subsequent investigation of the *B*-structure has given a repeat period along the fibre axis materially different from that obtained by Rundle and his colleagues. However, Kreger (1946, 1947, 1951) chose to interpret the patterns in a rather different manner—he realized that two glucose units in the chair conformation could not possibly be stretched to 10.6 Å, and therefore proposed that the fibre period contained three glucose units in a helical form, the unit cell being hexagonal. Subsequently, Spark (1952) also suggested a helical form, but with approximately four glucose residues per turn.

Various computer calculations have also been carried out in order to provide a model for the *B*-type X-ray spectrum. Rao and Sundararajan (1968) derived a helical model using potential energy calculations, in which there were four glucose units per helical turn, as postulated earlier by Spark (1952). They suggested the most probable form to be a right-handed helix, with the hydroxyl groups of the constituent glucose units pointing away from the helix axis to facilitate hydrogen bonding between, rather than along, chains. It is necessary to introduce some stabilizing factor such as hydrogen bonding, because the postulated model occurs in a region of the potential energy diagram in which each residue has approximately 8 kcal/mol more energy than the minimum value.

Blackwell, Sarko and Marchessault (1969) carried out an analysis of all possible chain conformations by computing the non-bonded potential energies for a maltose residue. Their calculations showed that the most probable form of *B*-amylose was a left-handed single helix with six glucose units per helical turn. The model also postulated the presence of relatively large amounts of water of crystallization—the fibre repeat period increased from the value of 8 Å typical of the *V*-form (see below) to 10.4 Å as a result of the insertion of a water molecule between glucose units in adjacent helical turns. It has long been recognized that water is of considerable importance in the crystallization of amylose, and various workers (Sair and Fetzer 1944; Guilbot, Charbonnière and Drapron 1961) have noted that there is an apparent discontinuity in physical properties when the amount of water present reaches approximately 10 per cent, corresponding to one molecule of water per anhydroglucose unit, but the *B*-pattern can be obtained at much lower amounts of water of crystallization than that invoked by Blackwell, Sarko and Marchessault (1969) for their model. However, there is an additional complicating factor—because any polymer material can at best be only semi-crystalline, the water molecules in amylose will be distributed between the crystalline and amorphous regions, and the exact proportion found in each of these regions cannot be easily measured.

Goebel, Dimpfl and Brant (1970) used the fibre repeat period (10.4 Å) and six-fold helical structure suggested by Blackwell, Sarko and Marchessault (1969) as the criteria for defining a computed model. Their calculations, based on geometries derived from cyclohexa-amylose, showed that both right- and left-handed six-fold helixes of the desired pitch were possible, but that the four-fold helical structure suggested by Rao and Sundararajan (1968) was somewhat unlikely. However, as we have noted earlier (see section 4.2.4), the choice of monomer geometry is all-important in computer model building. For example, Goebel, Dimpfl and Brant (1970) admit that their postulated model for *B*-amylose is improbable if the geometries derived from methyl *β*-maltoside monohydrate are used in the computations; conversely, the seven-fold helix postulated by Zaslow (1963) for the amylose–*t*-butanol complex is attainable using the geometry of the methyl maltoside, but not if that derived from cyclohexa-amylose is employed. It is necessary to emphasize again that agreement between an experimentally observed parameter and a computer derived model may be largely illusory, having been occasioned by the subjective choice of geometries used in the calculations.

Whilst the unit-cell dimensions for starch, or amylose, giving the *B*-pattern appear to have achieved general acceptance, although the interpretation of these measurements differ, the little information that exists regarding the *A*-pattern is contradictory. Thus Bear and French (1941) suggested that the *A*-type spectrum of starch was a comparatively minor modification of the *B*-type (for the *A*-pattern: $a_0=15.4$ Å, $b_0=8.87$ Å, and $c_0=6.18$ Å; whilst for the *B*-pattern: $a_0=16.0$ Å, $b=10.6$ Å, and $c_0=9.2$ Å), although sufficient to give a triclinic cell rather than the orthorhombic cell of the *B*-pattern. Senti and Witnauer (1946) prepared amylose fibres giving the *A*-pattern by treating amylose acetate with 2 per cent caustic potash in 75 per cent methanol (or ethanol) and exposing the resultant alkali amylose to high humidity (80 per cent) for several days.* This work yielded a fibre period of 10.5 Å, a value quite inconsistent with any cell dimension suggested by Bear and French (1941) but not very different from the fibre period (10.6 Å) accepted for amylose in the *B*-form.

The above brief summary indicates that the arrangement of chains giving either the *A*- or *B*-patterns is not understood. French (1972), in formally retracting the model of Rundle, Daasch and French (1944) for *B*-amylose, has emphasized that none of the alternative models provide a structure capable of explaining the experimental X-ray observations without distorting bond lengths and bond angles, or introducing into the unit cell unacceptably high amounts of water of crystallization. In fact, he has suggested that because none of the single-stranded helical models give a suitable model for the experimental results, attention be given to double-stranded helices of the form shown in figure 6.2.

French (1972) envisages that two types of double-helical structure could be formed, either from separate chains (see figure 6.2(a)), or by a process of a single chain folding back on itself (see figure 6.2(b)). It is, as French emphasizes, quite easy to construct space-filling amylose models that correspond to the structures shown in figure 6.2, although this fact does not provide evidence for the existence of such a structure at a molecular level. The model does explain, for example, the inability of retrograded amylose to take up iodine. This phenomenon is usually ascribed to the internal diameter of the helix of *B*-amylose being too small to accommodate the complexing species, but in the double helix model the amylose would be effectively complexed

* By exposing the fibre to saturated water vapour, a change from the *A*-form to the *B*-form was detected; complete conversion was achieved by boiling the vapour-treated filaments in water.

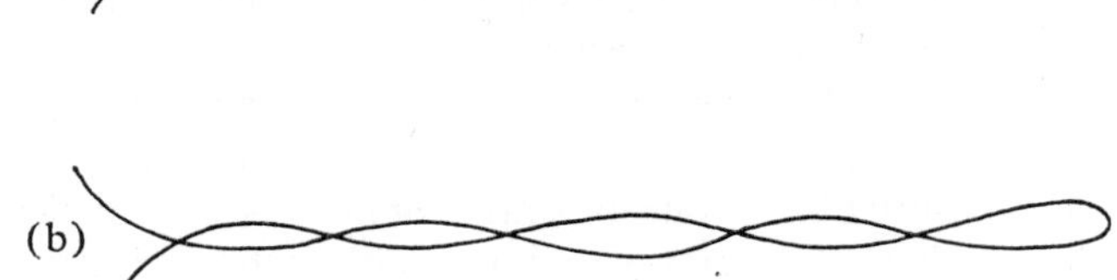


*Figure* 6.2. Symbolic representation of double helices in starch: (a) separate chains are involved, and they may be parallel or anti-parallel; (b) double helix formed by chain folding, chain must be anti-parallel (after French 1972).

with itself and thus unable to bind iodine.

The structure of only one crystalline form of amylose—the *V*-form—is widely accepted. The only references to a *naturally occurring* *V*-form are a personal communication from Zobel, quoted by Sterling (1968), according to which it is present in a mixed crystalline form in the granules of certain varieties of maize, and the observation that such starches have a lower gelatinization temperature than those exhibiting *A*-, *B*-, or *C*-patterns (Zobel, Cotton and Senti 1964). Whilst it was Katz (1937) who first noted the existence of *V*-amylose, the identification of this form as a helical inclusion complex resulted from the work of Rundle and his collaborators (Rundle and French 1943a, b; Rundle and Edwards 1943; Rundle 1947; Mikus, Hixon and Rundle 1946). The unit cell was shown to be orthorhombic, with dimensions $a_0=13.7$ Å, $b_0=23.8$ Å, and $c_0=8.05$ Å, obtained using the amylose–butan-1-ol complex; the diameter of the helix is given by $a_0$, the fibre repeat period by $c_0$, and each helical turn contained six glucose residues. Some confusion exists regarding the helix diameter, because it was originally reported that the anhydrous amylose–iodine complex had a smaller diameter than did the 'dried' butan-1-ol complex (Rundle and Edwards 1943), but later work (Rundle 1947) showed that more efficient drying of the latter complex reduced the diameter to 13.0 Å, the same value as obtained using the iodine complex. The completely dried form is now referred to as the anhydrous *V*-complex (or $V_a$-amylose), but even this is something of a misnomer in that it has been claimed (Valletta, Germino, Lang and Moshy 1964) that complexes dried to constant weight under stringent conditions may contain up to one mole of water and one-quarter mole of complexing agent per helical turn. The other form is referred to as the hydrated *V*-complex (or $V_h$-amylose). The size of the unit cell appears to depend on the degree of hydration, for Zobel, French and Hinkle (1967) showed the dimensions of $V_a$-amylose to be $a_0=13.0$ Å, $b_0=22.5$ Å, and $c_0=7.90$ Å, whilst for $V_h$-amylose $a_0=13.65$ Å, $b_0=23.70$ Å and $c_0=8.05$ Å, and obtained a series of intermediate values for the helix diameter during the process of drying.

Complexes of *V*-amylose involving more than six glucose units per helical turn also exist. Bear (1944) noted that the complexes of amylose with *t*-butanol, and other branched alcohols, yielded diffraction patterns that were too complex to be indexed. He suggested that the helix diameter was probably greater than the value of 13.7 Å obtained for the amylose–butan-1-ol complex. Zaslow (1963) later investigated the amylose–*t*-butanol complex in detail, and suggested a helical structure with a diameter of 14.94 Å containing *seven* glucose units per helical turn. The existence of the seven-fold helix for complexes between amylose and the branched alcohols was subsequently confirmed by Yamashita and Hirai (1966). Using α-naphthol as the complexing agent, Yamashita and Monobe (1971) obtained a helical structure with a diameter of 16.2 Å, equivalent to *eight* glucose residues per helical turn.

Little direct evidence exists regarding the chirality of the *V*-amylose helix. On the basis that amylose triacetate forms a left-handed helix and that *V*- or *B*-forms may be prepared from the derivative (Sarko and Marchessault 1967), it has been argued that *V*- and *B*-amylose also form left-handed helices (Jackobs, Bumb and Zaslow 1968; Blackwell, Sarko and Marchessault 1969). However, Zaslow and Miller (1961) showed that there was an intervening amorphous phase in the conversion of *V*-amylose to *B*-amylose, so that a reversal of the helical screw sense is not impossible although it may be rather unlikely from energetic considerations (Geobel, Dimpfl and Brant 1970). Thus even in the case of the crystalline modification of amylose subjected to the most detailed examination, there are certain aspects of the structure that are not completely understood.

### **6.2.2.** Molecular Orientation within the Granule.

Starch granules show birefringence in polarized light, giving the typical 'Maltese cross' pattern shown in plate 1 for potato starch granules. Starch granules are both semi-crystalline and birefringent, and some confusion exists because these two factors are not necessarily related. For example, a film of atactic polystyrene may be warmed, stretched, and rapidly cooled to give a material that is birefringent (because the long axes of the chains tend to be oriented in the direction of the applied stress) but not crystalline (because the basic irregularity of the atactic structure precludes any three-dimensional ordering). Conversely, a piece of paper is semi-crystalline and the crystallites themselves are birefringent, but because the arrangement of the crystalline structures is haphazard the paper as a whole is not birefringent. Thus the fact that starch is birefringent implies only that there is a high degree of molecular orientation within the granule, without reference to any crystalline form.

It must not be thought that the birefringent starch granule is in any way a unique structure. Most synthetic semi-crystalline polymers form spherulitic structures, both on growth from the molten state or from concentrated solution. These spherulites expand until, in the case of the solid material, they impinge, but at an early stage in growth they are not unlike starch granules, as may be seen in plate 2. In our opinion, many of the factors shown to be of importance in polymeric spherulites may also be applicable to the starch granule.

To elucidate the structure of a spherulite, it is necessary to assume that the polymer chains are optically positive, i.e. the highest index of refraction occurs along the axis of the chain, rather than at right-angles to it. The polarizing microscope is used to find whether the spherulite is positively or negatively birefringent, i.e. whether the chain axis is to be found radially or tangentially arranged, by using a unit retardation plate to produce the first order red between crossed polars in white light. A small reduction in relative retardation gives first order yellow, whereas a small increase in relative retardation changes the first order red to second order blue. By examining the colours of the spherulite as the position of the retardation plate is changed, it is possible to deduce whether the tangential, or radial, refractive index is the greater. For starch granules, the radial refractive index is the larger, so that they are *positively birefringent*.

French (1972) produced 'optical maps' of various irregularly shaped starch granules by photographing them in a series of orientations using the polarizing microscope. The predominant orientation of the chains in each photograph can be deduced from the fact that at the position of the 'dark cross' the molecular axis of the chains is either parallel or perpendicular to the plane of polarization. For these non-spherical granules the molecular arrangement of chains is such that their long axes are perpendicular to the surface of the granule.

The development of the technique of solid-state light-scattering has provided another tool for investigating molecular orientation in the starch granule. It was initially developed by Stein and his co-workers (Stein and Keane 1956; Stein and Rhodes 1960; Stein, Keedy and Rhodes 1962; Stein and Rhodes 1962, 1963; Stein 1964) for texture studies of spherulites of synthetic polymers, but was subsequently applied to single starch granules from shoti, potato and tapioca (Borch, Sarko and Marchessault 1969).

Light is scattered as a result of fluctuations in anisotropy or density within the granule. To distinguish between these two possibilities plane-polarized incident light is used, and the scattered radiation recorded after passage through an analyser. If only density fluctuations are present in an isotropic superstructure, the scattered waves will oscillate in the same plane as the incoming beam, and will be unable to penetrate an analyser oriented at right angles to the polarizer; conversely if a scattering pattern is obtained then the superstructure is anisotropic. The pattern obtained with a horizontally oriented analyser and a vertically polarized incident beam is designated the $H_v$-pattern, and is the one most commonly employed in examining spherulitic structures (the $V_v$ pattern, which is obtained when both the analyser and the incident beam are vertically polarized, records the scattering due to both density and anisotropy fluctuations).

Borch, Sarko and Marchessault (1969) showed that the $H_v$-patterns for tapioca and potato starches yielded the 'clover leaf' forms expected for spherulitic structures. Indeed, the scattering envelope for a single granule of the former starch was in close agreement with the predictions of the Stein–Rhodes equation, and provided the first unambiguous verification of that theory. However, deviations from the theory become obvious in the case of potato starch when the relative intensities of higher order maxima were considered, and led to the suggestion that potato starch granules contained an isotropic centre in an otherwise anisotropic structure. The fact that it is necessary to assume the diameter of this isotropic region to be some 50 per cent of the granule diameter does cause one to regard the model with some dubiety. With shoti starch, non-spherulitic patterns were obtained. These observations reflect the morphology of the granules, for tapioca granules are spherical, potato granules tend to be ellipsoidal, and the shoti granules are elongated, with the hilum situated at the extreme end (Schoch and Maywald 1968).

In a subsequent study of the anisotropic light scattering by the asymmetric starch granules of shoti starch, Mencik, Marchessault and Sarko (1971) provided a model for the observed $H_v$-patterns. They suggested that the scattering envelope could be explained by a lamellar model, in which the lamellae were at right angles to the axis of symmetry of the granule. This lamellar model is in fact quite similar to the layered structures observed in the optical microscope (see section 6.2.3), and is consistent with the growth of the granule by apposition.

The concept of a layered granule structure for tapioca and potato starches was considered by Finkelstein and Sarko (1972). In the case of tapioca starch the layer thickness, if present at all, was less than 0.5 $\mu$m. On the other hand, to account for the scattering envelope of potato starch it was necessary to assume that the granule contained only a small number of layers, the best agreement between theory and experiment being obtained with a model containing only three layers. The existence of clearly defined layers within the granule of potato starch has, of course, long been recognized from studies using the optical microscope, and their spacing (4–7 $\mu$m) is such that the model delineated by Finkelstein and Sarko (1972) may indeed be correct. Nevertheless, as the authors admit, the agreement between experiment and theory may be somewhat fortuitous, because the potato starch granule is unlikely to possess the spherical symmetry assumed by these authors.

The application of solid-state light-scattering techniques to the starch granule has not greatly increased our knowledge of its structure. It has, however, been invaluable in demonstrating that the starch granule may be treated as a polymer spherulite, with the proviso that its shape and nucleation centre are governed by the plastid in which it is formed so that spherical symmetry cannot always be attained. Notwithstanding this limitation, it should be possible to apply the theories of spherulitic growth elucidated for synthetic polymers to the starch granule.

**6.2.3.** Granule Morphology.

When an aqueous suspension of potato starch granules is examined by optical microscopy, perhaps the most conspicuous feature is the presence of layering within the granule (see plate 3). For many years it was supposed that these layers represented the difference in the mechanism of starch deposition between day and night, but growing potatoes in a regime of constant light still yielded the layers (Roberts and Proctor 1954; Bünning and Hess 1954; Mes and Menge 1954). Treatment of the potato starch granule with dilute mineral acid, as in the process of Lintnerization, accentuates these layers and, further, suggests that each is composed of fine lamellae approximately 1000 Å in breadth (Frey-Wyssling and Buttrose 1961). However, when dried potato starch granules are examined in pentane, which has the same refractive index as water but is incapable of swelling the granule, the layers are not visible (Hanssen, Dodt and Niemann 1953), nor can they be detected in thin sections of the native granules by electron microscopy (Buttrose 1963).

In cereal starch granules, layering is not obvious unless the granules have been exposed to the action of mineral acid or *alpha*-amylase. Thin sections of maize starch granules, examined by electron microscopy, have variously been reported to exhibit, or not to exhibit, layering (see Badenhuizen 1959). The ever-present danger in working with thin sections is that their mode of preparation may induce artefact formation. In the particular case of maize starch, it is not clear whether the absence, or presence, of such layers constitutes the artefact. In contradistinction to the case of potato, however, the layering in the starch granules of cereals is apparently related to environmental effects, because granules produced by plants grown under constant light do not exhibit layers when treated with acid or *alpha*-amylase.

The significance of granular layering remains uncertain, especially in view of the fact that the layers are not obvious in the dried granule, or prior to degradation of the granule. It can be concluded only that certain treatments accentuate the degree of anisotropy of the units of starch granule formation. The postulate of Czaja (1954), namely that the layers could be due to the alternate deposition of amylose and amylopectin, has little experimental basis. Indeed, inasmuch as the granules of waxy-maize starch also exhibit layers after the appropriate treatment with mineral acid (Badenhuizen 1955), the evidence would appear to negate this hypothesis.

It is noteworthy that in the crystallization of synthetic high polymers, ringed structures somewhat analogous to the layers observed in starch granules are fairly commonplace. These ring phenomena are thought to be due to the twisting of crystalline fibrils during the growth of the spherulite (Price 1959; Keller 1959) from primary nuclei, by the accretion of material at the expanding surface. Material that cannot be accommodated within the crystal structure is pushed out into the melt or solution. As the crystal expands it tends to branch, and the non-crystalline material is deposited between these branches. Thus the spherulite is semi-crystalline, in that it contains both crystalline and amorphous material.

Borch, Sarko and Marchessault (1969) have unambiguously demonstrated that the tapioca starch granule can be considered as a true spherulitic structure, i.e. it possesses the same characteristics as the bodies grown in polymer melts. Probably the structure of all starch granules is approximately similar, but for the fact that the granule is formed within a plastid which controls both its shape and the point at which nucleation occurs, i.e. the hilum.

Potato starch granules are ovoid (see plate 3), and the very obvious growth rings originate from the hilum, which is the point at which the arms of the Maltese cross, observed on examining the granules in polarized light, meet. The vast majority of potato starch granules have a hilum which is not central, but is displaced towards one end. This effect is even more obvious in other types of starch, for example, in the bullet-shaped granules of shoti starch the hilum occurs virtually at the pointed end.

In addition to the *simple* type of starch granule, in which a plastid gives rise to a single nucleus and consequently one granule, a second type of granule is also found. This is the *compound* granule, in the formation of which one plastid gives rise to a multiplicity of nuclei, and subsequently granules, each of which exhibits its distinct birefringence pattern. Although potato starch is not usually thought to have compound granules, a close inspection of published photomicrographs (Schoch 1961; Seidemann 1966) clearly shows their existence. A survey of various cultivars yielded one (Pentland Dell) particularly rich in such granules, which are shown in plates 4 and 5 (Banks, Greenwood and Muir 1973b). These photomicrographs show a triplet, i.e. three nuclei have developed within one plastid to produce three individual birefringence patterns. It is also noteworthy that the subdivisions between the three segments of the granule are straight lines. In the case of the spherulites of synthetic polymers this effect is taken to imply that the nuclei were created at the same instant, and that subsequent growth occurred at the same rate.

In certain starches, there is some dispute as to whether the granules are truly complex or, as a result of the forces produced by dehydration, are merely fissured. For example, in the case of wrinkled-seeded pea starch, Badenhuizen (1959) has insisted that the granules are simple but with pronounced radial cracks, in contradistinction to earlier work (Reichert 1913) in which they had been classified as compound. The scanning electron microscope provides a more detailed picture of the starch granule, and this instrument has been used to examine the granules of wrinkled-seeded pea starch as shown in plates 6–8 (Banks, Greenwood and Muir 1974c). These photomicrographs demonstrate quite clearly that the majority of granules are complex in nature, especially those granules exhibiting a central cavity. Again, birefringence patterns obtained with these granules (shown in plate 9) in no way resemble those from simple starch granules such as the potato (cf. plate 5). Each granular segment apparently possesses its own birefringence pattern, which of course would be expected if the granule were truly compound. Furthermore, by applying high rates of shear to aqueous suspensions of the granules it proved possible to fracture them, and the dislodged segments were found to retain their individual birefringence patterns.

Birefringence patterns obtained from the compound granules of potato starch indicate that a number of quite distinct nuclei have been formed which have subsequently grown with spherical symmetry until mutual impingement occurred. The complex patterns exhibited by the pea starch are much more difficult to interpret, however, and we suggest that a number (from three to ten) of nuclei are formed near the centre of the amyloplast, and subsequently grow in a sectorial fashion. As in the case of the compound granules of potato starch, the linear interface between segments indicates that the nuclei are born simultaneously, and that crystallization proceeds at the same rate in all segments.

Varieties of three common cultivars—maize, pea, and barley—have been found to have abnormally high amylose contents, because of the expression of a recessive gene. In each case, the shape of the starch granule also changes: in amylomaize the granules become irregular in shape (see plates 10–12) and often show non-birefringent regions; in wrinkled-seeded pea the complex granules described above are the dominant type when the pea reaches maturity; in barley of high amylose content the granules are smaller and more irregular than those of the parent normal barley. Whilst generally some correlation between amylose content and the morphology of the starch granules exists, there are exceptions. For example, we have found a variety of pea having a starch with some 80 per cent amylose (i.e. appreciably higher than the value of 65 per cent normally associated with pea starches of high amylose content), in which complex granules are entirely absent. This observed separation of the genes which control enhanced amylose content and abnormal shape may be of considerable practical significance, because geneticists have tended to judge the back-crossing experiments involving starches of high amylose content solely on the basis of granule morphology.

**6.2.4.** Susceptibility to Degradation by Acids and Enzymes.

*Acid Degradation Studies.* The action of mineral acid on starch granules accentuates certain characteristics, notably the presence of layers of apparently different degrees of anisotropy. Chemical changes resulting from this treatment have consequently been studied in the hope of elucidating the architecture of the granule (Meyer and Menzi 1953; Kerr 1952; Cowie and Greenwood 1957c; Arbuckle and Greenwood 1958b). Acid degradation of potato starch (Cowie and Greenwood 1957c), and of wheat starch (Arbuckle and Greenwood 1958b), appeared to indicate that the rate of hydrolysis of amylopectin was greater than that of amylose. These authors were aware of the shortcomings in their mathematical treatment and of the potential errors involved in using sedimentation velocity, rather than a method based on colligative properties, to follow the changes in the molecular weight of the amylopectin. Experimental results showed that the number of bonds broken per initial molecule per unit time was greater for amylopectin. Since the use of the sedimentation coefficient would underestimate the degradation of the branched component, it was concluded that the amylopectin fraction was hydrolysed preferentially. Unfortunately, further consideration shows that this conclusion is not valid because of the large difference in the initial molecular weights of the two components. For example, if a value of 100000 glucose residues is ascribed to the degree of polymerization of the amylopectin, and 1000 residues to that of the amylose, there will be a ratio of one amylopectin molecule to thirty amylose molecules (i.e. 23 per cent amylose by weight); if, in unit time, one bond in 1000 is hydrolysed randomly in both fractions, then 99 bonds per initial molecule will be broken in the case of amylopectin, but only one per initial molecule for the amylose. Thus, despite the fact that both fractions are equally susceptible to acid hydrolysis, the rate, expressed in terms of bonds broken per initial molecule, is apparently two orders of magnitude greater for the amylopectin than for the amylose. In fact, to establish that preferential hydrolysis of either component does occur would require much more detailed studies than any carried out to date.

It has long been recognized that treatment of starch granules with mineral acids greatly affects their ability to swell in hot water. After fairly brief exposure to acid, the granules apparently dissolve completely in hot water: indeed, the Lintner process for producing soluble starch is based on this concept. However, even after prolonged exposure of the granules some acid-resistant polysaccharide remains: the so-called *Nägeli amylodextrin*. Kainuma and French (1971) have recently summarized the properties of *potato* amylodextrin as follows: the material dissolves easily in hot water, and can be recrystallized from aqueous solution; the crystalline structure depends on the conditions (temperature, concentration, and presence of salts), with *A*-, *B*-, or *C*-patterns being obtained; after protracted exposure to acid the residual material retains the external form of the parent starch, exhibiting both birefringence and X-ray diffraction patterns characteristic of the starting material; the number-average degree of polymerization is approximately 15–30 glucose units; on $\beta$-amylolysis, approximately 80 per cent of the polysaccharide is converted to maltose, the incomplete conversion being due to branch points; the native amylodextrin gives little or no colour with iodine, but once it has been dissolved the solution stains red or purple; the specific rotation in water is $+195°$; and the amylodextrin is very resistant to further hydrolysis by the acid at room temperature.

Kainuma and French (1971) then prepared amylodextrins from various starches and a commercial amylose preparation. The amounts of the different polysaccharides solubilized by this treatment are recorded in table 6.1. It is perhaps surprising that the granules are so resistant to mineral acid: rather less than 50 per cent of the polysaccharide material is solubilized in the most susceptible starch. The other starches are much less susceptible, especially those of high amylose content. The low relative solubilization of potato starch granules is unexpected, as they gelatinize so easily.

*Table* 6.1. The properties of the Nägeli amylodextrins[1] prepared from various sources (Kainuma and French 1971).

| source | material solubilized (%) | type of X-ray pattern | solubility in boiling water |
|---|---|---|---|
| waxy maize | 48.4 | *A* | soluble |
| potato | 25.4 | *B* | soluble |
| wrinkled-seeded pea | 22.8 | *B* | insoluble |
| amylomaize | 18.2 | *B* | insoluble |
| amylose | 17.0 | amorphous | insoluble |

[1] Samples suspended in 16% sulphuric acid for 90 days at 22–25°C.

The complexity of these acid degradation studies, and the care needed in interpreting the results, is emphasized by comparing the work of Kainuma and French (1971) with that of Buttrose (1963, 1966). There is some indication from the former work that starches with an *A*-type spectrum are more readily solubilized by acid than are those with a *B*-type diffraction pattern, whilst the latter work shows the reverse. These differences probably arise from the different conditions (type of acid, strength of acid, and temperature) which have been used in the two studies, and the most important of these variables is the temperature, in that different types of granule swell to various extents as their aqueous suspensions are heated.

It is also surprising that the commercial amylose sample examined by Kainuma and French (1971) was so resistant to hydrolysis, as it gave initially an amorphous X-ray diffraction pattern. Our experience of similarly prepared commercial samples is that they have very low solubility in boiling water, and are not dissolved even by the action of 1M caustic alkali at room temperature. These observations are in accord with the properties of retrograded amylose, for which one would expect a *B*-type diffraction pattern. The fact that no such pattern is obtained suggests that the degree of three-dimensional order is low; nevertheless, the intermolecular forces are such as to stabilize the amylose against both solution and acid hydrolysis.

Assuming that starch samples are composed of local regions of order and disorder, it would be expected that the main hydrolytic action of the acid would occur in the disordered zones as, for example, in the case of cellulose exposed to mineral acid. (Here the resistant particles represent the regions of high order, i.e. the crystallites.) A similar approach has also been made in studying the structure of semi-crystalline synthetic polymers (Palmer and Cobbold 1964); when solid polyethylene was exposed to the action of fuming nitric acid it was found that the residual material was as crystalline as the parent polymer, despite the fact that the number-average degree of polymerization had been reduced to approximately 200 carbon atoms. The length of this material (260 Å) corresponded to the fold period of the polyethylene single crystal. Palmer and Cobbold (1964) therefore concluded that the folding which characterizes the crystallization of the polymer from dilute solution is to be found also in the material crystallized from the melt (the actual folds are not part of the crystalline assembly, and therefore are accessible to hydrolysis). By analogy, the amylodextrins would be expected to represent the fundamental building units in the starch granule. In this case, however, there exists the difficulty that the samples (amylomaize, wrinkled-seeded pea, and amylose) that are most poorly crystalline (as defined by an X-ray diffraction pattern) are the most resistant to acid degradation. Kainuma and French (1971) therefore concluded that it cannot be the crystalline organization alone that protects the starch chains from attack, but rather that whatever organization protects starch chains against acidic hydrolysis also gives them a high tendency to aggregate.

French (1972) fractionated the amylodextrin from waxy-maize starch by gel permeation chromatography, and found the elution diagram to show three distinct peaks: that of highest molecular weight contained a multiplicity of branch points; that of medium range a single branch-point; and the lowest molecular weight fraction contained a series of linear chains (degree of polymerization 10–20 glucose units). French (1972) also postulated that the branch points could serve as the initiation points for the formation of double helices, of the type shown in figure 6.3.

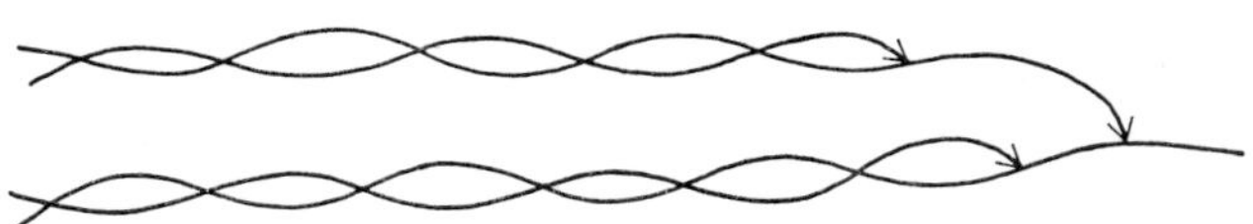

*Figure* 6.3. Symbolic representation of a double helix originating from branch-points in starch (after French 1972).

*Enzyme Degradation Studies.* Amylases (see chapter 5) have been widely used to investigate the structure of starch granules. The early observation that *beta*-amylase had no effect on intact starch granules (Sandstedt, Blish, Meecham and Bode 1937) is still valid. Other early studies established several important points, namely that the susceptibility of granules to degradation by *alpha*-amylase depended on both the source of the starch and of the *alpha*-amylase (Stamberg and Bailey 1939), and that potato starch granules were particularly resistant to hydrolysis by most common *alpha*-amylases (Balls and Schwimmer 1944).

In a detailed study of the action of several amylases on granular starch, Leach and Schoch (1961) confirmed the resistant nature of potato starch. Bacterial, malted barley, fungal, and pancreatic *alpha*-amylases were allowed to act on maize and potato starches for 24 hours; the bacterial and pancreatic enzymes produced reasonable amounts of degradation of the potato starch only at 50°C, whereas the maize starch was hydrolysed by all the amylases (other than the fungal) at 30°C. The comparative susceptibilities of various starches to the bacterial enzyme were then compared, with the results shown in table 6.2. Waxy-maize starch is the most susceptible to degradation by the bacterial *alpha*-amylase, and amylomaize the least. Heat–moisture treatment of potato starch reduced the degree of amylolysis only a little. Leach and Schoch (1961) claimed that neither maize nor potato starch adsorbed any *alpha*-amylase from solution, but their observation, being based on a spectrophotometric measurement of the amount of protein in solution, is vitiated by their use of a preparation containing non-amylolytic protein.

*Table* 6.2. The solubilization of granular starches by bacterial *alpha*-amylase (Leach and Schoch 1961).

| starch | % solubilized[1] |
|---|---|
| waxy maize | 57.8 |
| tapioca | 55.7 |
| waxy sorghum | 55.7 |
| sorghum | 50.2 |
| maize | 49.5 |
| wheat | 48.5 |
| rice | 32.6 |
| sago | 27.0 |
| arrowroot | 20.8 |
| potato | 18.4 |
| heat–moisture treated potato | 15.5 |
| amylomaize | 14.0 |

[1] 24 hours at 50°C.

These authors noted two distinct patterns of *alpha*-amylolytic attack. In type *A* extensive granule erosion and fragmentation occurred (e.g. for maize, sorghum, and their waxy counterparts), whilst in type *B* there was apparently a selective granule-by-granule degradation (e.g. for the remaining starches in table 6.2, with the exception of that from wheat, which was hydrolysed by a mechanism intermediate between types *A* and *B*). Granules degraded by process *A* showed initial attack at the hilum, and birefringence first disappeared from this region, in agreement with the earlier studies of Schwimmer (1945) and Sandstedt (1955). According to Leach and Schoch (1961) the potato starch granules show no loss of birefringence even after approximately 50 per cent solubilization, but occasional 'ghost granules' are seen undergoing rapid dissolution. From this observation, they concluded that the type *B* process was of an 'all or nothing' nature, i.e. the granules were resistant to attack, but once they had succumbed to an initial hydrolysis they dissolved quite readily. In view of the fact that the granules of waxy-maize starch can lose some 50 per cent of their contents without any change in shape or in optical properties during exposure to mineral acid, it is perhaps worthwhile to consider an alternative explanation for the type *B* process, namely that the potato starch granule can retain its shape and birefringence even when some 50 per cent of its contents have been removed. The occasional 'ghost granules' would then correspond to a very advanced state of hydrolysis.

The majority of starches examined by Leach and Schoch (1961) showed a slight increase in iodine-binding capacity after treatment with the bacterial *alpha*-amylase. Nordin and Kim (1960) also noted that exposure of sorghum starch to *alpha*-amylase *in vitro* led initially to an increase in the iodine-binding capacity. *In vivo* studies of malted barley (Greenwood and Thomson 1959, 1961a) led to a similar conclusion, interpreted as showing the α-amylolysis of amylose and the β-amylolysis of amylopectin. In fact, these experimental observations may well be artefacts, due to the difficulty of obtaining a true molecular dispersion of undegraded starch granules in caustic alkali (see table 3.1).

The action of different *alpha*-amylases (human salivary, pig pancreatic, *B. subtilis*, and *A. oryzae*) on a variety of starch granules (protozoal, rice, waxy maize, maize, and potato) at 35–40°C was also examined by Walker and Hope (1963). Salivary *alpha*-amylase was found to degrade maize starch rapidly, and was adsorbed onto the surface of the granule, especially at low temperature; porcine pancreatic *alpha*-amylase behaved in a similar fashion, but *B. subtilis* enzyme hydrolysed the maize starch granules only slowly, and was adsorbed onto the granules at 0°C but not at 35°C; *alpha*-amylase from *A. oryzae* was without action on the maize starch granules, and none of the enzymes were capable of degrading potato starch under the conditions investigated. Again, the extreme resistance of potato starch granules to α-amylolysis was confirmed.

This work also showed that the efficiency of adsorption of the *alpha*-mylase was proportional to the surface area of the granules, the adsorption obeying the Freundlich relation $E_a = kAE^n$, where $E_a$ is the amount of enzyme adsorbed, $E$ is the equilibrium concentration of the enzyme, $A$ is the surface area of the starch granules, and $k$ and $n$ are constants. The significance of enzyme adsorption in heterogeneous reactions has been dealt with by McLaren (1963) and McLaren and Packer (1970), and a suitable kinetic model has been developed.

Studies of partially hydrolysed wheat starch granules have been made using light microscopy (Sandstedt 1955), transmission electron microscopy (Buttrose 1960, 1962), and scanning electron microscopy (Evers and McDermott 1970). The first and third of these techniques effectively give the same type of information, but the depth of focus available to the scanning electron microscope confers on it distinct advantages. Evers and McDermott (1970) studied the effect of wheat *alpha*-amylase on the large lenticular granules of wheat starch (see plates 13 and 14). Whilst it is obvious that major changes are induced in the morphology of wheat starch granules by this *alpha*-amylase, it is difficult to relate the observed changes with any particular structure. In some granules the equatorial plane seems to constitute a point of weakness in the structure, and $\alpha$-amylolysis occurs there preferentially. In other granules, penetration of the enzyme occurs in a haphazard fashion at points on the surface which then become deep pores. The radial cracks, which are often in evidence in the study of wheat starch granules by light microscopy, are not seen using the scanning electron microscope. Their presence should lead to relatively easy penetration of the granule surface, and consequent destruction of the granule from the inside. The results of Evers and McDermott (1970) also confirm the long-held view that the centre of the granule is much less resistant to *alpha*-amylolysis than is any other region. Badenhuizen (1938) had noted that this type of granule could be cleaved at the equatorial plane by means of micro-manipulators. Evers and McDermott (1970) obtained similar cleavage by suspending the granules in a liquid medium on a microscope slide and applying a shear action by the movement of a coverslip. These observations suggest that the granule may be composed of two saucer-shaped halves.

Scanning electron microscopy was also used by Sterling (1971) to examine the fractured faces of native and acid-treated (Lintnerized) potato starch granules. These faces exhibited a definite structure, not merely the random fracture face of an amorphous material, and indicated a fibrillar organization of material, arranged in a radial fashion. Some of these fibrils appeared to be up to 14 $\mu$m in length, and thus must pass through many of the fine (50–100 nm wide), and a few of the coarse (2–7 $\mu$m wide), lamellations found in the potato starch granule. These observations were consistent with earlier work (Hess, Mahl and Gütter 1955; Sterling and Pangborn 1960).

The scanning electron micrographs of Sterling (1971) confirm the presence of radial order within the starch granule, but provide little or no evidence of the tangential lamellae that are so obvious in light microscopy. Indeed Hall and Sayre (1970), whilst showing radial fractions in the potato starch granule (by scanning electron microscopy), stated that there was no radial layering, i.e. no concentric lamellations. In a study of the effects of ultrasound on potato starch granules by scanning electron microscopy Gallant, Degrois, Sterling and Guilbot (1972) also noted that there was no separation of concentric tangential lamellae: all the changes brought about by ultrasonic radiation occurred in the radial sense. Thus certain techniques apparently show the granule to possess tangential order, whilst others suggest a radial orientation of structural elements.

Transmission electron microscopy has again added little to our understanding of the structure of the wheat starch granule. Sections of granules partially degraded by *alpha*-amylase show a layered structure, with the layers being much closer together at the edge of the granule (Buttrose 1960, 1962). In an attempt to improve the information available from this technique, Gallant *et al.* (1969, 1972) treated the granules with periodic acid and then formed a silver complex to increase the contrast. With this technique the authors obtained the usual lamellar structures for the granules of potato and maize starches, and were able to conclude that those rings most susceptible to the action of bacterial *alpha*-amylase were also those most readily oxidized by periodate. When the granules were treated consecutively with mineral acid, *alpha*-amylase, and periodate, followed by formation of the silver complex, evidence was obtained of the microfibrillar structures reported by other workers (Sterling and Pangborn 1960; Heyn 1959). However, it was suggested that the microfibrils were oriented radially in the layers close to the hilum, but tangentially in the outer layers. In potato starch granules subjected to $\gamma$-radiation the microfibrils were again observed, but in this case the dominant orientation was in the radial sense even in the outer layers. Gallant and Guilbot (1969) suggested that these structural elements might well have been oriented tangentially in the native granule, but have subsequently assumed the radial form as a result of water penetrating the granule. This explanation does not, in our view, appear to be very satisfactory.

Degradation of six starches (potato, wheat, manioc, maize, waxy maize and amylomaize) by porcine pancreatic *alpha*-amylase was studied by means of both scanning and transmission electron microscopy by Gallant, Derrien, Aumaitre and Guilbot (1973). No two starches were found to be degraded by the enzyme in exactly the same way. For example, potato starch was hydrolysed by an exocorrosion mechanism, i.e. the polysaccharide was removed from the granule surface, whilst at the other extreme only a few craters appeared on the surface of the amylomaize starch granule, through which the hydrolysis of the interior took place.

*Summary*. There can be no doubt that the study of the degradation of starch granules has provided a vast number of observations. Correlating these observations so that they provide a coherent model for the structure of the granule is difficult: so many of the optical observations are mutually contradictory. For potato starch, some workers find only radial structural elements, others only concentric layers, and others concentric layers in which structural elements may be either radial or tangential. It appears likely that much of the confusion in the literature arises because of artefacts introduced in sample preparation, or to the description of atypical events.

### 6.2.5. Gelatinization.

Dried starch granules swell when they are suspended in water, e.g. the diameter of the lenticular granule of wheat increases by approximately 30 per cent, and the major and minor axes of the potato starch granule increase by 47 and 29 per cent respectively (Hanssen, Dodt and Niemann 1955).

On heating the suspension the granules continue to slowly expand but retain their optical properties, i.e. this swelling is reversible up to a point at which the granules start to lose their birefringence and, in many cases, start to swell hugely. Over a relatively narrow temperature range (approximately 5–10 deg. C), all the granules in the population show these marked irreversible changes in properties and are said to have undergone *gelatinization.*

Gelatinization curves, i.e. the fraction of granules gelatinized as a function of temperature, can be readily obtained using the hot-stage microscope (Schoch and Maywald 1956). The method is fairly sensitive, reproducing quite small differences in the gelatinization temperatures of different populations. For example, figure 6.4 shows the gelatinization curves for potato starch obtained from the tuber, the sprouted tuber, and the sprouts themselves (Banks and Greenwood 1959b). In any one population the larger granules appear to be the most susceptible to gelatinization, whilst the small granules are more resistant. Also, according to Savage, Deatherage, MacMasters and Senti (1958), an increase in amylose content tends to raise the gelatinization temperature. For the starches shown in figure 6.4 the number-average diameter of the granules from the tuber was 42 $\mu m$ and that from the sprouts 14 $\mu m$; the amylose contents were 20 and 17 per cent respectively.

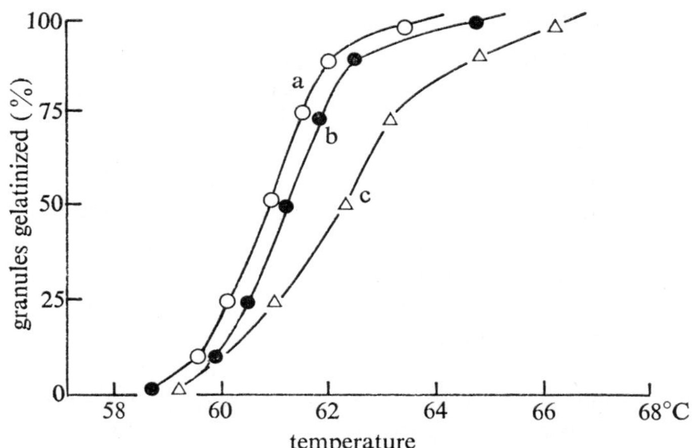


*Figure* 6.4. Percentage of potato starch granules gelatinized in water as a function of temperature: (a) starch from original tubers; (b) starch from shoots; (c) starch from sprouted tubers (Banks and Greenwood 1959b).

The combination of the effects of lower granule size (tending to increase the gelatinization temperature) and lower amylose content (tending to decrease gelatinization temperature) may be sufficient to account for the tuber and sprout starches having virtually the same gelatinization curve. However, the phenomenon of gelatinization is exceedingly complex, and factors other than amylose content and granule size are important.

An extremely interesting observation has recently been reported by Gough and Pybus (1971), who kept an aqueous suspension of wheat starch at a temperature just below that at which gelatinization occurs (50°C) for a period of 72 hours. The properties of the granules were greatly altered by this treatment, and they became swollen and more permeable to the entry of fairly large dyestuff molecules. On the other hand, the granules retained their optical properties (although some granules appeared less anisotropic) and *A*-type X-ray diffraction patterns, and their susceptibility to bacterial *alpha*-amylase was increased only marginally, suggesting that much of the original structure survived the treatment in hot water. The most interesting changes, however, were encountered in the gelatinization characteristics recorded in table 6.3. The temperature range over which gelatinization occurs is narrowed, and moved to a higher value. Furthermore, the temperature range over which individual granules gelatinize was reduced from 0.5–1.5 deg. C for the parent starch to less than 0.2 deg. C for the heat-treated material. It was also observed that the increased susceptibility of the granules to staining with dyestuffs was not directly related to the 'sharpening' of the gelatinization temperature range.

*Table* 6.3. The gelatinization characteristics of starch before and after treatment with water at 50°C for 72 hours (Gough and Pybus 1971).

| sample | gelatinization temperature range (deg. C) | |
|---|---|---|
| | population | individual granules |
| 1, untreated | 52–61 | 0.5–1.5 |
| 1, treated | 65.4–65.8 | 0.2 |
| 2, untreated | 56–61 | 0.5–1.5 |
| 2, treated | 65.4–65.6 | 0.2 |

An explanation of this phenomenon may be found in the physics of semi-crystalline polymers. It is recognized that the melting points of such polymers are a function of the temperature at which crystallization took place. Generally, as the temperature of crystallization increases, so does the melting point. Indeed, it is possible to choose conditions so that melting and crystallization occur at the same temperature. We suggest that this type of behaviour occurs during the heat-treatment of the starch granules—the smaller, or less perfect, crystallites are melted by the high temperature, and more perfect structures are formed. Such a process would be time- and temperature-dependent, as indeed is the case. The result, by analogy with synthetic polymers, would be an increase in the melting temperature and a narrowing of the temperature range over which melting occurred. Since in the process of gelatinization the crystalline order of the granule is destroyed, an analogy can be made between melting and gelatinization. The second effect of the heat treatment, namely the swelling of the granules, has no direct counterpart in the model we are considering, and the results of Gough and Pybus (1971) suggest that some degree of irreversibility of swelling below the gelatinization temperature may be introduced if exposure to warm water is prolonged. Since the granules do swell in water considerable forces must be applied to the granular architecture, and it is possible that some hours are required before the necessary molecular relaxation processes can produce a stable structure.

The gelatinization of starch may be followed in other ways, e.g. gelatinization is effectively a melting process, and therefore calorimetric methods may be employed. Differential scanning calorimetry has been used to measure the endothermic heat $\Delta H$ associated with gelatinization (Stevens and Elton 1971). Typical gelatinization endotherms are shown in figure 6.5 for starches from three genotypes of maize (waxy, normal, and high amylose content);

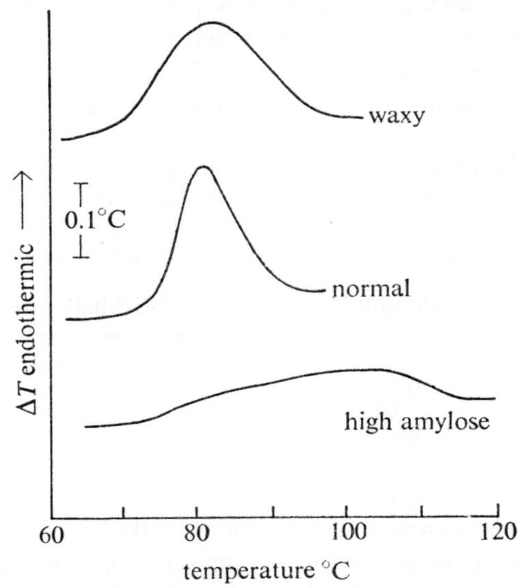


*Figure* 6.5. Gelatinization endotherms of starches from normal maize, waxy maize, and amylomaize (Stevens and Elton 1971).

the heats of gelatinization and the details of the gelatinization process for various starches are recorded in table 6.4. It is especially interesting that within the maize starches the effect of increasing amylose content is to decrease the heat of gelatinization. It has long been recognized that waxy-maize starch possesses at least the same degree of order (as indicated by X-ray diffraction pattern and birefringence) as does normal maize starch (Baker and Whelan 1950), whilst the degree of order within amylomaize starch is much lower. Stevens and Elton (1971) were unable to obtain an accurate measure of the heat of gelatinization of amylomaize starch because the very low peak extended into the region beyond 100°C, where experimentation difficulties are encountered in systems containing water. They conclude, however, that the heat of gelatinization of amylomaize starch is low in comparison to the other types of maize starch.

Using the same technique, a value of 2.3 cal/g was obtained for wheat starch accorded the hot water treatment of Gough and Pybus (1971). This value is somewhat lower than that recorded for untreated wheat starch (see table 6.4) suggesting that the crystalline content of the starch has decreased as a result of the hot water treatment.

*Table* 6.4. Heats of gelatinization and the endotherm characteristics (obtained by differential scanning calorimetry) of various starches (adapted from Stevens and Elton 1971).

| starch | $\Delta H$ cal/g | endotherm temperatures: onset | peak | conclusion |
|---|---|---|---|---|
| wheat | 2.7 | 54 | 69 | 86 |
| rice | 3.4 | 66 | 82 | 100 |
| maize | 3.6 | 68 | 79 | 96 |
| tapioca | 4.0 | 66 | 78 | 100 |
| arrowroot | 4.6 | 73 | 84 | 106 |
| waxy maize | 4.9 | 62 | 81 | 98 |
| potato | 5.3 | 58 | 71 | 95 |

Another commonly used method of investigating the gelatinization process is to measure the viscosity of the system, for the granules swell to an extent determined by the source of the starch and thus are capable of increasing the viscosity of the suspension to varying extents. The measurement of viscosity is easily carried out by means of the 'Brabender' visco-amylograph. Typical results obtained by this technique are shown in figure 6.6 for waxy maize, normal maize, and amylomaize starches, with amylose contents of 0, 28, and 55 per cent respectively.

As the starch suspension is heated the granules swell and impinge upon each other, so increasing the viscosity of the paste, but eventually the cohesive forces in the original granular structure become weakened and the viscosity decreases as the integrity of the granule is lost. Thus waxy-maize starch swells comparatively readily to produce a high peak viscosity, but the granular structure then collapses and the viscosity decreases. In contrast, the presence of the amylose fraction in normal maize starch restricts the degree of swelling, so that the peak viscosity is much less than that of waxy-maize starch, but it helps to maintain the integrity of the granule so that a relatively strong paste results. No increase in the viscosity of the suspension of amylomaize starch granules is recorded under these conditions, despite the fact that calorimetry indicates that some gelatinization occurs in the temperature range employed in this measurement of viscosity (see figure 6.5), and also despite the evidence of leaching studies which show that extraction with boiling water removes some 35 per cent of the polysaccharide material of this particular type of amylomaize starch (Banks, Greenwood and Muir 1971a).

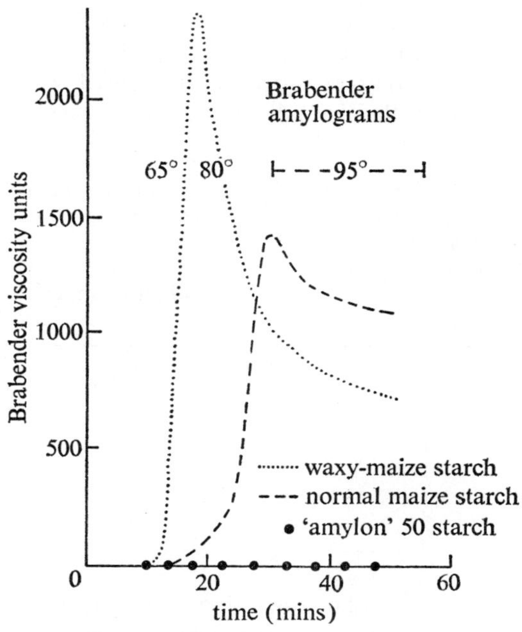


*Figure* 6.6. The pasting behaviour of various maize starches measured in a 'Brabender' visco-amylograph. Starch concentration was 40 g polysaccharide in 460 ml water; the heating programme was (a) start at 50°C, (b) heat at 1.5 deg. C/min. to 95°C, and (c) hold at 95°C for 30 min. (Banks, Greenwood and Muir 1973c).

The gelatinization of starch is also affected by the presence of salts, e.g. starch will not gelatinize on boiling in sodium sulphate solution but will gelatinize in sodium iodide solution at room temperature. It is generally accepted that if starch is exposed to a series of anions (all having the same cation), the resulting changes in the gelatinization temperature will occur in the same order as the anions occur in the lyotropic series. With cations having a common anion the situation is rather more complex, and only recently has an explanation been provided for these diverse phenomena. Gough and Pybus (1973) followed the gelatinization of wheat starch in solutions of various chlorides (chloride being chosen because it is the least effective of any

---

*Figure* 6.7. Diagrammatic representation of starch granules, viewed between crossed polars, undergoing various types of gelatinization (Gough and Pybus 1973).

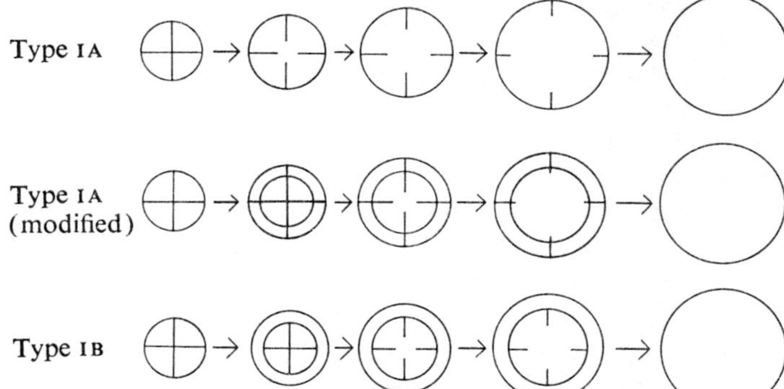


Type I involves tangential swelling with simultaneous loss of birefringence from hilum to periphery. Type IA: isotropic material spreads from hilum to periphery; type IA (modified): as in IA, but outer layers retain their anisotropy; type IB: outer layers swell with loss of birefringence.

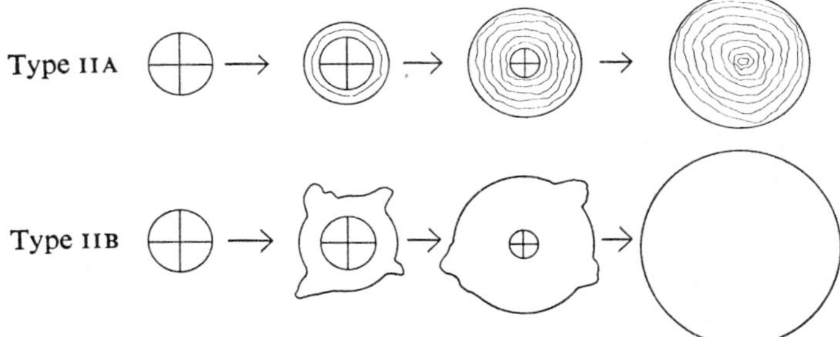


Type II involves both swelling and loss of birefringence from periphery to hilum. Type IIA: retains concentric layers under peripheral swelling; type IIB: swelling in outer layers more pronounced to give irregular outline.

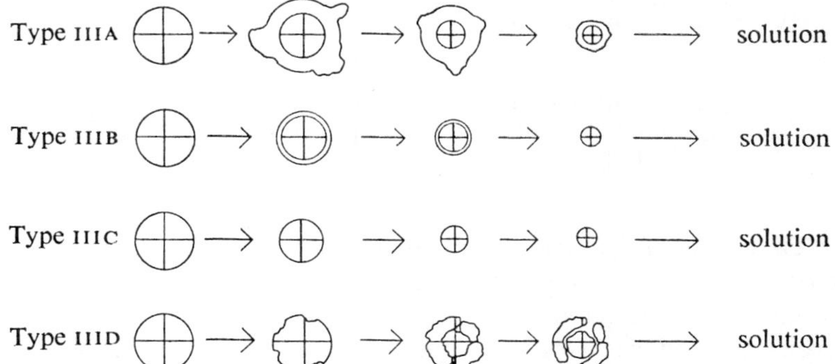


Type III involves apparent dispersion of the granule. Type IIIA: irregular structure formed; type IIIB: swelling and dissolution proceed at same rate; type IIIC: shrinkage from outside without loss of birefringence; type IIID: no loss of birefringence, but formation of radial channels.

anion for modifying the gelatinization behaviour of starch), using hot-stage microscopy. These authors established that gelatinization could take place in three distinct forms, types I, II, and III, with further subdivisions within each type as shown diagrammatically in figure 6.7.

Gough and Pybus (1973) found that solutions of certain metal chlorides exhibited all three types of gelatinization pattern as the salt concentration increased. Typical curves relating the gelatinization temperature to the concentrations of three chlorides (calcium, cupric, and aluminium) are shown in figure 6.8; the type of gelatinization at each concentration is also indicated. The curves for the different cations are broadly similar, and the gelatinization temperature initially increases with increasing salt concentration, reaches a maximum value, decreases to pass through a minimum, and finally increases again. Type IA gelatinization is observed until the minimum is reached, and then type II followed by type III gelatinization. Also, prior to the minimum being reached the gelatinization temperature is independent of the rate of heating, whilst subsequently it is independent on that factor.

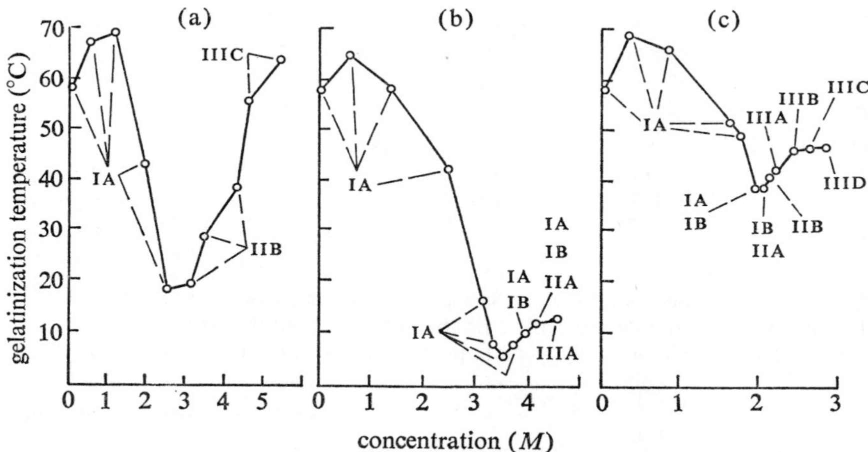


*Figure* 6.8. Relation between salt concentration and the type and temperature of gelatinization of wheat starch granules: (a) calcium chloride; (b) cupric chloride; (c) aluminium chloride (Gough and Pybus 1973).

The shapes of the curves were explained by postulating that at low concentrations of salt the active principle in the swelling of the granule was water; its activity being lowered by the presence of the salt, higher temperatures were necessary to induce gelatinization. However, as the salt concentration is increased, the water present is confined to the limited hydration of the ions which are responsible for gelatinizing the granule. The most fundamental change is observed at the minimum in the gelatinization-temperature–salt-concentration relation, when the mode of gelatinization changes from type I to type II. Gough and Pybus (1973) suggest that, at this point, attack on the granule is so rapid that the outer layers swell before the salt has penetrated the granule. Diffusion of the salt through these swollen layers is then so slow as to become the rate-limiting factor, and hence the gelatinization process would be expected to be dependent on the rate of heating, as is in fact observed. With further increase in the salt concentration, the swelling of the outer layer is so rapid that it precludes diffusion of the salt until the layer is removed by dissolution, so giving rise to type III gelatinization. This theory, which the authors support with data realting to the viscosity of dilute suspensions of wheat starch granules and to sedimentation volumes in the various salt solutions, does much to explain the many incongruities found in the literature regarding the effect of cations on gelatinization.

Dissolution of various starches in cold anhydrous dimethylsulphoxide was studied by Leach and Schoch (1962), who found that the granules dissolved without swelling, but at different rates depending on the type of starch. As in the case of degradation by *alpha*-amylases, maize, sorghum, and their waxy counterparts were the starches most rapidly solubilized, with potato starch being solubilized only slowly. However, the analogy between enzymic solubilization and that resulting from treatment with dimethylsulphoxide is by no means perfect, in that amylomaize starch is fairly readily dissolved by the latter treatment but is resistant to enzymic degradation. These studies were interpreted as providing supporting evidence for the postulate (Leach, McCowen and Schoch 1959) that in the maize starch granule there exist distinct areas of crystallinity and amorphous regions, whereas in potato starch the structure is more uniform.

However, in our opinion, the large number of painstaking studies of gelatinization phenomena have provided no evidence for a detailed model of the starch granule. In fact, like all the other methods used to examine the starch granule, this technique provides only general background information.

### **6.2.6.** Structure of the Starch Granule.

In this section we shall attempt to summarize the salient features of granular structure, but, of necessity, much of the discussion must be speculative.

The starch granule, like a polymer spherulite, is a semi-crystalline entity composed of crystalline and amorphous regions. By analogy with synthetic polymers we may theorize about the arrangement of these regions, and three models have been proposed. In the original concept of the *fringed micelle* the long polymer chains would pass through a crystallite, into an amorphous region, and then subsequently through other crystalline and amorphous regions. With the discovery of *chain folding* in polymer single crystals, the model was modified so that in the crystalline regions the chains could be folded rather than being outstretched, so that there is a fairly sharp division between crystalline and amorphous regions. In the alternative model, the *paracrystalline* concept, this sharp division disappears, and the spherulite is treated as a single crystal with large numbers of faults or defects.

The physical forces that cause a synthetic polymer to adopt the spherulitic mode as it crystallizes might well be operative during the crystallization of starch, and models devised for synthetic polymers can be considered as a suitable basis for explaining the structure of the starch granule. Indeed, Borch, Sarko and Marchessault (1969), using the technique of solid-state light scattering, have demonstrated a close similarity between the spherical granules of tapioca starch and spherulites of synthetic polymers (see section 6.2.2).

Moving from the general spherulitic model to a more descriptive one presents several difficulties. First, we would agree with French (1972) that the structure of the chains in the crystalline assembly giving either *A*- or *B*-type *X*-ray diffraction patterns is unknown, and with his suggestion that some type of double helix may be involved. Second, there is some dispute as to the proportion of glucose units to be found within the crystalline regions in a starch granule, for estimates have ranged from zero to 60 per cent (Sterling 1968). Third, and perhaps most baffling, is the fact that the overwhelming mass of evidence indicates that the principal crystalline component is amylopectin, the branched fraction. We have noted that the birefringence and X-ray diffraction pattern of waxy-maize starch with zero amylose content resemble those of normal maize starch, whilst the amylomaize starches exhibit poorer crystallinity and are much less birefringent. Similar results are obtained when comparing barley starches from waxy, normal, and high amylose genotypes, and also from normal and high amylose pea starches. Additional evidence comes from leaching experiments, for when starches are leached at or slightly above their gelatinization temperatures a considerable fraction of the total amylose component is released into the supernatant, whilst the amylopectin is retained in the swollen granule. Indeed, by a suitable pre-treatment of the granule, Montgomery and Senti (1958) were able to extract all the amylose without granule rupture and with retention of crystallinity. The only reasonable conclusion from such observations is that amylose is not, under normal conditions, concerned in the crystalline regions of the granule.

The conclusion has very important implications. In the crystallization of synthetic polymers a multiply-branched structure, containing 4–5 per cent of branch points, would certainly *not* be predicted to constitute the crystalline entity in preference to a virtually linear component. In fact, branching is accepted as being detrimental to crystallization and to the physical properties dependent on crystallization. Thus high-pressure polyethylene, which is branched,* is only 50 per cent crystalline and softens at 90°C, whereas low-

*In high-pressure polyethylene there is a mixture of long-chain and short-chain branching, so that the apparent number-average chain length is about 20 methylene units. However, the branching is predominantly short-chain, so that it would correspond to the Staudinger 'herringbone' structure—a model invalidated for amylopectin.

pressure polyethylene is essentially linear, has a crystallinity of 90 per cent and a softening point of 120°C.

Accepting that amylopectin is the principal component of the crystalline regions in the granules raises another problem, namely that it is almost impossible to obtain the isolated component in a crystalline form. Doi and Doi (1969) obtained polycrystalline granules of amylopectin in gelatin, but these assemblies were not birefringent. They did, however, exhibit X-ray diffraction patterns of the *A*- and *B*-types, the form of the pattern depending on the temperature of crystallization. The staling of bread is probably due largely to the crystallization of the amylopectin component, but the relative concentrations of starch and water in this system are such that the gelatinized granule probable retains some 'memory' of its original structure, and this acts as a 'seed' for the crystallization process. The addition of iodine–iodide to gelatinized wheat starch granules is sufficient to restore a weak positive birefringence, and overnight storage of the swollen granules produces the same effect. We have repeatedly endeavoured to produce some form of retrogradation in amylopectin but have met with success only in those fractions obtained from starches having abnormally high amylose contents, which contain short-chain linear material in a size range ideal for retrogradation.

Similar attempts were made by Vasko and Koenig (1972) to prepare crystalline amylopectin by precipitation from ethylene diamine solution with methanol, and by casting as films from solutions in water and dimethylsulphoxide, the films being allowed to swell by exposure to water vapour at 35°C and 50°C, and being aged at room temperature. In all cases, the X-ray diffraction patterns indicated that crystalline ordering of the polysaccharide did not exist. However, studies using infra-red absorption showed, by analogy with earlier results obtained from amylose (Koenig and Vasko 1970), the presence of chain folding. The intensity of the band associated with folding increased as the sample aged for both amylose and amylopectin, and it was concluded that even although amylopectin could not crystallize it formed some kind of structure involving folding of the chains.

It is of some interest that a paracrystalline form of glycogen was obtained by De Wulf and Hers (1972) by various procedures involving the freezing and thawing of aqueous solutions of the polysaccharide. French and Kikumoto (1973) found that repeated freezing (at −20°C) and thawing (at 4°C) did not produce a crystalline material, whereas a single freezing at −20°C, followed by storage at −80°C for 30 days and thawing at 4°C, yielded a dense white precipitate. On separation and drying (at 30 per cent relative humidity), the isolated material yielded a *B*-type X-ray spectrum. Such a treatment might also produce a crystalline form of amylopectin.

Any model of the starch granule must take into account that the order within the crystalline regions is such that, by a proper choice of growth conditions, it is often possible to favour either the *A*- or *B*-type X-ray diffraction pattern. Again, the *B*-type pattern of potato starch can also be converted to the *A*-type by heat–moisture treatment, although in many ways potato starch presents something of an anomaly. The intensity of the X-ray diffraction pattern, the high birefringence, and the comparatively high heat of gelatinization (see table 6.4), allied with the fact that the diffraction pattern is of the *B*-type (the same as that obtained from retrograded amylose), suggest that the granule should be difficult to disperse in water. In fact, potato starch is exceedingly easy to disperse in water, and the pre-treatments that are necessary to ensure the complete dispersal of cereal starches (see section 2.2.2) are not required. But in contrast, despite the ease with which the potato starch granule swells in water, it is exceedingly resistant to attack by *alpha*-amylase.

Other starches, notably those of high amylose content, also exhibit *B*-type X-ray diffraction patterns (although somewhat weak ones) and are very resistant to α-amylolysis, but they have only a very limited ability to swell. We suggest that in these starches the amylose component is to be found in the crystalline regions. This conclusion arises from the fact that amylomaize starch of very high amylose content contains so little 'normal' amylopectin, that the amylose must of necessity take part in the crystalline assembly. In this respect, it is interesting to note that the X-ray diffraction pattern of these amylomaize starches cannot be changed from the *B*- to the *A*-type by a heat–moisture treatment (Charbonnière, Mercier, Tollier and Guilbot 1968); neither can that of retrograded amylose. It is commonly assumed that the major difference between the *A*- and *B*-type structures is that the latter contains more water of crystallization, for it has been postulated that a molecule of water provides the bridging hydrogen bonds between successive helical turns in *B*-amylose, whilst in *A*-amylose the hydrogen bond is directly between the glucose residues in adjoining helical turns (Blackwell, Sarko and Marchessault 1969).

We suggest that these observations are combined with those of French (1972) regarding the possibility of double helix formation. Thus, we envisage the *B*-pattern of potato starch arises from an array of double helices with water providing the bridging hydrogen bond between each duplex; heat–moisture treatment dehydrates the system so that water is eliminated and hydrogen bonding occurs directly between each duplex. As water is further eliminated from the amorphous regions, the strains set up by removal of this plasticizing agent are sufficient to disrupt the crystalline array, and the sample becomes amorphous because three-dimensional order is lacking. (This hypothesis explains the well-known fact that dry starch tends to show an amorphous X-ray spectrum.) If hydrogen bonds are to be easily broken in the case of the change *B*-spectrum→*A*-spectrum→amorphous, then each duplex may be linked to its neighbours by only a few such bonds.

If amylopectin is the crystalline entity then only comparatively short lengths of chain need be involved in the structure with correspondingly few hydrogen bonds. On the other hand, when amylose forms the duplex a large number of hydrogen bonds will be involved in the crystal lattice, and the structure will be resistant to dehydration by heat–moisture treatment, as is observed in the case of amylomaize starch. This concept requires the rider that chain folding does not occur when amylose adopts the *B*-form (otherwise the length of helix involved in hydrogen bonding would be comparable to that entailed in the crystallization of amylopectin, and would therefore be equally susceptible to dehydration), in pleasing agreement with experiment (Vasko and Koenig 1972).

On the basis of this model, the amylopectin component is the dominant factor in crystallization for the majority of starches. However, the ease of swelling of waxy starches in cold water shows that these starches possess a considerable proportion of amorphous material. But this observation is to be expected. Although the external chains of amylopectin have been suggested as forming the crystalline entities, it is difficult to see even the external chains of the Meyer structure having sufficient order to form the necessary crystalline array, even if the branch points can act as nuclei for the formation of double helices as suggested by French (1972). The alternative suggestion advanced by French (1972), that we must seek new models for the amylopectin macromolecule, is more probably correct. He suggested as possibilities a modified trichitic structure and also a racemose structure; both are shown in figure 6.9.

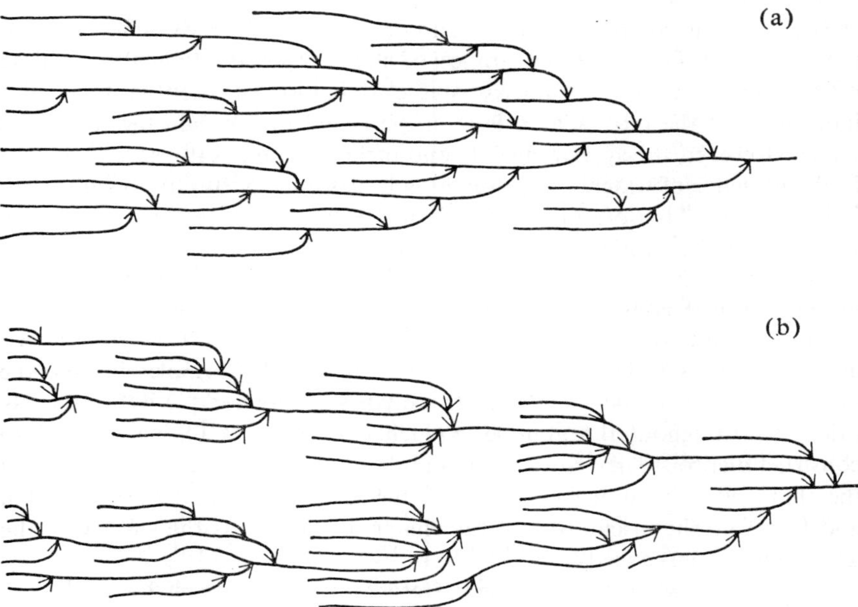


*Figure* 6.9. Proposed models for the structure of amylopectin: (a) modified trichitic structure; (b) racemose structure (after French 1972).

The two models would appear to account adequately for most of the known properties of amylopectin—susceptibility to enzymes, the production of double-branched dextrins on $\alpha$-amylolysis, and the hydrodynamic behaviour of the macromolecule. On the other hand, an irregularly-branched Meyer structure would also account for the above properties. The advantage of the trichitic, and more especially the racemose, models is that one can imagine them participating fairly readily in a crystalline structure (Robin *et al.* 1974).

Whilst it is easy to imagine either the trichitic or racemose structures being involved in a crystalline array within the granule, it is difficult to understand why the separated component does not form a crystal structure. One is tempted to postulate that glucose units are added at the surface of the granule to chains that may already be involved in a crystalline array, but this concept will be dealt with later (see section 6.3).

We must also consider that starch granules, including the waxy types, are both crystalline *and* birefringent. The sign of birefringence is such that the chains in the amylopectin are most probably oriented with their long axes radially inclined (or, more correctly, at right angles to the surface). Labelling starch granules by bombardment with tritium atoms indicated that the chains were so arranged that non-reducing end-units occurred at the granule surface (Nordin, Moser, Rao, Giri and Liang 1970). Similarly, the erosion pattern resulting from the action of glucoamylase on wheat starch granules was found to be quite different from that of *alpha*-amylase (Evers, Gough and Pybus 1971), and quite compatible with the non-reducing chain-ends being found at the granule surface. The appearance of stress cracking along the radii, so often observed during the microscopic examination of dried granules, also indicates that the molecular axis is in the radial rather than the tangential plane. However, the appearance of concentric rings in potato starch granules in the presence of water, and in cereal granules on exposure to mineral acid, indicates some kind of tangential ordering. Why these layers are not seen in the dry granule is rather difficult to explain.

So far, we have discussed the disposition of amylopectin in the starch granule without considering the amylose fraction. In most normal starches, i.e. those with amylose contents of 15–35 per cent, this polysaccharide apparently adds little to the crystalline nature of the granule. Indeed, the evidence does not indicate that the amylose component is even strongly oriented throughout the granule. French (1972), using iodine staining to construct dichroism maps of starch granules, found that radial orientation of the chains was favoured in the core of the granule, but that the dichroism, and therefore the orientation of the chains, decreased as the surface of the granule was approached.

The form in which the amylose is present must be a matter of speculation. Many starches, particularly those derived from cereal sources, contain lipid material that cannot be removed by extraction with fat-solvents such as chloroform, but require that the starch be swollen by the extracting solvent.

In such circumstances, it is difficult to envisage the lipid being present anywhere other than in the core of the amylose helix. On the other hand, the *V*-type X-ray diffraction pattern is not observed in cereal starches until they have been gelatinized, and then only a very weak pattern is seen. Thus there would appear to be little ordering of any amylose *V*-helices present. It is most unlikely that the *V*-type helix is formed in the absence of a complexing agent, which would suggest that most of the amylose is present as an amorphous form. In agreement with this conclusion is the fact that starch granules will stain with iodine vapour only in the presence of water, which acts as a molecular lubricant to allow the amylose chains to adopt the conformation necessary to give the blue complex. Some of the amylose could be present as a double helix for, if the helices were not aligned, this arrangement would not affect the X-ray diffraction pattern.

*Summary.* We will summarize our model by considering three specific starches—potato, maize, and amylomaize. In the *potato starch granule* there is a relatively high degree of crystallinity (contributed solely by the amylopectin), and the amylose is found in the amorphous state. Below the gelatinization temperature, the crystallinity is sufficient to give a compact structure that does not allow penetration by *alpha*-amylase; the strong radial orientation of the molecules provides an unreceptive surface for the enzyme, and hence $\alpha$-amylolysis is slow. On gelatinization, the amylose contributes nothing to the stability of the granule which therefore swells until it bursts. We envisage the heat treatment causing two effects: (a) the dehydration that changes the X-ray diffraction pattern to the *A*-type; and (b) the conversion of a fraction of the amorphous amylose to a helical form. Being less soluble than the amorphous form, the helical regions would act as weak centres of crystallinity when the granule gelatinized, preventing it from bursting.

In the *maize starch granule* the amylopectin may again constitute the crystalline skeleton, but the overall degree of crystallinity and orientation of the macromolecules would be lower than in the case of potato. The surface of the granule would then be more susceptible to the action of *alpha*-amylase, and the poorer orientation of the molecular chains would allow the enzyme easier access to the interior of the molecule. Much of the amylose must be present as a complex of fat, and as such is probably in the *V*-form. Again, we believe that the limited solubility of this form would be sufficient to set up a weak crystalline structure that would retard granule swelling.

In the *amylomaize starch granules*, amylose in the retrograded form would form the crystalline structure. Retrograded amylose is quite resistant to the action of *alpha*-amylase, hence one would expect the hydrolysis of the granule to be very slow. Although only a small fraction of the amylose molecules need be involved in the long regions of double helix, the stability of such a structure might well be expected to prevent the granule swelling to any extent in hot water.

## 6.3. Biosynthesis of Starch.

In this section we are concerned mainly with the chemistry and biochemistry of the synthesis of the granule components, rather than with the botanical connotations. This latter aspect has been dealt with more than adequately by Badenhuizen and his collaborators (Badenhuizen 1958, 1959, 1969, 1971; Salema and Badenhuizen 1967).

### 6.3.1. Properties of Starch as a Function of Plant Growth.

The properties of starch granules from a given botanical source are not static, but vary during the growth of the plant concerned. In particular, a quite dramatic increase in the iodine-binding capacity of the polysaccharide occurs in starches from many different sources: e.g. in those from wheat (Bice, MacMasters and Hilbert 1945); maize (Wolf, MacMasters, Hubbard and Rist 1948; Erlander 1960); barley (Harris and MacWilliam 1958; Banks, Greenwood and Muir 1973a, d); amylomaize (Mercier, Charbonnière, Gallant and Guilbot 1970); tobacco leaves (Matheson and Wheatley 1962); peas (Greenwood and Thomson 1962b); sweet potato (Ono 1969); and potato (Geddes, Greenwood and MacKenzie 1965). Furthermore, the morphology of the granule and certain molecular properties of the components may also change during growth. Studies of these changes have been carried out for peas (Greenwood and Thomson 1962b), barley (Banks, Greenwood and Muir 1973a, d), and the potato (Geddes, Greenwood and MacKenzie 1965).

*Growth Studies on Pea Starches.* As detailed earlier (see section 2.5.3), starch from the smooth-seeded pea contains, at maturity, some 35 per cent of amylose, whilst that from wrinkled-seeded varieties contains twice as much amylose. The results of growth studies carried out on the two types of pea are shown in table 6.5.

*Table* 6.5. Properties of pea starches as a function of development (Greenwood and Thomson 1962b).

| sample[1] | starch content (%) | average granule diameter (μm) | granule type | amylose (%) |
|---|---|---|---|---|
| s1 | 22 | 10 | simple | 15 |
| s2 | 26 | 15 | simple | 19 |
| s3 | 32 | 30 | simple | 29 |
| s4 | 40 | 40 | simple | 37 |
| w1 | 12 | 8 | simple | 23 |
| w2 | 16 | 12 | simple | 36 |
| w3 | 22 | 20 | simple+ compound | 59 |
| w4 | 31 | 30 | simple+ compound | 69 |

[1] s is starch from smooth-seeded pea, w is starch from wrinkled-seeded pea.

For both genotypes, as the starch content increases so too does the average diameter of the starch granule. However, while the type of granule (simple) remains the same throughout the development of the smooth-seeded pea, there is a change in the starch from the wrinkled-seeded pea: initially the granules are simple, but as the starch content increases a second population of compound granules appears. Interestingly enough, the compound granules become apparent only when the amylose content of the developing starch exceeds approximately 35 per cent, which is the limiting value associated with the dominant genotype (the smooth-seeded pea).

The properties of the amylose components isolated from the starches of both pea genotypes are shown in table 6.6. Each amylose fraction possesses a high iodine-binding capacity, and thus the calculated amylose contents of the starches must be real rather than apparent. These results are totally opposed to the suggestion of Matheson (1971) that the iodine-binding capacity of the amylose fraction itself increases during development. We suggest, in fact, that the results for wheat starch (Wood 1960) and rice starch (Briones, Magbanua and Juliano 1968), according to which the iodine-binding capacities of the isolated amylose components increase with maturity, are due to difficulties in fractionating the starch rather than to any aspect of the biosynthetic process.

In table 6.6, the complete conversion to maltose recorded for the concurrent action of *beta*-amylase and *Z*-enzyme shows that amylopectin is absent from these fractions, but the extent of conversion to maltose by *beta*-amylase acting alone decreases as the growing season progresses. Thus,

*Table* 6.6. Properties of the amylose components isolated from pea starches during growth (Greenwood and Thomson, 1962b).

| sample[1] | $[\beta]$[2] (%) | $[\beta+Z]$[3] (%) | IBC[4] (%) | $[\eta]$[5] (ml/g) |
|---|---|---|---|---|
| s1 | 90 | 100 | 18.7 | 145 |
| s2 | 86 | 101 | 19.0 | 160 |
| s3 | 79 | 100 | 19.1 | 195 |
| s4 | 74 | 100 | 18.9 | 210 |
| w1 | 90 | 100 | 19.3 | 120 |
| w2 | 82 | 100 | 19.1 | 130 |
| w3 | 80 | 100 | 18.9 | 135 |
| w4 | 74 | 101 | 19.1 | 150 |

[1] As in table 6.5.
[2] $[\beta]$ is the extent of conversion into maltose obtained using purified *beta*-amylase in this and subsequent tables.
[3] $[\beta+Z]$ is the extent of conversion into maltose obtained using *beta*-amylase preparations contaminated with *Z*-enzyme in this and subsequent tables.
[4] IBC is the iodine-binding capacity.
[5] $[\eta]$ is the limiting viscosity measured in 0.5 M KOH.

according to the model we developed in section 2.3.5, the extent of long-chain branching in the amylose fraction must increase as the starch content increases. The limiting viscosity number also increases, and therefore there must be an even more pronounced increase in the molecular weight of the amylose component (as the extent of branching increases, the limiting viscosity number becomes insensitive to the change in molecular weight).

The properties of the amylopectin fractions were found to vary but little as growth progressed. Those from the wrinkled-seeded pea generally exhibited higher iodine-binding capacities than those from the smooth-seeded genotype. This, of course, is in agreement with the earliest studies on the starches (Potter, Silveira, McCready and Owens 1953), but we have already shown (see section 2.5.3) that the polysaccharide is a mixture of normal amylopectin and short-chain material, some of which is linear and some lightly branched, rather than being an amylopectin with abnormally long outer chains as was postulated in that early work.

*Growth Studies on Barley Starches.* A similar comparison, between starches of normal and high amylose contents as a function of growth, has recently been reported for barley (Banks, Greenwood and Muir 1973a, d). In this case there is the additional complication that as the barley endosperm matures, starch granules of two quite distinct sizes and types are deposited (a similar phenomenon is encountered in the endosperm of wheat). Arbitrarily, we may define the large granules as having a diameter greater than 10 $\mu$m (the majority attain a diameter between 15 and 35 $\mu$m), and the small granules as having a diameter of less than 10 $\mu$m, the average value being approximately 5 $\mu$m. It is generally accepted that the large granules appear during the initial stages of starch deposition, whereas the small ones are laid down at a later stage of growth, in interstices between the large granules (Buttrose 1963). The distinction between small and large granules is much less apparent in the case of the barley genotype giving starch of high amylose content, because at maturity the average size of the granule is much less than that of the normal genotype (Merritt and Walker 1969a, b).

This observation is supported by measurements of the granule size distributions during growth. The frequency–size and weight–size distributions for granules of normal barley are shown in figure 6.10(a) and (b) respectively (the weight–size distribution was obtained by assuming the granules to be oblate ellipsoids with the major axis twice as great as the minor axis, and to have a common density). In the period to 20 days from anthesis there is a single distribution of granule sizes, although the overall size does increase. Subsequently, however, a discontinuity in the frequency–size distribution becomes apparent. At maturity, the small granules (diameter less than 10 $\mu$m) account for approximately 80 per cent by number of the total population, but only 10 per cent by weight.

The starches isolated from the high-amylose barley have, at an equivalent stage of development, markedly smaller granules than those from the

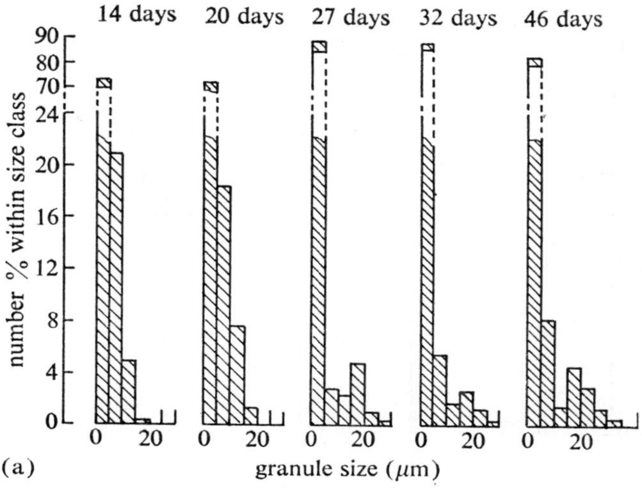


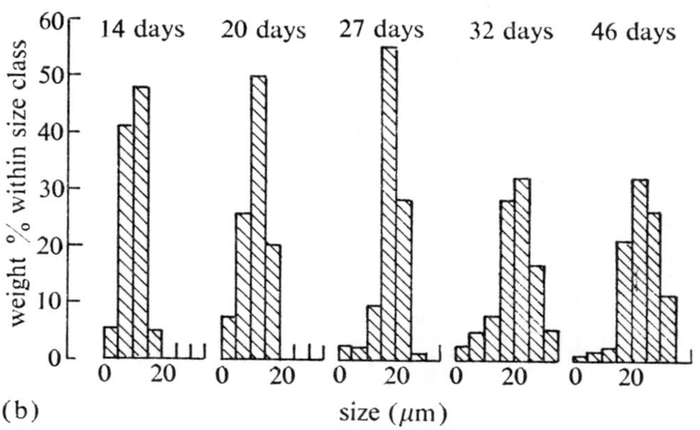


*Figure* 6.10. Granule size distributions for starches from normal barley during growth: (a) number/size distribution; (b) weight/size distribution (Banks, Greenwood and Muir 1973d).

dominant genotype, and moreover there is no indication of a bimodal distribution of granule sizes (see figure 6.11(a) and (b)). Even on a weight basis the proportion of small granules at maturity is high, for the weight-average granule diameter is 13 μm, whereas the corresponding value for the starch from normal barley is 24 μm.

The difference in granule size distribution between the starches from the two barley genotypes is much too pronounced to have its origins in any experimental artefact. It is perhaps best emphasized by comparing plates 15 and 16, showing scanning electron photomicrographs of the mature starch granules.

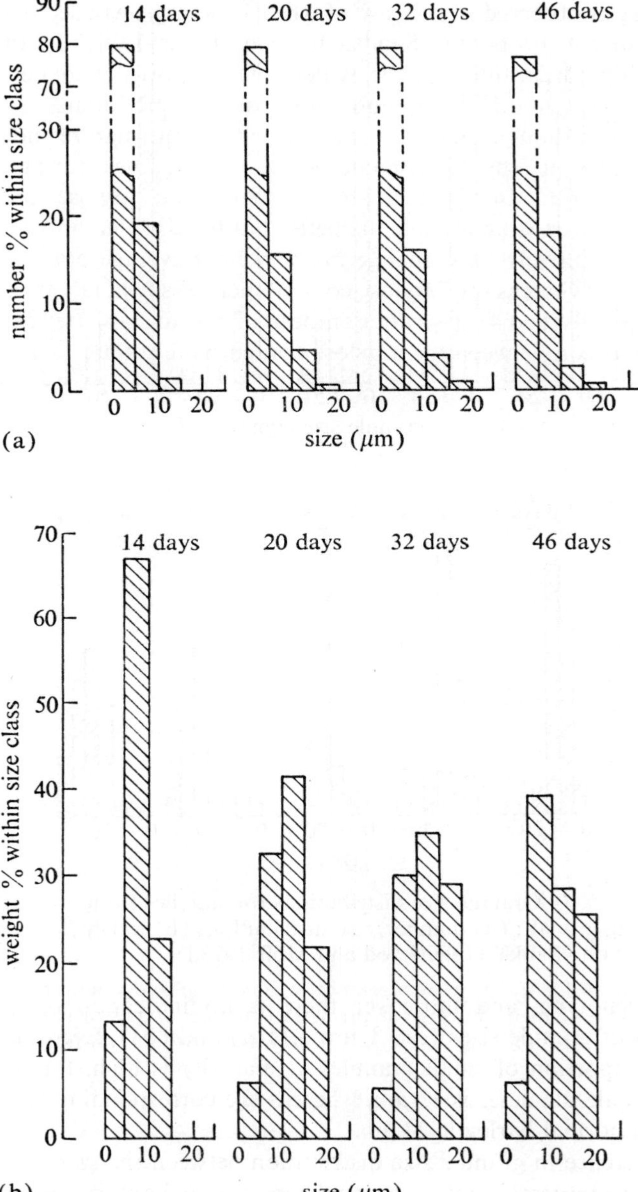


*Figure* 6.11. Granule size distributions for starches from high-amylose barley during growth: (a) number/size distribution; (b) weight/size distribution (Banks, Greenwood and Muir 1973d).

In the early stages of development of barley, many irregularly-shaped granules are found (May and Buttrose 1959; Badenhuizen 1969). The same phenomenon is observed in the case of wheat, and a growth sequence for the granule has been proposed (Sandstedt 1946; Evers 1971). In the case of normal barley starch, there is little evidence to support this mode of granule deposition. The starch from the immature plant has predominantly the same granule shape, although smaller, as the starch from the mature endosperm.

The increase in amylose content of the two starches during growth is shown in figure 6.12. In the case of normal barley the amylose content again increases with increasing granule diameter, but for the high-amylose samples there is comparatively little change in the weight-average granule diameter during growth, whereas the amylose content increases by a factor of 1.5 in the period between 14 and 46 days after anthesis. Thus, in this latter case, there is no simple relation between the amylose content and granule size.

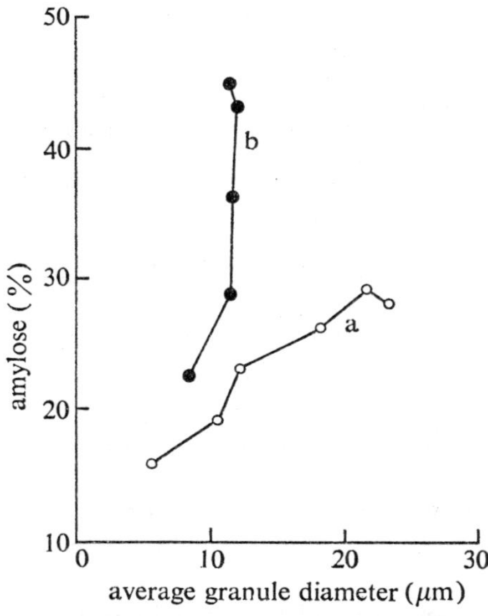


*Figure* 6.12. Relation between amylose content and weight-average granule diameter during the growth of two genotypes of barley: (a) normal; (b) high-amylose (Banks, Greenwood and Muir 1973d).

In figure 6.13 the amylose-contents of the two starches are graphed as a function of the relative weight of starch present (i.e. the amount of starch present in the ear during development as a percentage of that present at maturity). In normal barley starch, the major changes in amylose-content take place during the deposition of the first 30 per cent of the starch,*

*In the initial stages of development, the pericarp plays a significant role in starch metabolism (Banks, Evers and Muir 1972). The above results

whilst in the starch from the high-amylose barley significant changes in the amylose content occur throughout the period of starch deposition.

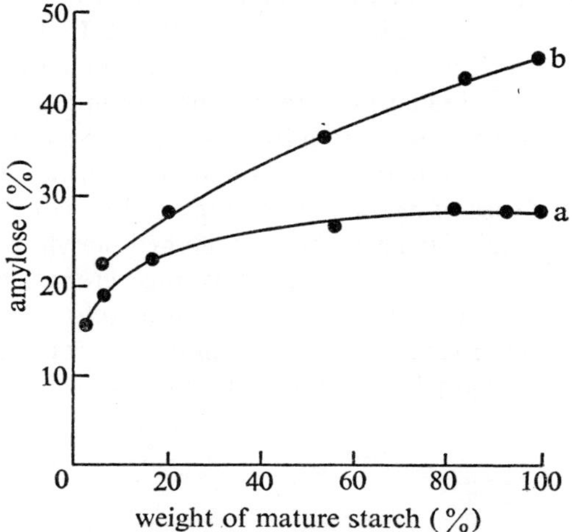


*Figure* 6.13. Amylose content of barley starches at various stages of maturity. The stage of development is expressed in terms of the amount of starch present as a percentage of the final weight of mature starch: (a) normal; (b) high-amylose (Evers, Greenwood, Muir and Venables 1974).

Mature starches of the barley genotypes, and also of wheat, were fractionated according to granule size (Evers, Greenwood, Muir and Venables 1974), using both differential sedimentation in water (Geddes, Greenwood and MacKenzie 1965) and a sieving technique (Evers 1973). The properties of the resultant fractions are shown in table 6.7. It would appear that in wheat and barley starches of normal amylose content there is no significant difference in the amylose–amylopectin ratio between starch granules of different sizes. Only in the starch of abnormally high amylose content does the fraction of smallest diameter (less than 10 $\mu$m) contain more amylose than the fractions of higher size-ranges.

There is a tendency for a genetically-determined increase in amylose content to be accompanied by a change in granular form: in pea starch we have noted the appearance of compound granules, in barley starch the genotype of high amylose content has much smaller granules than the parent cultivar, and irregularly-shaped granules are manifest in amylomaize starch. Thus, there is some evidence to support our claim that the amylose content of a starch and its granular form are determined by independent control mechanisms.

---

refer to the *total* starch present in the developing ear, i.e. from both endosperm and pericarp.

*Table* 6.7. The amylose contents of mature starch granules of various sizes (Evers, Greenwood, Muir and Venables 1974).

| | amylose content (%) for granule size: | | | |
|---|---|---|---|---|
| | 10 μm | 10–15 μm | 15–20 μm | >20 μm |
| barley, normal | 30 | 29 | 30 | 28 |
| barley, high amylose | 53 | 45 | 44 | i.s.[1] |
| wheat | 25 | i.s.[1] | 27 | 28 |

[1] i.s. = insufficient sample.

*Table* 6.8. Properties of the amylose fractions from the starches of normal barley, as growth progresses (Banks, Greenwood and Muir 1973a).

| maturity[1] (days) | [β] (%) | [β+Z] (%) | IBC (%) | [η][2] (ml/g) |
|---|---|---|---|---|
| 14 | 100 | 101 | 19.4 | 215 |
| 20 | 100 | 99 | 19.5 | 250 |
| 27 | 97 | 100 | 19.6 | 270 |
| 32 | 91 | 101 | 19.3 | 405 |
| 46 | 83 | 101 | 19.6 | 390 |

[1] Elapsed time from the average date of anthesis.
[2] Measured in 0.15 M KOH.

*Table* 6.9. Properties of the amylose fractions from the starches of high-amylose barley, as growth progresses (Banks, Greenwood and Muir 1973a).

| maturity (days) | [β] (%) | [β+Z] (%) | IBC (%) | [η][1] (ml/g) |
|---|---|---|---|---|
| 14 | 100 | 98 | 20.1 | n.d.[2] |
| 20 | 98 | 101 | 19.2 | 200 |
| 27 | 95 | 100 | 20.0 | 240 |
| 32 | 90 | 100 | 19.5 | n.d. |
| 46 | 92 | 99 | n.d. | 370 |

[1] Measured in 0.15 M KOH. [2] n.d. = not determined.

The molecular properties of the amylose and amylopectin fractions isolated from the two genotypes of barley starch have also been determined as a function of plant growth. Tables 6.8 and 6.9 contain the relevant information for the amylose components of these starches. All fractions are free from amylopectin, as shown by the high iodine-binding capacities and the complete conversion to maltose achieved by the concurrent action of *beta*-amylase and

*Z*-enzyme. In both genotypes the extent of conversion to maltose by the action of pure *beta*-amylase decreases as development proceeds, which indicates a growing accumulation of branched amylose. The limiting viscosity numbers of the amylose fractions again increase as the growing season progresses.

*Table* 6.10. Properties of the amylopectin fractions from the starches of normal barley, as growth progresses (Banks, Greenwood and Muir 1973a).

| maturity(1) (days) | $[\beta]$ (%) | $[\beta + Pu]^{(2)}$ (%) | $[\eta]^{(3)}$ (ml/g) | $\bar{M}_w \times 10^{-6}$ | IBC (%) | average length of unit chain |
|---|---|---|---|---|---|---|
| 14 | 57 | 102 | 150 | 50 | 0.69 | 20.2 |
| 20 | 58 | 101 | 140 | 100 | 0.99 | 21.5 |
| 27 | 56 | 99 | 150 | 250 | 1.10 | 20.2 |
| 32 | 56 | 100 | 145 | 250 | 1.06 | 20.7 |
| 46 | 57 | 101 | 140 | 250 | 0.42 | 20.8 |

[1] As in table 6.8.
[2] $[\beta + Pu]$ is the extent of conversion into maltose obtained by the concurrent action of *beta*-amylase and pullulanase.
[3] Measured in dimethylsulphoxide.

*Table* 6.11. Properties of the amylopectin fractions from the starches of high-amylose barley, as growth progresses (Banks, Greenwood and Muir 1973a)

| maturity(1) (days) | $[\beta]$ (%) | $[\beta + Pu]$ (%) | $\bar{M}_w \times 10^{-6}$ | IBC (%) | average length of unit chain |
|---|---|---|---|---|---|
| 20 | 62.0 | 99 | n.d. | n.d. | 24.7 |
| 27 | 59.5 | 100 | n.d. | 0.94 | 25.0 |
| 32 | 62.5 | 100 | 150 | 0.53 | 24.1 |
| 46 | 62.0 | 101 | 250 | 0.47 | 24.8 |

[1] As in table 6.8.

Tables 6.10 and 6.11 show the properties of the branched materials. Just as in the amylose fractions, the molecular weight of the amylopectin increases as starch deposition proceeds. In each series, the $\beta$-amylolysis limit and the chain length are constant throughout the period of growth. However, the values of these parameters are greater in the branched fractions isolated from the starches of high amylose content. Although the polysaccharide from normal barley binds more iodine than that from the high-amylose barley, the form of the binding curve in the former case suggested contamination with amylose of high molecular weight, whereas the curves obtained from the latter polysaccharide were abnormal in form. Hence the anomalous properties of this amylopectin from high-amylose barley could be due to it having either an abnormal external chain-length, or being a mixture of short-chain material and normal amylopectin.

The growth studies yielded insufficient polysaccharide to distinguish between these possibilities, but the structure of the amylopectin fraction of starches isolated from mature forms of both starches has been established (Banks, Greenwood and Walker 1971). It was found that the material sedimented from aqueous solutions of the two amylopectin fractions were identical with respect to iodine-binding capacity, $\beta$-amylolysis limit, and chain length. However, the supernatant from the high-amylose starch contained much more polysaccharide material than that from the normal genotype. Subsequent investigation of this material has shown it to be a mixture of linear and lightly-branched molecules, with a number-average chain length of approximately 100 anhydroglucose residues. Thus it is very similar in structure to the fractions isolated from the starches of high amylose content of both pea and amylomaize.

*Growth Studies on Amylomaize.* A growth study on amylomaize has also been reported (Mercier, Charbonnière, Gallant and Guilbot 1970). Measurement of iodine-binding capacity at 20°C showed an apparent increase in amylose content as growth progressed. However, the same measurement carried out at 2°C showed the highest amylose content to be associated with the youngest tissue. These results were interpreted as being due to the deposition of a large amount of anomalous material (see section 2.5.3) in the initial stages, and its subsequent dilution with normal amylose and amylopectin. Measurements on the isolated components were not reported.

*Growth Studies on Potato.* These detailed studies on the changes in properties of the components of pea and barley starches show that certain biosynthetic processes are common to both. However, both are seed starches, and it is therefore of some interest to consider the development of starch in some other storage organ. For this reason, we have included growth studies on the potato tuber. Whilst it is possible to relate the growth process of a seed to an absolute point in time (anthesis, for example), the same is not true of tuber starches, and it is necessary to employ a somewhat arbitrary standard, namely the size of the tuber. The gradation of tubers harvested at different times according to their size introduces the additional complication that the results may be biased by differences in growth rate rather than in age. However, extraction and examination of starch from tubers graded by this system showed the same general pattern as in the case of pea and barley starches, as may be seen from table 6.12. Thus as tuber size increases, so too do the starch content, average granule diameter, and the amylose content.

Molecular properties of the amylose and amylopectin fractions obtained from these potato starches are recorded in table 6.13. The values for the iodine-binding capacities and the susceptibility of the amylose to the

combined actions of *beta*-amylase and Z-enzyme again show that the increases in the iodine uptake by starch as growth proceeds reflect a real increase in amylose content, and not a change in the structure of this polysaccharide. Susceptibility of the amylose to β-amylolysis decreases during development, and, apart from the sample from the smallest tubers, the limiting viscosity number shows a steady increase. The amylopectin β-amylolysis limit shows a slight decrease as growth proceeds, but the molecular weight of this component again increases markedly.

*Table* 6.12. Properties of potato starch granules as a function of tuber size (Geddes, Greenwood and MacKenzie 1965).

| sample | tuber size (cm) | starch content (% wet weight) | average granule diameter (μm) | amylose content (%) |
|---|---|---|---|---|
| 1 | 0–1 | 5.0 | 10 | 12.5 |
| 2 | 2–3 | 6.4 | 20 | 13.9 |
| 3 | 4–5 | 9.2 | 25 | 16.4 |
| 4 | 5–6 | 11.0 | 28 | 17.2 |
| 5 | 6–7 | 13.4 | 32 | 18.5 |
| 6 | 8–9 | 17.5 | 43 | 19.0 |
| 7 | 10–11 | 18.0 | 45 | 19.8 |
| 8 | 15 | 18.5 | 48 | 20.0 |

*Table* 6.13. Properties of the amylose and amylopectin components of potato starches extracted from tubers of different sizes (Geddes, Greenwood and MacKenzie 1965).

| sample[1] | amylose | | | | amylopectin | |
|---|---|---|---|---|---|---|
| | [β] (%) | [β+Z] (%) | IBC (%) | [η][2] (ml/g) | [β] (%) | $\bar{M}_w \times 10^{-6}$ |
| 1 | 92 | 101 | 19.5 | 305 | 56 | 9 |
| 2 | 90 | 99 | 19.5 | 185 | 55 | 14 |
| 3 | 88 | 101 | 19.5 | 275 | 54 | 24 |
| 4 | 87 | 101 | 19.5 | 345 | 53 | 30 |
| 5 | 85 | 101 | 19.5 | 415 | 52 | 35 |
| 6 | 84 | 100 | 19.5 | 450 | 51 | 35 |
| 7 | 83 | 101 | 19.5 | 505 | 52 | 76 |
| 8 | 81 | 99 | 19.5 | 530 | 52 | 120 |

[1] As in table 6.12. [2] Measured in 0.5 M KOH.

The granules of potato starch show no change in morphology during development, but merely increase in average size. At any stage in the growth cycle there is a wide distribution of granule sizes, and these may be fractionated into a series of narrower size ranges. The starch granules of sample 4 (i.e. the starch isolated from tubers having a diameter of 5–6 cm) were fractionated by differential sedimentation in water; the properties of the resultant granular starches and their components are shown in table 6.14.

As the granule size increases, the proportion of amylose also increases. Similarly, the properties of the amylose and amylopectin components of these sub-fractions of starch granules vary in the same way as they did during development, and, for both polysaccharides, the $\beta$-amylolysis limit decreases and the molecular weight increases. Moreover, if the parameters recorded in table 6.14 for the starch sub-fractions are compared with those obtained for granules of a similar diameter (see tables 6.12 and 6.13), a close similarity in their values is observed.

*Table* 6.14. Fractionation of potato starch according to granule size: properties of the granular starch sub-fractions and of their amylose and amylopectin components (Geddes, Greenwood and MacKenzie 1965).

| sample[1] | starch | | amylose | | | amylopectin | |
|---|---|---|---|---|---|---|---|
| | average granule diameter ($\mu$m) | amylose content (%) | $[\beta]$ (%) | $[\beta+Z]$ (%) | $[\eta]^{(2)}$ (ml/g) | $[\beta]$ (%) | $\bar{M}_w \times 10^{-6}$ |
| 4.1 | 37 | 19.5 | 85 | 99 | 400 | 52 | 40 |
| 4.2 | 28 | 18.0 | 86 | 101 | 335 | 55 | 32 |
| 4.3 | 16 | 16.7 | 92 | 100 | 275 | 54 | 22 |
| 4.4 | 10 | 16.0 | 92 | 100 | 210 | 57 | 15 |
| 4.5 | 7 | 14.4 | 91 | 101 | 310 | 59 | 10 |

[1] Samples were obtained by the sub-fractionation of the starch of sample 4 of table 6.12.
[2] Measured in 0.5 M KOH.

It can be concluded, therefore, that in potato starch the properties of the granule and of its components are directly related to the size of the granule. The additional fact that the proportion of amylose varies during growth implies either (a) that once a granule is laid down during growth no further increase in size occurs, so that the small granules found in the starch from a large tuber are those formed early in growth, or (b) that the properties of the granule alter in a highly specialized manner as its diameter increases, so that the biosynthetic process is controlled by a feedback mechanism related to granule diameter.

The former possibility might appear much more likely. However, Badenhuizen and Dutton (1956) showed that, after exposure to $^{14}CO_2$, autoradiography of the potato tuber starch indicated that there was no correlation between the degree of radioactivity and the granule size. The activity of individual tubers varied over a wide range, some being inactive whilst others were highly active. This work showed that in the main the activity was concentrated at the periphery of the granules, constituting proof that granules grew by apposition rather than by intersusception; but it is not clear from the description given by Badenhuizen and Dutton (1956) how long the plants were exposed to the $^{14}CO_2$. Because the time was probably fairly short, as the tubers were from a very narrow size range (0.4–3.4 cm; i.e. comparable only with samples 1 and 2 of our work, see table 6.12), this study cannot be used to distinguish between the two possibilities that we have offered to account for the relation between granule properties and size found in potato starch.

*Summary.* We would draw attention to what we believe to be the salient points in the development of granular starch. There can be little doubt that in the majority of cases the proportion of amylose increases as growth proceeds, and the molecular size of both the amylose and amylopectin components increases in this period. Also, the susceptibility of the amylose fraction to the action of *beta*-amylase decreases, showing that the extent of long-chain branching in amylose increases as the plant develops. However, the manner in which these changes influence the fine structure of the granule remains obscure. For example, the starches of normal amylose content isolated from pea, barley, and potato all exhibit the above variations during development. Nevertheless, the granules are so constructed that, during growth, the gelatinization temperature of pea starch increases, that of potato starch decreases, and that of barley starch remains constant!

Whilst most starches show certain common features in their development, there are biosynthetic processes which do not appear to be widespread. Thus the form of the granule may change, as in wrinkled-seeded pea starch, or remain the same (but increasing in size) as happens in potato starch. In the latter starch the amylose content apparently increases with granule size, whilst in the mature wheat and barley starch it is independent of granule size. But, despite this difference, all three starches exhibit an increase in amylose content throughout their development.

We have suggested that in the case of potato starch the small granules present in a mixed population obtained from a large tuber are those laid down at a relatively early stage of development. In the case of the cereal starches, we would speculate that each granule is nucleated at approximately the same time, and then grows throughout the development period; the necessary corollary is that within the cereal starch granule there is an increase in the proportion of amylose as the periphery is approached. Thus, the development of starch granules in the potato may represent a 'single granule' mechanism (once nucleated, granules grow to their limiting sizes relatively rapidly and before others are nucleated), whereas a 'multi-granule' mechanism might be the preferred mode in some cereals (all granules are nucleated at much the same time, and continue to grow throughout the period of starch deposition).

Although these growth studies have not contributed extensively to our understanding of the structure of the starch granule, they have demonstrated that granular properties change during the development of the plant, and that some at least of these changes follow a set pattern. The conclusion to be drawn must necessarily be that the enzymes involved in the biosynthetic process also exhibit regular changes in their activities. It is also of some interest that when genetic manipulation allows an increase in the amylose content of starch, certain other characteristic changes are noted. For example, the starch contents of pea, barley, and maize decrease on going from the normal genotypes to the mutants, giving granules abnormally rich in amylose; on the other hand, the mutant recessive waxy varieties of barley, maize, and sorghum possess the same proportion of starch as do the corresponding dominant genotypes. Thus it would appear that partial suppression of the enzyme system producing amylopectin leads to an overall loss in the amount of storage polysaccharide, but complete suppression of the enzyme system responsible for amylose production can be achieved without any change in the total amount of starch laid down. Furthermore, in each of the three cultivars (pea, barley, and maize), the expression of the gene giving an enhanced amylose content results in the accumulation of a polysaccharide of comparatively low degree of polymerization, i.e. a mixture of linear and lightly-branched molecules. The appearance of this anomalous polysaccharide in all three cases might suggest that a common denominator has been introduced into the biosynthetic pathway. Identification of the enzyme systems so modified would help to elucidate the biosynthetic mechanism by which starch is formed and deposited.

**6.3.2.** Starch-Synthesizing Enzymes.

The synthesis of amylose and amylopectin may require three types of enzyme systems, i.e. those necessary to (a) form a primer, (b) add glucose units by α-1:4-bonds to that primer, and (c) introduce α-1:6-branch points.

*Formation of Primer*. Elongation of the amylosic chain and the introduction of branch points has been considered to require a priming molecule, although this concept is currently the subject of some debate. Disproportionating enzyme (*D*-enzyme), isolated first from potato (Peat, Whelan and Rees 1953) and subsequently from sweet corn (Lee, Marshall and Whelan 1971), has been claimed to perform this function. It catalyses a reaction of the type

$$2 \text{ maltotriose} \rightleftharpoons \text{maltopentaose} + \text{glucose}$$

or

$$2 \text{ maltotetraose} \rightleftharpoons \text{maltoheptaose} + \text{glucose}.$$

By removal of the glucose the equilibrium can be displaced, and linear chains containing approximately 50 glucose units can be formed (Walker and Whelan 1959). Whelan (1971) has suggested that the necessary substrate oligomers could be formed by hydrolytic carbohydrases assuming a synthetic rôle. Thus *alpha*-glucosidase acting on glucose could form maltose, which in turn could be condensed by *alpha*-amylase to give maltotetraose. This hypothesis is based on the observation of Hehre, Okada and Genghof (1969) that *alpha*- and *beta*-amylase will rapidly dimerize the same anomeric forms of maltose that they liberate from starch (see section 5.5). However, such condensation reactions demand high concentrations of sugars of the appropriate anomeric form, and whether or not such conditions could be achieved *in vivo* is debatable.

*D*-enzyme appears redundant in the synthesis of primer, for maltotriose, the smallest substrate for the disproportionating reaction, is itself capable of functioning as a primer for plant phosphorylase (albeit somewhat inefficiently), and the synthetase system can use maltose as a primer, at least in the case of glycogen formation (Goldemberg 1962; Leloir 1964). Whelan (1958) suggested that *D*-enzyme might also be involved in the breakdown of starch because phosphorylase cannot hydrolyse a substrate having four or fewer glucose units, and therefore some 20 per cent of the amylopectin molecule would not be available to the combined action of phosphorylase and *R*-enzyme; the disproportionating reaction would ensure that the polysaccharide was degraded virtually completely to glucose and glucose-1-phosphate (one molecule of maltotetraose would be the only other product). This system is, in fact, envisaged as being analogous to the degradation of glycogen, but in the latter case the transferase and the de-branching enzyme are found in a single enzyme molecule (see section 5.4). This hypothesis assumes that phosphorylase plays an important rôle in the *in vivo* degradation of the starch granule, an assumption that appears to have little foundation.

*Formation of* α-1:4-*Bonds*. For some twenty years after its discovery phosphorylase was thought to be the only enzyme capable of elongating amylosic chains in both plants and mammals. The enzymes from these two sources differ in several respects—the mammalian enzyme is much more active when glycogen is the substrate than is the plant enzyme, and the two differ in molecular structure (Fukui and Kamogawa 1969). Thus phosphorylase from animal tissue exists in two forms, phosphorylase *a* (a tetramer) and phosphorylase *b* (a dimer). Conversion of phosphorylase *b* to *a* occurs by phosphorylation of the protein by reaction with ATP, mediated by phosphorylase kinase; the reverse conversion is mediated by a specific phosphatase. The *b*-form is active only in the presence of adenosine 3′:5′-cyclic monophosphate (i.e. cyclic AMP), whereas the *a*-form is active in the absence of this nucleotide. In phosphorylase of plant origin, on the other hand, the *a*- and *b*-forms are unknown, and the enzyme is active in the absence of cyclic AMP.

Phosphorylase catalyses reactions of the type

$$\text{G-1-P} + \text{G}_n \rightleftharpoons \text{G}_{n+1} + \text{P}(\text{inorganic})$$

where $\text{G}_n$ is a primer molecule. The enzyme will, depending on the conditions, catalyse the reaction in either direction, and may therefore be classified as either a degradative or synthesizing enzyme. The position of the equilibrium is affected by the ratio of glucose-1-phosphate to inorganic phosphate, and this in turn varies with pH. For example, a rise in pH from 5 to 7 causes the equilibrium ratio of glucose-1-phosphate to inorganic phosphate to decrease from 10.8 to 3.1, i.e. the rise in pH favours the synthetic reaction. However, even under the best conditions the equilibrium ratio is such that the molar concentration of substrate glucose-1-phosphate is some three times greater than that of the product, inorganic phosphate, which suggests that *in vitro* phosphorylase is not a highly efficient synthesizing system. Of course, the *in vivo* reaction, occurring in the more complex environment of the cell, may be more efficient because of the simultaneous removal of inorganic phosphate from the system.

Plant phosphorylase has been claimed to be capable of the *de novo* synthesis of polysaccharide (Green and Stumpf 1942; Dyar 1950; Badenhuizen, Kreutzer and Zar 1957; Lee 1960), but the unprimed reaction has also been claimed for muscle phosphorylase (Illingworth, Brown and Cori 1961; Brown, Illingworth and Cori 1961). However, Kamogawa, Fukui and Nikuni (1968) showed that although crystalline potato phosphorylase was apparently capable of the *de novo* synthesis of polysaccharide, this capacity was removed by pre-incubation of the enzyme preparation with glucoamylase, indicating that the phosphorylase was itself contaminated with primer; when this material was removed the enzyme was incapable of synthesis. More recently, it has been suggested that one or more of the phosphorylase isoenzymes found in maize endosperm (Tsai and Nelson 1969a) and in potato tubers (Slabnik and Frydman 1970; Frydman and Slabnik 1973) is capable of the *de novo* synthesis of polysaccharide. Similarly, one of the two isozymic forms found in the blue-green and the green algae (Frederick 1964) is thought to function without a primer; it is a glycoprotein, the glucosyl moiety of which can be removed only by treatment with *alpha*-amylase (Frederick 1971). The concept that the enzyme molecule should contain its own primer is interesting, but difficult to distinguish from the alternative possibility that a small amount of primer is present in the phosphorylase preparation or in the glucose-1-phosphate merely as a physical contaminant (Abdullah, Fischer, Qureshi, Slessor and Whelan 1965).

The discovery of an alternative pathway to the synthesis of α-1:4-bonds, involving sugar nucleotides as the glucosyl donors, is of more recent origin (De Fekete, Leloir and Cardini 1960; Leloir, De Fekete and Cardini 1961). In contrast to the phosphorylase reaction, the action of the synthesizing enzyme in this case (starch synthetase) is irreversible. The enzyme was found to be bound to the starch granule, and to catalyse the reaction

$$\text{UDPG} + G_n \rightarrow G_{n+1} + \text{UDP} + P_i$$

where UDPG is uridine-5′-(α-D-glucopyranosyl pyrophosphate), and $P_i$ is pyrophosphate. Subsequent studies on the bound synthetase showed, in fact, that adenosine-5′-(α-D-glucopyranosyl pyrophosphate), i.e. ADPG, is a much more efficient glucosyl donor than was the uridine derivative (Recondo and Leloir 1961). Similar conclusions were reached for the developing rice grain, where the bound synthetase could use either ADPG or UDPG but preferred the former nucleotide sugar (Murata, Sugiyama and Akazawa 1964; Akazawa, Minamikawa and Murata 1964). In all the above studies, the granules had been isolated from plant organs in which the starch constituted a reserve polysaccharide; studies on leaf starches, which have but a transitory existence, showed that the bound enzyme could use ADPG but not UDPG as the glucosyl donor (Murata and Akazawa 1964). The starch in the soya bean, which might be regarded as a source of reserve polysaccharide, is apparently anomalous as it contains synthetase incapable of utilizing UDPG, and thus is rather more akin to the leaf starches (Frydman and Cardini 1967). However, the starch in the soya bean is itself anomalous, being laid down not on the development of the seed but rather on its germination (Yin and Sun 1947).

The starch granules of waxy maize were reported to be deficient in the bound synthetase (Nelson and Rines 1962), an observation subsequently confirmed by Frydman (1963). Similarly, the glutinous variety of rice was found to contain very little granule-bound synthetase activity (Murata, Sugiyama and Akazawa 1965). In fact, the enzyme system utilizing ADPG in these waxy varieties occurs not as a granule-bound form, but rather as a soluble form. This soluble form of starch synthetase was first reported in extracts of sweet corn (Frydman and Cardini 1964a), and then in the extracts of tobacco leaves and potato tubers (Frydman and Cardini 1964b). It was later obtained from the chloroplasts of spinach (Doi, Doi and Nikuni 1964; Ghosh and Preiss 1965a, b), and from extracts of glutinous rice grains (Murata, Sugiyama and Akazawa 1965). In all cases, the soluble form will incorporate into a suitable primer the glucosyl moiety only of ADPG, as it cannot make use of UDPG as a glucosyl donor.

The relation between the bound and soluble forms of starch synthetase is still not entirely clear. Chandorkar and Badenhuizen (1966) showed that grinding potato starch granules led to an increase in the synthetase activity capable of utilizing ADPG, with a concomitant decrease in that using UDPG. These observations were extended by Frydman and Cardini (1967) using a number of starches, and destroying their structures by grinding, α-amylolysis, and exposure to concentrated solutions (7 and 8 M) of urea. This work confirmed that any attempt to solubilize the granule-bound starch synthetase led to an increase in the activity of the enzyme incorporating in a primer the glucosyl moiety from ADPG, but a disappearance of that utilizing UDPG. The fact that, in the higher plants, the starch granules of the glutinous varieties of maize and rice are deficient in granule-bound synthetase activity is generally taken to imply that this enzyme must be associated with the amylose fraction (Akazawa and Murata 1965; Murata and Akazawa 1966; Frydman, De Souza and Cardini 1966). Further proof is provided by the ease with which amylose will bind the soluble form of starch synthetase (Akazawa and Murata 1965). However, the bound form in this case still retains its absolute specificity towards ADPG, which leaves the granule-bound synthetase capable of utilizing UDPG as a glucosyl donor in a somewhat equivocal position—is the enzyme quite distinct from that making use of ADPG, or has the binding within the starch granule physically modified the synthetase utilizing ADPG so that it can also accept UDPG as a donor substrate? Frydman and Cardini (1967) concluded that the differences in specificity between the synthetase bound to the granule and the soluble synthetase might be due to changes in the conformation of the same enzyme. We believe that their conclusion is still valid, and the fact that when the soluble form is bound by amylose it retains its specificity for ADPG does not preclude the possibility that the synthetase bound to amylose in the starch granule, with the restrictions necessarily imposed by the structure of that granule, could have its structure modified in such a way as to accept UDPG as a donor substrate.

Although concerned here mainly with the biosynthesis of starch, we note that the glucosyl nucleotides are also involved in the synthesis of α-1:4-linkages in both mammalian tissue and in bacteria. Glycogen synthesis in the former source utilizes UDPG as the glucosyl donor (Leloir and Cardini 1957; Leloir, Olavarria, Goldemberg and Carminatti 1959); bacteria, on the other hand, utilize only ADPG of the sugar nucleosides in the synthesis of glycogen (Greenberg and Preiss 1964, 1965; Preiss and Greenberg 1965; Preiss, Ozbun, Hawker, Greenberg and Lammel 1973).

We have emphasized the fact that the absence of granule-bound synthetase is correlated with the absence of amylose only in the higher plants. The red alga, *Serraticordic maxima*, which produces Floridean starch—an amylopectin-type polysaccharide (Greenwood and Thomson 1961b)—contains the bound synthetase in forms capable of utilizing both ADPG and UDPG as glucosyl donors (Nagashima, Nakamura, Nisizawa and Hori 1971). In the case of this particular red alga, therefore, the starch synthetase must perforce be bound by the amylopectin.

It must be emphasized that the binding of starch synthetase by non-glutinous starch granules is very strong indeed. For example, the enzyme could not be liberated by treating the granules of rice starch with reagents such as 0.1M Tris buffer, 4.0M urea, or 6.0M urea (Tanaka, Minagawa and Akazawa 1967); only when the treatment was sufficient to disrupt the crystalline structure of the granule did a sharp decline in synthetase activity occur.

The earliest work showed that starch synthetase was incapable of synthesizing polysaccharide *de novo* but required a primer, although even maltose could serve this purpose (Frydman 1963). Just as in the case of phosphorylase, however, an increasing number of claims are being made that one of the known isozyme forms of starch synthetase is capable of the *de novo* production of polysaccharide. Thus Gahan and Conrad (1968) described a synthetase present in extracts of *Aerobacter aerogenes* that synthesized glycogen without any primer requirement. These authors also suggested that the *de novo* synthesis differed from the normal process by proceeding from the reducing end of the growing polysaccharide molecule. Subsequently, one isozyme form of the synthetase from each of the following sources has been found to function in the absence of added primer: spinach (Ozbun, Hawker and Preiss 1971a; Ozbun, Hawker and Preiss 1972), maize (Ozbun, Hawker and Preiss 1971b), potato tubers (Hawker, Ozbun and Preiss 1972), *Escherichia coli* (Fox, Kennedy, Hawker, Ozbun, Greenberg, Lammel and Preiss 1973), and rat liver (Krisman 1972, 1973). The unprimed, but not the primer-dependent, reaction has been found in many instances to be stimulated by the addition of high concentrations of certain salts (e.g. citrate or EDTA) and protein. Also, the *de novo* synthesis apparently survives pre-treatment of both enzyme and substrates with glucoamylase in most cases, suggesting either that the synthetase is truly independent of primer, or that the enzyme molecule itself possessed a portion that could function as a primer but was unavailable to hydrolysis by glucoamylase. However, Schiefer (1973) found that treatment of the starch synthetase from sweet corn with glucoamylase and *alpha*-amylase abolished the primer-independent reaction, even in the presence of high concentrations of salts. More recently, on the other hand, Preiss and his co-workers (Hawker, Ozbun, Ozaki, Greenberg and Preiss 1974) have re-affirmed their belief in *de novo* synthesis by the synthetase of spinach leaf, although they suggest that the unprimed reaction may be induced by the presence in the enzyme of a small amount of endogenous glucan which is resistant to hydrolysis by both glucoamylase and *alpha*-amylase, but nevertheless constitutes a built-in primer. As in the case of the corresponding primer-independent phosphorylase isozymes, it is still not possible to reach an unambiguous conclusion concerning the biological significance of these *de novo* syntheses.

*Formation of* $\alpha$-1:6-*Bonds*. The enzymes considered so far are capable of producing only a linear chain of $\alpha$-1:4-linked glucosyl residues, i.e. amylose. To produce amylopectin or glycogen it is necessary to have a branching enzyme capable of introducing $\alpha$-1:6-linkages into the linear glucose chain. Such an enzyme (*Q*-enzyme) was first identified in the potato (Haworth, Peat and Bourne 1944; Barker, Bourne and Peat 1949a, b), and later in the broad bean (Hobson, Whelan and Peat 1950). Early studies also confirmed that this enzyme was not restricted to the higher plants, but was also found in lower organisms such as the flagellate *Polytomella cocca* (Bebbington, Bourne, Stacey and Wilkinson 1952). As originally defined, *Q*-enzyme catalysed the conversion of amylose to an amylopectin-type molecule but could not cause further branching to form glycogen. However, a second type of branching enzyme was isolated from yeast (Gunja, Manners and Maung 1960) that was active with amylopectin as substrate, producing a polysaccharide very similar to glycogen in its limiting viscosity number, chain length, and reaction with iodine–potassium iodide (Gunja and Manners 1959). Thus these early studies suggested that two quite distinct branching enzyme systems existed, one capable of converting amylose or amylopectin to a glycogen-like molecule and the other catalysing the conversion of amylose to an amylopectin-type structure, but unable to further branch amylopectin.

Later work has tended to support these conclusions. Lavintman (1966) found that her branching enzyme preparation from sweet corn appeared to have two distinct activities, corresponding to those discussed above. Subsequent separation of the two activities was reported (Manners, Rowe and Rowe 1968). Frederick (1971) reported that the branching activities isolated from *Oscillatoria princeps* and from *Spirogyra setiformis* were quite different, that from the former source producing a glycogen-like polysaccharide from amylose, whereas that from the latter could synthesize only amylopectin. The enzymes producing these quite distinct polymers were found to be charge isomers, differing merely by an amino acid residue. The evolutionary pattern for the development of the different forms of branching enzyme has recently been analysed by Frederick (1973). He points out that the isozyme patterns in the *Thallophyte* are correlated with the type of storage glucan formed, the most elementary forms producing a highly branched glycogen-type polysaccharide, progressing to an amylopectin-type polyglucan (Floridean starch), and finally to a mixture of amylose and amylopectin as the evolutionary ladder is ascended.

It is of some interest that the capacity to produce a glycogen-like polymer is retained in the higher plants. Thus sweet corn, in addition to the normal reserve starch, synthesizes large quantities of a glycogen-type molecule (phytoglycogen), and therefore possesses the capacity to produce two quite distinct polysaccharide types—an observation in full agreement with the conclusions of Lavintman (1966) regarding the specificities of the branching enzymes present in sweet corn. More singular, however, is the fact that the enzyme branching-activity known to produce phytoglycogen in sweet corn is also found in two other maize genotypes, in neither of which phytoglycogen has been detected (Black, Loerch, McArdle and Creech 1966). Hodges, Creech and Loerch (1969) concluded that there were indeed two branching enzymes in sweet corn, one producing phytoglycogen and the other amylopectin, but both having amylose as their substrate, i.e. the enzyme giving the branching pattern characteristic of phytoglycogen was claimed to have no action on amylopectin.

However, Drummond, Smith and Whelan (1972), in a detailed study of the purification and properties of potato *Q*-enzyme, have altered the classical definition of the enzyme activity by showing that it can further branch amylopectin. The introduction of these new branch points leads to the formation of side chains containing six glucose residues and an increase in the extent of hydrolysis of the polysaccharide by pullulanase, but the wavelength of maximum absorbance of the iodine complex decreases only comparatively slightly (from 530 nm to 510 nm), and the change in the extent of conversion on $\beta$-amylolysis is similarly small (52 per cent for the parent amylopectin and 50 per cent after it had been treated with *Q*-enzyme). Thus the product of the action of this branching enzyme on amylopectin is closer in structure to the parent polysaccharide than it is to glycogen. This amended concept is somewhat similar to that advanced by Larner (1953), who claimed that *Q*-enzyme had the capacity to introduce branch points in the outer chains of amylopectin having chain lengths of at least 14 glucose units.

The branching enzyme acts as a transglycosylase, transferring a portion of an amylose chain to a suitable acceptor molecule. Whelan (1971) has indicated that inter-chain transfer occurs, i.e. when a portion of the amylose chain is severed it is transferred to another molecule rather than being attached to the residual portion of the same chain. Early work (Nussenbaum and Hassid 1952; Peat, Whelan and Bailey 1953) suggested that a minimum chain-length of 40 glucose residues was required before *Q*-enzyme could exert its activity. It has recently been suggested (Drummond, Smith and Whelan 1972) that the minimum segment length of a donor chain remaining, after *Q*-enzyme acting on amylopectin has transferred the outer section of the chain, is six glucose units. This calculation is based on two untested assumptions, namely that the ability of the enzyme to split any bond may depend on the proximity of the branch point, and that the length of the donor chain is probably shorter than the length required for rapid branching. Thus the need to remove as large a segment as possible, whilst not approaching too close to the branch point, is postulated to account for the significant proportion of maltohexaosyl side-chains found in amylopectin exposed to *Q*-enzyme.

Some doubt exists regarding the mode of action of *Q*-enzyme. Thus Griffin and Wu (1968, 1971) claim that the activities responsible for the scission of the amylose chain and the subsequent synthesis of an $\alpha$-1:6-branch point are not found in the same enzyme molecule, but rather are the result of two distinct (and separable) enzymes. They have produced evidence to the effect that the two activities (isolated from maize, waxy maize, and potato tubers) can be separated on DEAE-cellulose. One, with a molecular weight of approximately 70000, hydrolyses amylose but its action on amylopectin depends on its source, for the action of the potato enzyme is restricted to the external chains whereas those of waxy maize and maize can bypass branch points; the other component (molecular weight approximately 20000) introduces barriers to the action of *beta*-amylase into amylose. Drummond, Smith and Whelan (1972), on the other hand, deny that hydrolysis occurs, i.e. the enzyme at no stage uses water as an acceptor molecule. These two views of the mode of action of *Q*-enzyme are quite incompatible; the decision as to which is correct must await more detailed investigation.

We have dealt here very briefly with the enzymes thought to be directly involved in starch synthesis. The next consideration is the way in which the enzymes must be manipulated in order to yield both amylopectin and amylose.

**6.3.3.** Models of Starch Biosynthesis.

The amyloplast has the ability to simultaneously synthesize polysaccharides which differ very greatly in their degree of branching, but attempts to reproduce this phenomenon *in vitro* have been totally unsuccessful. Again, the relative importance of the nucleotide and phosphorylase pathways in the *in vivo* synthesis of $\alpha$-1:4-linkages is a matter of debate, but there is no doubt that much of the starch-synthesizing tissue of higher plants contains sufficient phosphorylase and synthetase so that either enzyme could account for all the $\alpha$-1:4-linkages synthesized.

It must be stressed that studies involving tissue *homogenates* can give a totally misleading picture of events at the cell level. For example, the developing barley kernel has been shown to contain high *alpha*-amylase activity, which increases to a maximum some 10–15 days after anthesis and then falls to very low levels as the kernel develops (Duffus 1969; MacGregor, La Berge and Meredith 1971; La Berge, MacGregor and Meredith 1971). Thus at a period when starch is being deposited in the kernel there is an apparent sharp increase in the level of *alpha*-amylase, an enzyme associated with degradation rather than synthesis. This anomaly was explained when it was found that in both barley (MacGregor, Gordon, Meredith and Lacroix 1972) and wheat (Banks, Evers and Muir 1972) the *alpha*-amylase activity was restricted to the testa–pericarp fraction of the kernel, whereas starch deposition was taking place in the endosperm. Thus homogenates of the developing kernel gave the misleading impression that *alpha*-amylase could be associated with the deposition of starch.

*The Synthetic Role of Phosphorylase.* Views that the synthetases are responsible for the formation of the $\alpha$-1:4-bonds in starch tend to categorize phosphorylase solely as a degradative enzyme. However, only one enzyme of plant origin, *alpha*-amylase, has been shown to hydrolyse intact starch granules, and during the germination of cereals, for example, this enzyme is produced in very large amounts. There is no evidence that, under these circumstances, phosphorylase plays a significant rôle in the degradative process. Indeed, Kiribuchi and Nakamura (1973) have recently shown that the endosperm of germinating barley is virtually devoid of phosphorylase whereas the developing endosperm is rich in this enzyme, an observation which supports the hypothesis that phosphorylase acts as a synthesizing enzyme.

Because *alpha*-amylase is inefficient in the heterogeneous reaction involved in granule hydrolysis and consequently is produced in large amounts, we doubt very much whether any potential substrate for the degradative action of phosphorylase could have more than a transient expression. Furthermore, in barley large amounts of *beta*-amylase are released on germination, and hence phosphorylase must find itself in competition with both amylases for potential substrate; a competition for which it is particularly ill-suited, bearing in mind that it is inhibited by the presence of *beta*-amylase (Porter 1950).

Of course, *beta*-amylase might itself be involved in the *in vivo* degradation of starch, for this enzyme is present in large quantities in the dormant soya bean, which contains no reserve starch. The same observation, however, applies equally to modern sugar-cane hybrids (predominantly *Saccharum officinarum*), which possess both phosphorylase and *beta*-amylase activities but do not lay down starch as a reserve carbohydrate, preferring sucrose for this purpose (Alexander 1973). Indeed, Alexander (1973) has shown that sugar-cane also contains both *Q*-enzyme and starch synthetase, so that the lack of starch cannot be due to the absence of enzymes capable of synthesizing starch. In all tissue in which phosphorylase was found to be present there was also an excess of *beta*-amylase, and it is probable that the presence of this latter enzyme prevents starch synthesis, and the plant therefore utilizes sucrose as a reserve carbohydrate. By analogy, the presence of *beta*-amylase in soya bean may be indicative not of the enzyme being superfluous, but rather as the reason for the soya bean being deficient in starch.

In our opinion phosphorylase acts in a synthetic rôle *in vivo*, and breakdown of starch occurs as a result of the action of *alpha*-amylase supported by *R*-enzyme, *beta*-amylase when present, and *alpha*-glucosidase.

*The Synthetic Rôle of Synthetase.* It is not known what proportion of α-1:4-linkages are produced by the synthetase relative to the action of phosphorylase. Many plants have the necessary supporting enzyme systems to enable the synthetase to function effectively. For example, the identification of the enzyme ADPG-pyrophosphorylase (Espada 1962; Ghosh and Preiss 1966; Tanaka, Minagawa and Akazawa 1967) which catalyses the reaction

$$\text{glucose-1-phosphate} + \text{ATP} \rightleftharpoons \text{ADPG} + \text{P}_i$$

explains the formation of ADPG in the plant; ADPG is then converted to starch by the action of starch synthetase and branching enzyme. In fact, two theoretical schemes have been given for the conversion of sucrose to starch. The first, evolved from studies on sweet corn (De Fekete and Cardini 1964; Frydman and Cardini 1964a), considers that UDPG is the natural substrate for the synthesis (and degradation) of sucrose whereas ADPG is the precursor of starch, so that the conversion of sucrose to starch takes the form

$$\text{sucrose} \xrightarrow[\text{sucrose synthetase}]{\text{UDP}} \text{UDPG} \xrightarrow[\text{UDPG-pyrophosphorylase}]{\text{P}} \text{G-1-P}$$

$$\text{G-1-P} \xrightarrow[\text{ADPG-pyrophosphorylase}]{\text{ATP}} \text{ADPG} \xrightarrow[\text{starch synthetase}]{} \text{starch}$$

The second theory, based on studies of the developing rice grain (Murata, Sugiyama and Akazawa 1964; Murata, Sugiyama, Minamikawa and Akazawa 1966), envisages a rather more direct conversion of the form

sucrose→ADPG (or UDPG)→starch

The first stage of the reaction involves ADP, or UDP, and sucrose synthetase, and the second stage is mediated by starch synthetase.

The fact that two quite distinct schemes exist for the sucrose-to-starch conversion within the *Gramineae* implies that the reactions involved are not really understood. Indeed, Okada and Hehre (1974) have recently pointed out that although both sets of authors claimed to reproduce the conversion process *in vitro* using [$^{14}$C]-sucrose, the actual amount of [$^{14}$C]-glucose incorporated in the polysaccharide was miniscule on both occasions. Okada and Hehre (1974) rightly emphasize that these studies (De Fekete and Cardini 1964; Frydman and Cardini 1964b; Murata, Sugiyama, Minikawa and Akazawa 1966) provide no evidence for the significant incorporation of the glucose moiety of sucrose into starch, but rather are consistent with the addition of a few glucosyl units to rare primer end-groups. We would agree with Okada and Hehre (1974) that the conversion of sucrose to starch in plants has not been adequately defined.

It should be noted that precisely the same pathway for the conversion of sucrose to starch cannot be operative in all plants. ADPG-pyrophosphorylase is found in cereal extracts (Espada 1962; Tanaka, Minagawa and Akazawa 1967) and in spinach leaves (Ghosh and Preiss 1966), but it is apparently absent from the potato tuber. In the tuber, therefore, it is necessary to postulate that the synthesis of ADPG is mediated by the action of sucrose synthetase.

Although it has not proved possible to discover whether phosphorylase or starch synthetase provides the major pathway to the synthesis of reserve polysaccharide in plants, it has been asserted that the glucosyl-nucleotide route is preferred in bacteria to the exclusion of all others (Preiss, Shen, Greenberg and Gentner 1966). Okada and Hehre (1974) draw attention to this assertion when discussing amylosucrase, a bacterial α-D-glucosylase that directly converts sucrose to a glycogen-like polysaccharide according to the reaction

$$n\text{-sucrose} + \text{acceptor} \rightleftharpoons (G_n)\text{-acceptor} + n\text{-fructose}.$$

Synthesis by this scheme obviously removes the need for the sugar-nucleotide pathway in some bacteria, e.g. *Neisseria perflava*. In contrast to the *in vitro* systems using glucosyl nucleotides as donors, amylosucrase incorporated large amounts (99.6 per cent) of [$^{14}$C]-glucose in the product α-glucan.

Thus, there are three routes by which an amylopectin- or glycogen-type molecule may be synthesized: (a) phosphorylase and branching enzyme, (b) starch synthetase and branching enzyme, and (c) amylosucrase. Each of these reactions can be categorized as a transglycosylation, the glucosyl moiety of the donor (glucose-1-phosphate, ADPG or UDPG, or sucrose) being transferred to an acceptor (primer) by the mediation of the appropriate enzyme. However, both phosphorylase and synthetase require the intervention of a separate branching enzyme to produce glycogen or amylopectin, whereas amylosucrase does not. Indeed, according to Okada and Hehre (1974) their results are consistent with the concept that the branched α-glucan is produced solely by the transfer of single glucosyl units, the reaction being catalysed by amylosucrase.

Although an enzyme of the amylosucrase type has been identified only in a limited number of bacteria and not in plants, one can speculate as to whether or not the biosynthetic pathway for starch formation is so well characterized as to preclude the presence of an enzyme system making direct use of the sugar of translocation, sucrose.

*Studies on Mutants.* A comparison of mutants of a single species, the mutants being such as to produce a change in the relative proportion of the amount of starch deposited or in the composition of that starch (or in both factors simultaneously), should prove useful. From this aspect, the large number of known mutant forms of maize (*Zea mays*) present an ideal subject for study (Akatsuka and Nelson 1969). The following gene substitutions are recognized: *waxy* (*wx*); *sugary*-1 ($su_1$) and *sugary*-2 ($su_2$); *shrunken*-1 ($sh_1$), *shrunken*-2 ($sh_2$) and *shrunken*-4 ($sh_4$); *brittle*-1 ($bt_1$), *brittle*-2 ($bt_2$) and *brittle*-4 ($bt_4$); *dull* (*du*); and *amylose-extender* (*ae*).* With the exception of *wx*, the expression of all these genes results in a low starch content relative to the value for normal dent corn.

Comparatively little success has been achieved in correlating the changes in starch content and type with any specific deficiencies in the synthesizing enzymes. The discovery that the starch granules of the *waxy* mutant were deficient in granule-bound starch synthetase (Nelson and Rines 1962) was taken as proof of the dominant rôle of phosphorylase in synthesis (Badenhuizen 1963); the subsequent isolation of the soluble form of the synthetase cast doubt on such a conclusion.

The $su_1$ mutant (i.e. sweet corn) is of especial interest for, although the starch content is very much lower than that of normal maize, phytoglycogen is also laid down as a reserve polysaccharide. This latter material has, as we have already noted, the chain length and solution properties of a glycogen-type macromolecule rather than those of an amylopectin (Peat, Whelan and Turvey 1956; Greenwood and Das Gupta 1958), although the iodine-binding properties of the phytoglycogen suggested that it was intermediate in structure between glycogen and amylopectin (Anderson and Greenwood 1955). It is interesting that the unit-chain profile obtained on de-branching phytoglycogen with *Cytophaga* isoamylase does not exhibit the single peak typical of

*The genes *ae* and *du* have also been characterized as *high-amylose*-1 ($ha_1$) and *high-amylose*-2 ($ha_2$) respectively; the present designation is due to Kramer, Bear and Zuber (1958).

glycogen but rather a 'bimodal' distribution (Gunja-Smith, Marshall, Mercier, Smith and Whelan 1970) more akin to that obtained from amylopectin (Mercier and Whelan 1970). Equally significant is the fact that the starch isolated from the $su_1$ genotype is identical to that of normal maize (Greenwood and Thomson 1961b).

The genetic interactions affecting the synthesis of phytoglycogen in maize have been dealt with in some detail (Black, Loerch, McArdle and Creech 1966). Various gene combinations, and the amounts of phytoglycogen they produce, are shown in table 6.15. It is noteworthy that although neither the *wx* nor *du* mutations alone produce any phytoglycogen, they possess the branching enzyme activity that allows the synthesis of that polysaccharide; in the combination *du wx* phytoglycogen is produced, however. Further, in the genotype *ae* $su_1$ there is complete suppression of the enzyme activity responsible for the branching of phytoglycogen and hence the polysaccharide is absent from this mutant, so that in this respect the gene *ae* is epistatic to $su_1$.

*Table* 6.15. The proportion of phytoglycogen in entire kernels of 15 maize genotypes at 24 days maturity related to the presence of phytoglycogen-branching enzyme (Black, Loerch, McArdle and Creech 1966).

| genotype (homozygous) | phytoglycogen (%) | phytoglycogen-branching enzyme[1] |
|---|---|---|
| normal | 0.0 | — |
| *ae* | 0.0 | — |
| *du* | 0.0 | + |
| $su_1$ | 24.9 | + |
| *wx* | 0.0 | + |
| *ae du* | 0.0 | n.d. |
| *ae* $su_1$ | 0.0 | — |
| *ae wx* | 0.0 | n.d. |
| *du* $su_1$ | 17.9 | + |
| *du wx* | 1.9 | + |
| $su_1$ *wx* | 19.0 | + |
| *ae du* $su_1$ | 6.6 | + |
| *ae du wx* | trace | + |
| *ae* $su_1$ *wx* | 7.3 | + |
| *du* $su_1$ *wx* | 27.9 | + |

[1] The symbol + signifies the presence of phytoglycogen-branching enzyme, — denotes its absence, and n.d. = not determined.

Phytoglycogen is, however, produced to varying extents in the triple mutant combinations involving *ae*, so that the suppressing effect of this gene is much less marked under these circumstances. The phytoglycogen isolated from each of the genotypes possessed comparable $\beta$-amylolysis limits and chain lengths, indicating a common synthetic pathway in the different genotypes (Black, Loerch, McArdle and Creech 1966).

The *ae*-mutant, which as we have noted earlier (section 2.5.3) contains little normal amylopectin but a high proportion of relatively short-chain material (degree of polymerization ~100 glucose residues), has not been shown to be deficient in any of the enzymes associated with the biosynthesis of starch. We suggest that there occurs in this mutant some interference in the action of *Q*-enzyme so that it continues to exert its ability to cleave a linear chain, but subsequently favours water rather than the primary hydroxyl group of a glucose residue as the acceptor molecule.

The effects of gene dosage at the *ae* locus on the amylose content of maize endosperm has been studied by Fergason, Helm and Zuber (1966). These authors found that each increase in the number of *ae* alleles above the nulliplex (*Ae Ae Ae*) increased the amylose content, although in the case of the simplex (*Ae Ae ae*) this increase was not statistically significant. However, for the duplex (*ae ae Ae*) the amylose content rose to 35 per cent, and for the triplex (*ae ae ae*) to 69 per cent.

The amylose contents of the starches from the genotypes *wx*, *du*, $su_1$, $su_2$ and *ae* were the subject of a study by Kramer, Pfahler and Whistler (1965). The amylose contents of all the double mutant combinations involving the *ae* allele were also measured, and the proportion of amylose found to be inversely related to that in the single mutants. From these results it was postulated that two quite distinct systems were operative for the synthesis of α-1:4-bonds, a view with which Creech (1965) concurred.

In their study of the mutant $sh_4$ Tsai and Nelson (1969b) reported that the level of phosphorylase was much lower than that of normal maize, as was the starch content. From this correlation, they suggested that phosphorylase was the enzyme responsible for the major portion of starch synthesis. However, it is now generally agreed that the mutation has resulted not in a specific decrease in phosphorylase activity but rather in a decrease in soluble protein, including starch synthetase (Lee and Braun 1973; Burr and Nelson 1973; Preiss, Ozbun, Hawker, Greenberg and Lammel 1973). It is not therefore possible to decide from a study of this particular mutant whether phosphorylase or starch synthetase plays the dominant rôle in the biosynthesis of starch.

The only examples in which there is an apparently unambiguous relation between a decrease in starch content and an abnormal enzyme level are found in the $sh_2$ and $bt_2$ mutants. In each case there is little or no ADPG-pyrophosphorylase activity (Tsai and Nelson 1966; Dickinson and Preiss 1969). The lack of this enzyme, allied with the low level of starch deposited in the endosperm, may indicate that starch synthetase is the major biosynthetic pathway for starch formation in maize.

In discussing the effect of different genes on starch formation in maize, it should be remembered that in addition to the gross changes in the content of starch, or the proportion of amylose in that starch, rather more subtle changes may occur in the structure of the granule as the result of the expression of a particular gene. For example, it has been shown that the single mutants *du*, *ae*, $su_2$, and mutant combinations *du ae* and *du* $su_2$, raise the amylose content of the starch relative to that of normal maize. However, on measuring the temperature at which birefringence in water is lost, it was found that $su_2$ and *du* $su_2$ lowered the value by approximately 14 deg. C, *du* and *du ae* had the same value (ca. 70°C) as did normal maize, and only *ae* increased the value by some 20 deg. C (Pfahler, Kramer and Whistler 1957). Such differences may be due merely to slight variations in the mode of deposition of the polysaccharide in the developing starch granule (i.e. a purely physical phenomenon), but they may be due to variations in the fine structures of the amylose and amylopectin components, such variations being enzymic in origin and therefore reflecting the path of starch biosynthesis.

To date, therefore, the study of maize mutants has proved somewhat disappointing, and only the lack of ADPG-pyrophosphorylase in the $sh_2$ and $bt_2$ mutants allied with their low starch contents can be considered as definitive evidence favouring one particular route (the glucosyl nucleotide pathway) for the biosynthesis of starch in maize.

*Studies of Single Strains.* A second approach to defining the relative roles of phosphorylase and starch synthesis has been adopted by Badenhuizen. Rather than employing mutants he grows a single strain at extremes of temperature, examines the starches formed in these conditions and relates their properties to the measured levels of synthesizing enzyme activity. This approach has been adopted with the alga *Polytoma uvella* and with higher plants such as the potato (Badenhuizen and Chandorkar 1965; Mangat and Badenhuizen 1971).

The amyloplast of *Polytoma uvella*, a heterotrophic unicellular alga, contains many starch granules (Lang 1963) possessing both amylose and amylopectin (Manners, Mercer, Stark and Ryley 1965). Relative to the higher plants, certain properties of both the starch and the starch synthetase are decidedly anomalous. For example, in the case of a shaken culture, although the starch content of the cells increases continually in the time of the experiment the amylose content passes through a maximum value (Mangat and Badenhuizen 1970). Also, whilst possessing both bound and soluble starch synthetase activity, the latter form of the enzyme was found to be able to utilize UDPG as a donor (McCracken and Badenhuizen 1970), in contradistinction to the soluble enzyme of higher plants which is specific for ADPG. Interestingly, both the soluble and granule-bound synthetases made more efficient use of ADPG than UDPG, the relative rate of utilization of ADPG to that of UDPG being approximately seven to eight in each case. Although it is possible that the soluble preparation contained an enzyme activity (and suitable substrate) to allow a conversion of UDPG to ADPG, the fact that the relative rates of incorporation of these two glycosyl nucleotides are the same for both soluble and granule-bound synthetase makes it likely that a soluble transferase capable of utilizing UDPG does exist in the plant kingdom.

Notwithstanding the somewhat anomalous situation of the starch and its synthesizing enzymes, *Polytoma uvella* has been employed to elucidate the biosynthetic pathway in starch (Mangat and Badenhuizen 1970). In the shaken culture grown at 22–25°C phosphorylase activity rose rapidly to a maximum and subsequently declined; the activity pattern of the soluble transferase followed virtually the same course. Concentrations of both ADPG and UDPG passed through a similar fairly broad maximum before declining rapidly, and at all stages the ratio UDPG:ADPG remained constant at a value of 3:1. *Q*-enzyme activity increased gradually in the same period to a constant value, so that the ratio of phosphorylase activity to that of *Q*-enzyme (i.e. the *P*/*Q* ratio) reached a maximum when amylose production reached a maximum. In view of the close similarity of the changes in both phosphorylase and synthetase activity, the significance of the coincidence of these maxima is uncertain and cannot be used to deduce the relative importance of the two enzymes in starch biosynthesis.

The starch from cells grown at the higher temperature of 30°C contained less amylose than did its counterpart from cells grown at 22–25°C (5 per cent amylose, as opposed to 13.5 per cent), whilst the starch content of the cells grown at the higher temperature was approximately 70 per cent of that obtained at 22–25°C. At 30°C the soluble synthetase possessed only some 16 per cent of the activity, whereas phosphorylase retained approximately 60 per cent of the activity, associated with cells grown at 22–25°C.

Further studies using a second culture medium fully substantiated the above results (Mangat and Badenhuizen 1971), and growth at high temperature again produced a much larger decrease in the activity of starch synthetase than it did in the activity of phosphorylase. A relation between the *P*/*Q* ratio and the amylose content was also noted for both the alga and developing maize endosperm—as the ratio increased, so too did the amylose content. In the developing potato tuber, the *P*/*Q* ratio remained constant over a series of growth temperatures and so did the amylose content; again, the activity of starch synthetase was found to fall as the growth temperature of the tuber was increased, although it did not disappear entirely as had earlier been reported (Badenhuizen and Chandorkar 1965).

These studies by Badenhuizen and his colleagues tend to imply that phosphorylase does play a major part in starch biosynthesis, but they are somewhat at variance with earlier conclusions. For example, Fuwa (1957) showed that the developing grains of maize and of waxy maize both had approximately the same *P*/*Q* ratios (and indeed the same activities for the two enzymes). This work has long been held to signify that phosphorylase was not seriously involved in the synthesis of starch, but now it is apparently contradicted by Badenhuizen's results.

In view of the many discrepant results in the literature, it is impossible to decide at this stage the relative importance of phosphorylase and starch synthetase in the formation of $\alpha$-1:4-bonds. But it should be noted that the developing maize grain in the period 8–28 days after pollination contains starch synthetase activity some two to eight times greater than the level necessary to account for the observed rate of starch deposition; the level of phosphorylase activity is forty times greater than that of the synthetase in the initial stages of development, and ten times greater in the later stages (Preiss, Ozbun, Hawker, Greenberg and Lammel 1973; Lee and Braun 1973). Why there should exist such an apparent overabundance of enzyme activity capable of synthesizing α-1:4-linkages to give a polyglucan cannot be explained yet.

It would appear to us that a *multipathway* system for starch biosynthesis is the most likely.

*Hypotheses of Starch Biosynthesis.* This subject has been reviewed recently (Manners 1968; Geddes 1969), and consequently we will deal with it only briefly.

In an ingenious hypothesis Erlander (1958) proposed that amylose and amylopectin were evolved from a common *glycogen* precursor: linear chains removed by an undiscovered de-branching enzyme were thought to be joined together to form amylose, whilst the residual de-branched precursor formed the amylopectin. The necessarily hypothetical nature of the de-branching enzyme has meant that this scheme has attracted little support, particularly as recent work has shown that glycogen cannot be de-branched in a way which will increase the ratio of *A*-chains to *B*-chains to the value found in amylopectin (Marshall and Whelan 1974).

Whelan (1963) postulated a model in which amylose was synthesized by phosphorylase, and amylopectin by a mixture of phosphorylase and *Q*-enzyme. The necessary separation of the two synthetic systems was achieved by supposing that the newly-synthesized amylopectin constituted a physical barrier to the diffusion of *Q*-enzyme.

Geddes and Greenwood (1969) have proposed the *multi-pathway* system shown in figure 6.14. These authors have suggested that the enzyme systems involved are adsorbed at the surface of the nucleating species in the amyloplast, and that this protein layer would remain associated with the expanding granule. Growth by apposition at the granule surface would then result in effective *stereochemical control* of the polysaccharide structures, particularly in the case of the branched macromolecules.

The various biosynthetic routes discussed above are incorporated into this scheme, but the main feature of the hypothesis is the formation of a linear dextrin ‘pool’. These dextrins are formed by either transglucosylase pathways (not shown for clarity), or by phosphorylase synthesis, or by degradation of amylose.

It was suggested that the intense synthesis of linear material at or near the surface will saturate *Q*-enzyme activity in that environment, and allow a proportion of the molecules to remain unbranched. Chain extension of branched material may proceed by a *P*-enzyme or starch synthetase pathway (the latter not shown for clarity). This hypothesis provides an explanation for the relative increase in amylose content compared to amylopectin which occurs with increase in granule size. As the surface area increases with increase in granule size more synthetase will be adsorbed from the interior of the amyloplast to increase the amount of linear material.

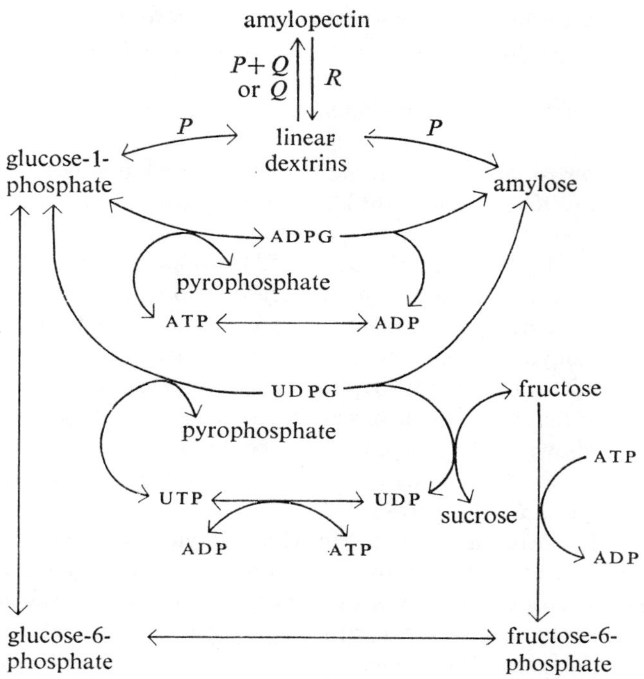


*Figure* 6.14. Scheme for starch biosynthesis (Geddes and Greenwood 1969).

French (1972) has also postulated that biosynthesis occurs at the granule surface, but that the amylose and amylopectin molecules are oriented in opposite senses; the amylopectin chains having their non-reducing ends at the granule surface whereas the reducing chain-ends of the amylose molecules are to be found there. Amylopectin then increases by the combined action of synthetase (or phosphorylase) and a branching enzyme; the amylose increases by accretion of glucose residues at the reducing end of the molecule, the reaction being catalysed by an as-yet undiscovered enzyme, and the polysaccharide is protected from the action of branching enzyme by the orientation of the molecule.

The concept of growth of the starch components occurring at the surface of the granule is at variance with the view of Badenhuizen (1965) that the polysaccharides are synthesized within the matrix of the amyloplast, and are subsequently transferred to the expanding granule. Evidence from the distribution of radioactivity in the starch of developing maize endosperm could be taken to support this latter view (Shannon, Creech and Loerch 1970; Shannon and Creech 1973).

More recently, Schiefer, Lee and Whelan (1973) have provided evidence that amylopectin is formed by a synthetase–branching enzyme complex, whereas amylose results from the action of a synthetase. The work also suggested that pre-formed amylose was less suitable as a substrate for the synthetase–branching enzyme complex than was a growing chain. This observation is immediately capable of explaining the separate existence of amylose and amylopectin in the starch granule—*Q*-enzyme is prevented from attacking pre-existing amylose by being complexed with the synthetase. The explanation is most attractive, but much more work is required before it can be accepted as the pathway for the biosynthesis of starch.

**6.3.4.** Summary.

The major routes involved in the biosynthesis of starch remain a subject of contention, for so much of the evidence offered in support of the various models is itself contradictory.

We have dealt with the problems raised by the apparent multiplicity of pathways available for the synthesis of starch, but have not yet considered reasons for the multiplicity of polysaccharides laid down as reserve carbohydrate. The reserve starch of most plants contains at least two polymers—amylose and amylopectin, together with a small amount of an ill-characterized material that can be either more or less highly branched than amylopectin. A third polyglucan, phytoglycogen, is present in sweet corn, and the profile of the chain-length distribution of this material is somewhat closer to that of amylopectin than to that of glycogen, although in its other properties (chain length, β-amylolysis limit, and iodine stain) it more nearly resembles the latter polysaccharide. However, at least one plant does have the ability to synthesize a true glycogen, for Rickson (1973) has reported that the polysaccharide present in the Müllerian body cells of *Cecropia peltata* is a glycogen. Thus inherent in the plant kingdom is the ability to synthesize a range of structures having a single purpose; the provision of reserve carbohydrate material.

The reason for the formation of a branched polysaccharide may be the obvious fact that the synthesizing enzymes are concerned with the concentration of non-reducing end-groups, and some form of multiply-branched structure is the most efficient way of providing such a substrate. In this respect it may be significant that the recessive genotypes of pea, barley, and maize that synthesize abnormally high proportions of amylose also contain less reserve starch than do the dominant genotypes. On the other hand, expression of the *waxy* gene does not cause an appreciable diminution in the level of reserve polysaccharide.

It is much more difficult to find a rational explanation for the purpose of amylose. The presence of 15–35 per cent of this material in the starch apparently confers some type of evolutionary advantage, since the majority of plants produce starch with an amylose content in this region. Amylose with a degree of polymerization of 50–70 anhydroglucose units has been found as a cell wall polysaccharide in the fungus *Hericium* (McCracken and Dodd 1971; Meeuse and Hall 1973), and also as extracellular capsular material in yeasts of the genus *Cryptococcus* (Hehre, Carlson and Hamilton 1949; Kooiman 1963). It is interesting to speculate that, at an early stage in evolution, amylose was developed to act as both structural and reserve material. Its subsequent incorporation within the starch granule must then have given such granules an advantage over the waxy type, either by changing their susceptibility to α-amylolysis, or by affecting the rate at which the granule could be solubilized.

The biosynthetic routes to starch, the reasons for the deposition of a multiplicity of polysaccharide structures, and the manner in which amylose and amylopectin can be simultaneously synthesized within the matrix of the plastid, remain matters for conjecture.

## Postscript

Many problems remain unsolved in the field of starch. Perhaps the largest of these is exemplified by the starch granule itself, and is typified by the scanning electron micrographs of granules of the maize genotypes (plates 10–12). The strangely-formed granules in the high amylose varieties cannot be explained, and their exact composition and mode of biosynthesis is unknown. In terms of the concept of individuality expressed in chapter 1, it is likely that the starch material in such granules is different, and progress will be limited until methods of separating and characterizing these granule types are found.

Some fundamental aspects of the subject are still in a state of flux, and we note, for example, that the long-held belief in the equality of numbers of *A* and *B* chains in amylopectin has now been abandoned (Marshall and Whelan 1974). The new model suggests the ratio of *A*:*B* chains has an average value of 2:1 for amylopectin, but retains the value 1:1 for glycogen.

The inability to apply broader perspectives, of which we complained in chapter 1, is exemplified by the recent suggestion of Marshall (1974) that those amylose molecules containing an α-1:6 linkage might still be linear. Whilst such a claim is valid from the enzymological standpoint, it can hardly be sustained by the known hydrodynamic behaviour of such fractions. Again, Marshall's claim (1974) that *Z*-enzyme is not an *alpha*-amylase is based on the erroneous assumption that the heat stability of a crude soya bean enzyme (Peat, Pirt and Whelan 1952b) should be the same as that of a purified sample (Greenwood, MacGregor and Milne 1965a).

In view of our comments in chapter 4, we should stress that the many hydrodynamic studies of *derivatives of amylose*—with which we have not had the space to deal—show these polymers to behave as random coils in solution with varying degrees of rigidity. However, it is of some interest that the field of classical polymer physics is now producing examples of polymers in which a helical conformation may be retained in solution. For example, poly(*n*-alkyl isocyanates) have long been known to be particularly rigid in solution (Burchard 1963a), but now this rigidity is being associated with a helical model. Interestingly, the increase in flexibility observed with increasing degree of polymerization is explained in terms of a slight probability of helix reversal (Tonelli 1974), essentially the same explanation advanced by Brant (see chapter 4) to account for the flexibility of amylose. Poly(β-hydroxybutyrate), a naturally-occurring polymer found in *Rhizobium*, has also been found to maintain a helical conformation in certain solvents, and even to undergo a sharp change in optical rotatory dispersion, reminiscent of a helix-coil transition, in some solvent mixtures (Marchessault, Okamura and Su 1970; Delsarte and Weill 1974).

We hope that the concepts which have been drawn together and developed in this monograph will help future research to fill in the gaps in the knowledge of this nutritionally- and industrially-important polymer.

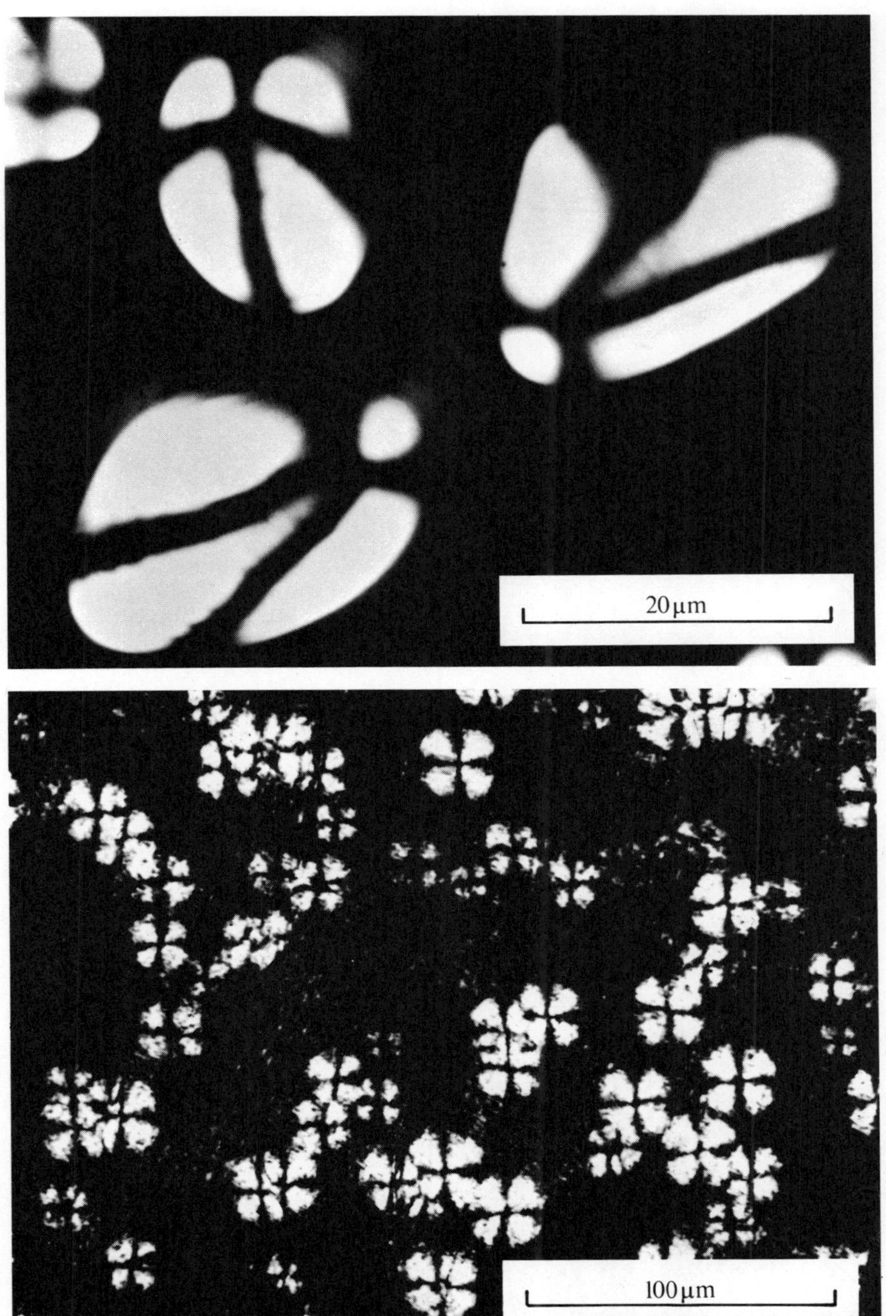


*Plate* 1. Birefringence patterns obtained from potato starch granules (Banks, Greenwood and Muir 1973b).

*Plate* 2. Polyethylene heated for 10 minutes at 250°C in the form of a film *ca*. 10 $\mu$m thick, between glass coverslips. Crystallized rapidly by cooling at room temperature. (By courtesy of Dr B. M. Gough.)

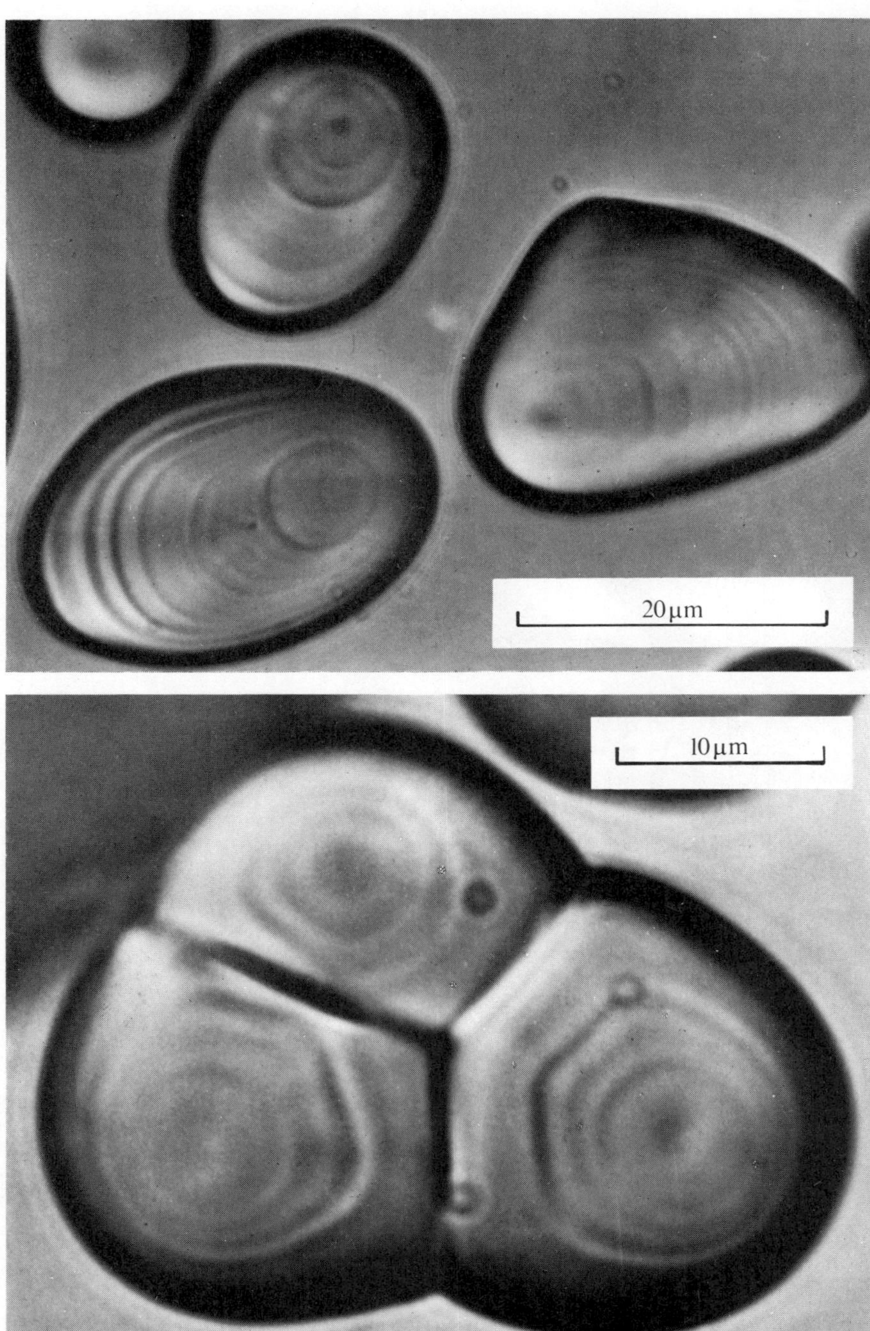


*Plate* 3. Potato starch granules showing layered structure (Banks, Greenwood and Muir 1973b).

*Plate* 4. Compound granule of potato starch (Banks, Greenwood and Muir 1973b).

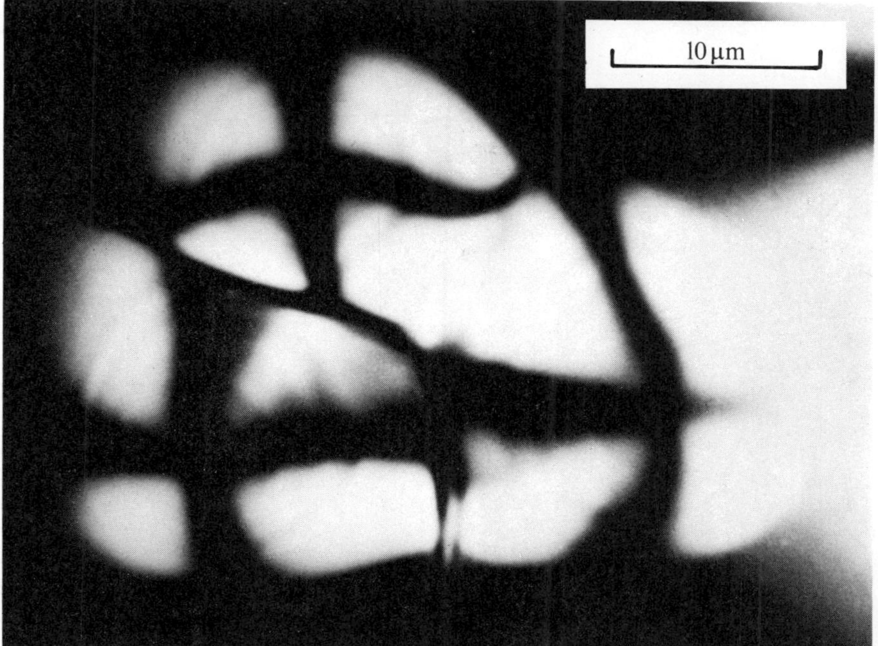

*Plate* 5. Compound granule of potato starch (as in plate 4) showing birefringence patterns (Banks, Greenwood and Muir 1973b).

*Plate* 6. Scanning electron micrograph of a granule from wrinkled-seeded pea starch showing surface fissures (Banks, Greenwood and Muir 1974c).

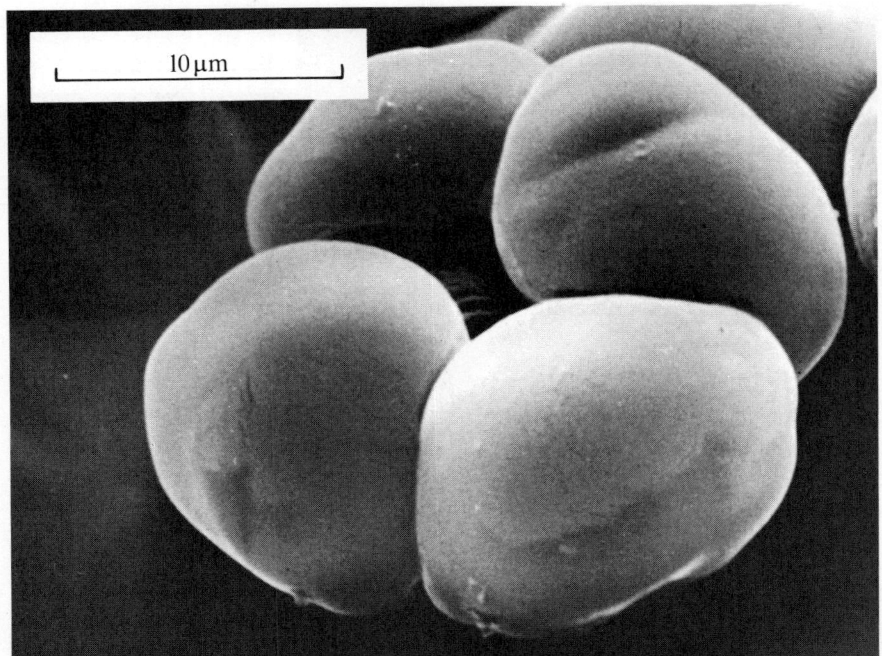


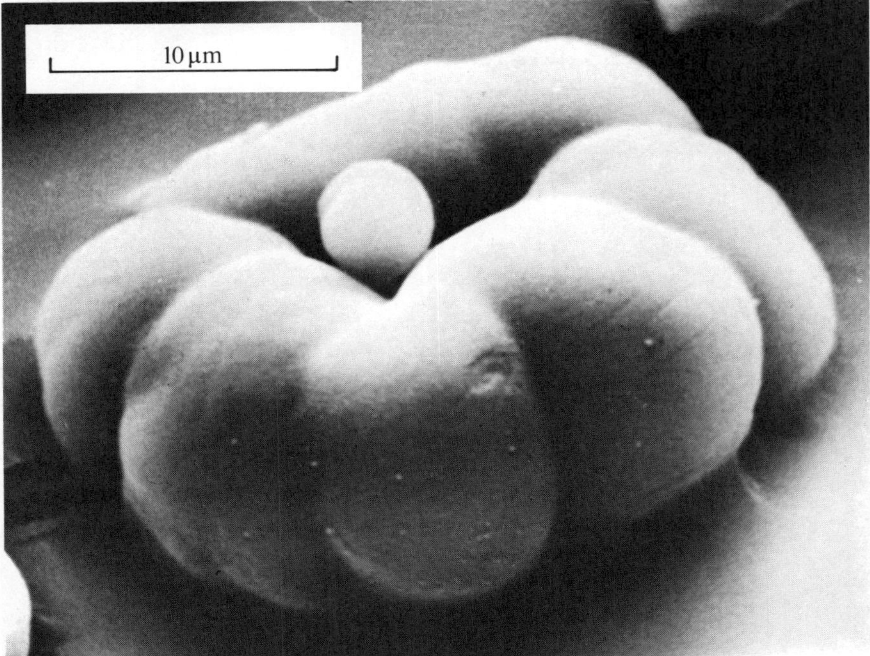


*Plate* 7. A typical compound granule of wrinkled-seeded pea starch with four segments and a deep central cavity (Banks, Greenwood and Muir 1974c).

*Plate* 8. A wrinkled-seeded pea starch granule with many segments, and with a small spherical starch granule resting in the depression (Banks, Greenwood and Muir 1974c).

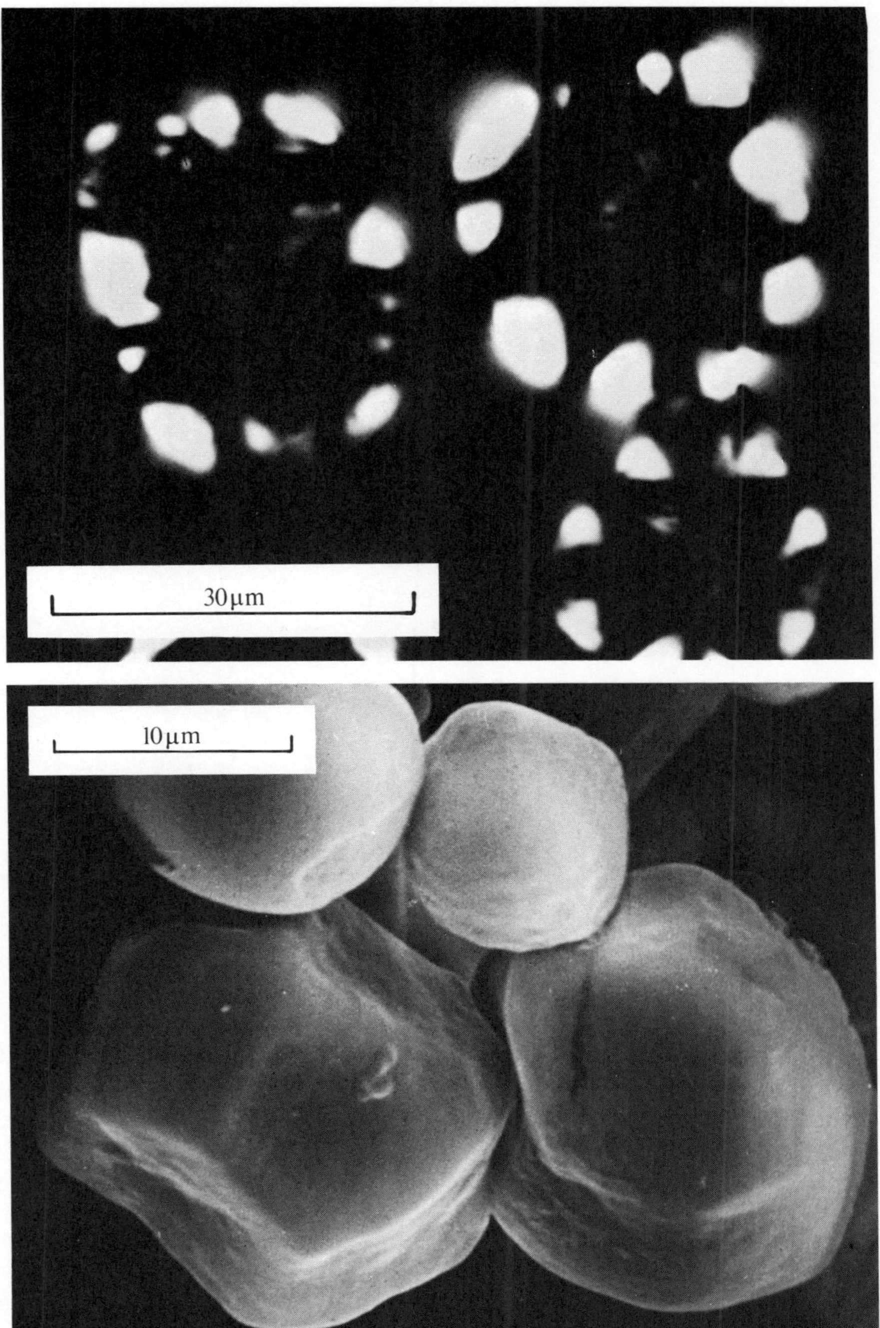


*Plate* 9. Granules of wrinkled-seeded pea starch viewed between crossed polars in the light microscope (Banks, Greenwood and Muir 1974c).

*Plate* 10. Scanning electron micrograph of granules of normal maize starch showing typical angular and rounded granules (Banks, Greenwood and Muir 1974a).

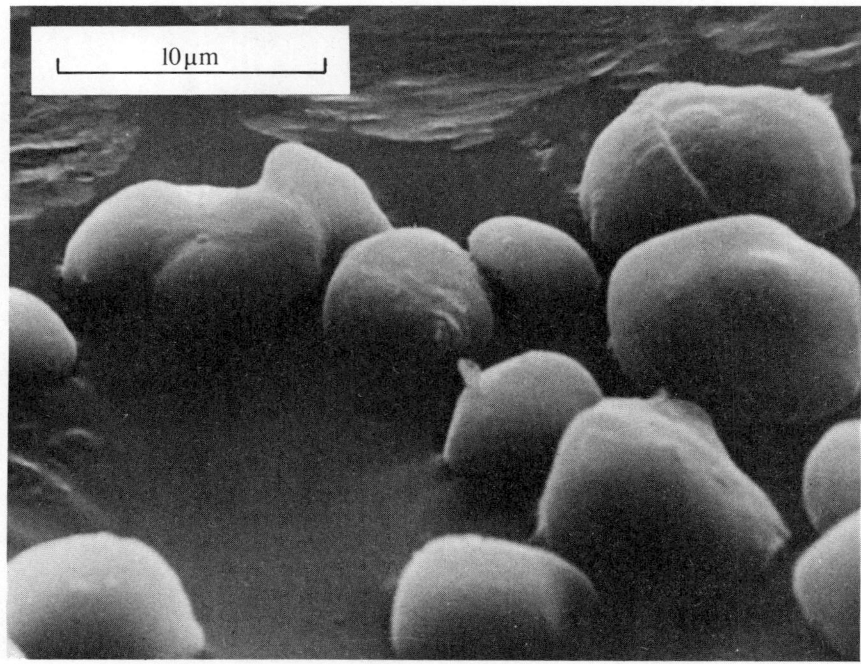


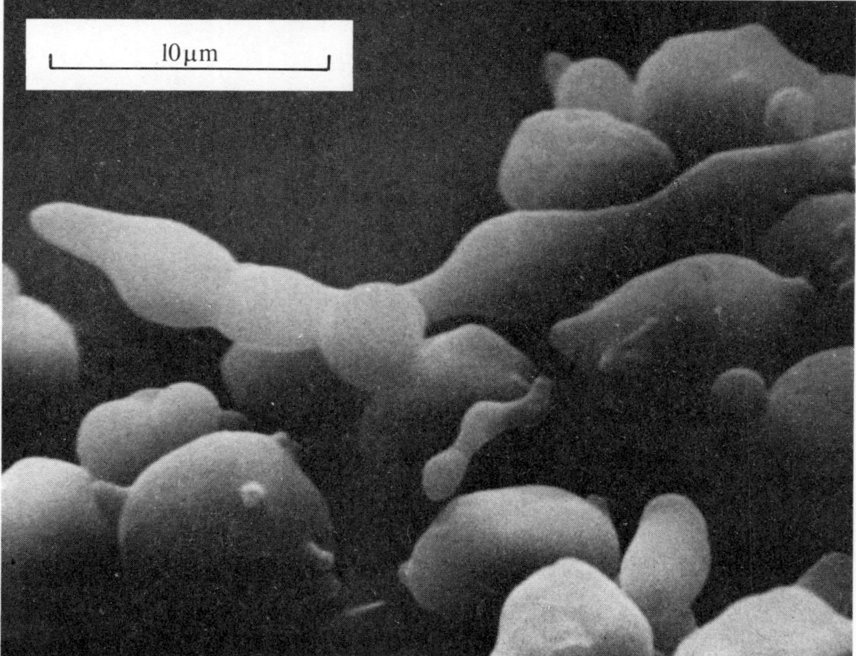


*Plate* 11. Scanning electron micrograph of granules of maize starch with a reputed amylose content of 50% (Amylon 50) showing unusual rounded granular forms (Banks, Greenwood and Muir 1974a).

*Plate* 12. Scanning electron micrograph of granules of maize starch with a reputed amylose content of 70% (Amylon 70) showing bizarre granular forms (Banks, Greenwood and Muir 1974a).

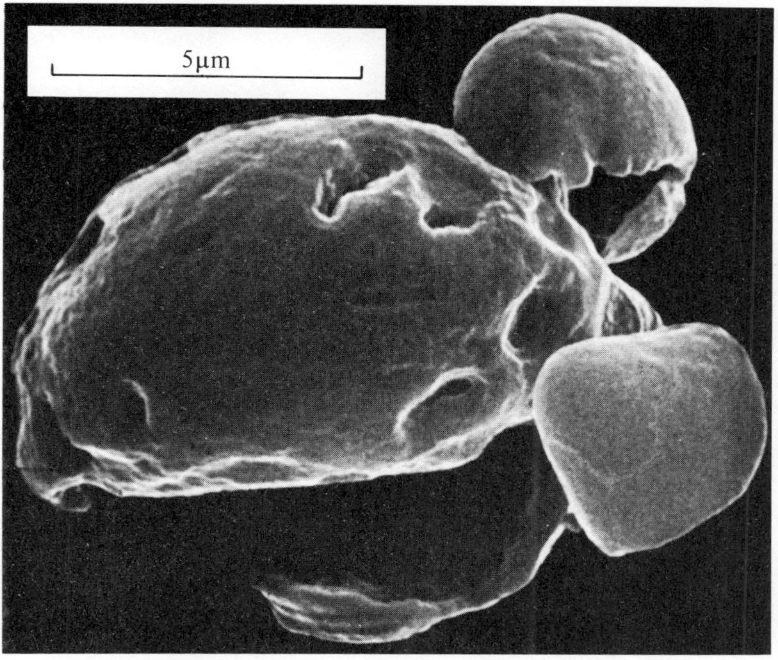


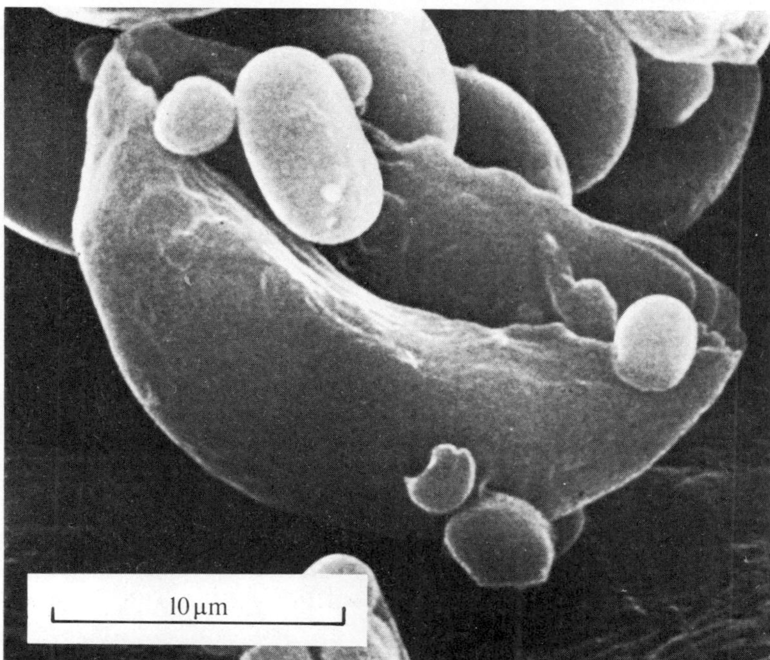


*Plate* 13. Fracture face profile of commercial wheat starch granule after 18 hours' incubation with wheat *alpha*-amylase (Evers and McDermott 1970).

*Plate* 14. Fracture face profile of commercial wheat starch granule after 18 hours' incubation with wheat *alpha*-amylase (Evers and McDermott 1970).

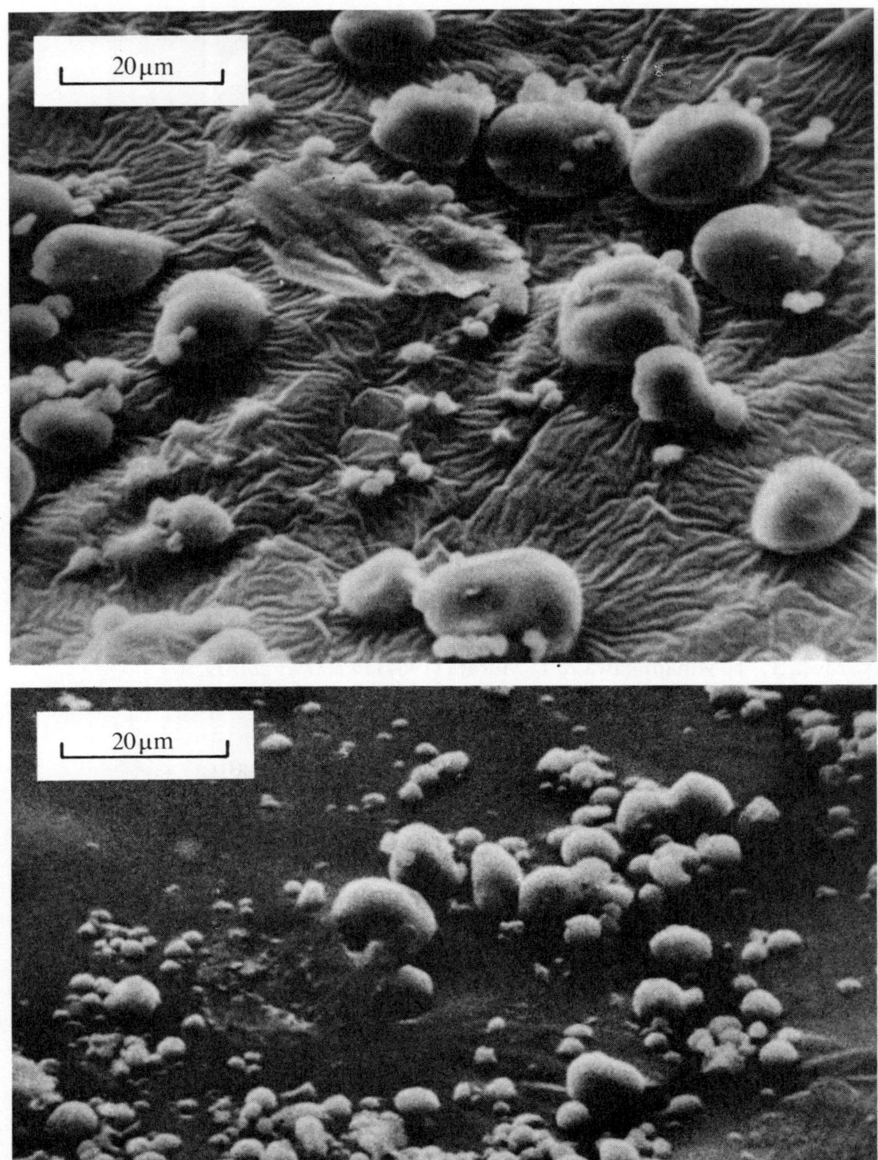


*Plate* 15. Scanning electron micrograph of granules from mature barley starch of normal amylose content (Banks, Greenwood and Muir 1973d).

*Plate* 16. Scanning electron micrograph of granules from mature barley starch of high amylose content (Banks, Greenwood and Muir 1973d).

## References

Abdullah, M., B. J. Catley, E. Y. C. Lee, J. F. Robyt, K. Wallenfels & W. J. Whelan (1966) *Cereal Chem. 43*, 111.
Abdullah, M., E. H. Fischer, M. Y. Qureshi, K. N. Slessor & W. J. Whelan (1965) *Biochem. J. 97*, 9P.
Abdullah, M. & D. French (1966) *Nature 210*, 200.
——— (1970) *Arch. Biochem. Biophys. 137*, 483.
Abdullah, M., D. French & J. F. Robyt (1966) *Arch. Biochem. Biophys. 114*, 595.
Abdullah, M. & W. J. Whelan (1963) *Nature 197*, 979.
Adkins, G. K., W. Banks & C. T. Greenwood (1966) *Carbohydrate Res. 2*, 502.
Adkins, G. K., W. Banks, C. T. Greenwood & A. W. MacGregor (1969) *Stärke 21*, 57.
Adkins, G. K. & C. T. Greenwood (1966a) *Stärke 18*, 213.
——— (1966b) *Stärke 18*, 171.
——— (1966c) *Stärke 18*, 237.
——— (1966d) *Carbohydrate Res. 3*, 81.
——— (1966e) *Carbohydrate Res. 3*, 152.
——— (1969a) *Stärke 21*, 153.
——— (1969b) *Carbohydrate Res. 11*, 217.
Adkins, G. K., C. T. Greenwood & D. J. Hourston (1970) *Cereal Chem. 47*, 13.
Akai, H., K. Yokobayashi, A. Misaki & T. Harada (1971) *Biochem. Biophys. Acta 237*, 422.
Akatsuka, T. & O. E. Nelson (1969) *J. Japanese Soc. Starch Sci. 17*, 99.
Akazawa, T., T. Minamikawa & T. Murata (1964) *Plant Physiol. 39*, 371.
Akazawa, T. & T. Murata (1965) *Biochem. Biophys. Res. Commun. 19*, 21.
Alexander, A. G. (1973) *Ann. N.Y. Acad. Sci. 210*, 64.
Allen, T. L. & R. M. Keefer (1965) *J. Amer. chem. Soc. 77*, 2957.
Anderson, D. M. W. & C. T. Greenwood (1955) *J. chem. Soc.*, 3016.
Arbuckle, A. W. & C. T. Greenwood (1958a) *J. chem. Soc.*, 2626.
——— (1958b) *J. chem. Soc.*, 2629.
Archibald, A. R., I. D. Fleming, A. M. Liddle, D. J. Manners, G. A. Mercer & A. Wright (1961) *J. chem. Soc.*, 1183.
Awtrey, A. D. & R. E. Connick (1951) *J. Amer. chem. Soc. 73*, 1842.

### B

Baba, A. & H. Kojima (1958) *Chem. Abs. 52*, 17337.
Babor, K., V. Kaláč & K. Tihlárik (1968) *Chem. Zvesti 22*, 321.
Badenhuizen, N. P. (1938) *Recueil Trav. Bot. Nederl. 35*, 559.
——— (1955) *Cereal Chem. 32*, 286.
——— (1958) *Encyclopaedia of Plant Physiol. 6*, 137.
——— (1959) *Protoplasmatologia II B*, 2, *b*, δ.
——— (1963) *Nature 197*, 464.
——— (1965) in *Starch Chemistry and Technology*, vol. 1 (eds R. L. Whistler & E. F. Paschall). New York: Academic Press Inc.
——— (1969) *The Biogenesis of Starch Granules in Higher Plants*. New York: Appleton-Century-Crafts.
——— (1971) in *Handbuch der Stärke in Einzeldarstellungen* (ed. M. Ulmann). Berlin: Paul Parey.
Badenhuizen, N. P. & K. R. Chandorkar (1965) *Cereal Chem. 42*, 44.
Badenhuizen, N. P. & R. W. Dutton (1956) *Protoplasma 47*, 156.

Badenhuizen, N. P., E. L. Kreutzer & L. Zar (1957) *South African J. Sci. 53*, 203.
Bailey, J. M. & D. French (1957) *J. biol. Chem. 226*, 1.
Bailey, J. M. & W. J. Whelan (1957) *Biochem. J. 67*, 540.
——— (1961) *J. biol. Chem. 236*, 969.
Baker, F. & W. J. Whelan (1950) *Nature 166*, 34.
Balls, A. K. & S. Schwimmer (1944) *J. biol. Chem. 156*, 203.
Banks, W. (1960) PhD thesis, Edinburgh University.
Banks, W., A. D. Evers & D. D. Muir (1972) *Chem. & Ind.*, 573.
Banks, W., R. Geddes, C. T. Greenwood & I. G. Jones (1972) *Stärke 24*, 245.
Banks, W. & C. T. Greenwood (1959a) *J. chem. Soc.*, 3436.
———(1959b) *Biochem. J. 73*, 237.
——— (1961) *Chem. & Ind.*, 714.
——— (1963a) *Makromol. Chem. 67*, 49.
——— (1963b) *Adv. Carbohydrate Chem. 18*, 357.
——— (1966) *Arch. Biochem. Biophys. 117*, 674.
——— (1967a) *Stärke 19*, 394.
——— (1967b) *Stärke 19*, 197.
——— (1967c) in *Conformation of Biopolymers* (ed. G. N. Ramachandran), 739. London: Academic Press.
——— (1968a) *Carbohydrate Res. 6*, 241.
——— (1968b) *Carbohydrate Res. 6*, 177.
——— (1968c) *Carbohydrate Res. 7*, 349.
——— (1968d) *Carbohydrate Res. 7*, 414.
——— (1968e) *Makromol. Chem. 114*, 245.
——— (1968f) *Carbohydrate Res. 6*, 171.
——— (1968g) *Stärke 20*, 315.
——— (1969a) *European Polymer J. 5*, 649.
——— (1969b) *Polymer 10*, 257.
——— (1970) *Arch. Biochem. Biophys. 136*, 320.
——— (1971a) *Stärke 23*, 222.
——— (1971b) *Stärke 23*, 300.
——— (1971c) *Makromol. Chem. 144*, 135.
——— (1971d) *Polymer 12*, 141.
——— (1972a) *Biopolymers 11*, 315.
——— (1972b) *Carbohydrate Res. 21*, 229.
——— (1973) *Ann. N.Y. Acad. Sci. 210*, 17.
Banks, W., C. T. Greenwood, D. J. Hourston & A. R. Procter (1971) *Polymer 12*, 452.
Banks, W., C. T. Greenwood & I. G. Jones (1960) *J. chem. Soc.*, 150.
Banks, W., C. T. Greenwood & K. M. Khan (1970a) *Stärke 22*, 292.
——— (1970b) *Stärke 22*, 333.
——— (1970c) *Carbohydrate Res. 12*, 79.
——— (1971a) *Carbohydrate Res. 17*, 25.
——— (1971b) *Carbohydrate Res. 19*, 252.
——— (1971c) *Carbohydrate Res. 20*, 233.
Banks, W., C. T. Greenwood & D. D. Muir (1971a) *Stärke 23*, 199.
——— (1971b) *Stärke 23*, 118.
——— (1972) *Stärke 24*, 181.
——— (1973a) *Stärke 23*, 153.
——— (1973*b*) *Stärke 25*, 331.
——— (1973c) in *Molecular Structure and Function of Food Carbohydrate* (ed. G. G. Birch & L. F. Green), 177. London: Applied Science.
——— (1973d) *Stärke 25*, 225.

Banks, W., C. T. Greenwood & D. D. Muir (1974a) *Stärke*, *26*, 289.
——— (1974b) *Stärke 26*, 73.
——— (1974c) *Stärke 26*, 46.
Banks, W., C. T. Greenwood & J. Sloss (1969) *Carbohydrate Res. 11*, 399.
Banks, W., C. T. Greenwood & J. Thomson (1959a) *Makromol. Chem. 31*, 197.
——— (1959b) *Chem. & Ind.*, 928.
Banks, W., C. T. Greenwood & J. T. Walker (1970) *Stärke 22*, 149.
——— (1971) *Stärke 23*, 12.
Banks, W., N. K. Mazumder & R. L. Spooner (1971) *Comp. Biochem. Physiol. 40 B*, 983.
Barker, S. A., E. J. Bourne & S. Peat (1949a) *J. chem. Soc.*, 1705.
——— (1949b) *J. chem. Soc.*, 1712.
Bates, F. L., D. French & R. E. Rundle (1943) *J. Amer. chem. Soc. 65*, 142.
Bathgate, G. N. & D. J. Manners (1966) *Biochem. J. 101*, 3c.
——— (1968) *Biochem. J. 107*, 443.
Baum, H. & G. A. Gilbert (1956) *J. colloid Sci. 11*, 428.
Baum, H., G. A. Gilbert & N. D. Scott (1956) *Nature 177*, 889.
Baumann, H. (1965) *J. polymer Sci. B3*, 1069.
Bear, R. S. (1944) *J. Amer. chem. Soc. 66*, 2122.
Bear, R. S. & D. French (1941) *J. Amer. chem. Soc. 63*, 2298.
Bebbington, A., E. J. Bourne, M. Stacey & I. A. Wilkinson (1952) *J. chem. Soc.*, 240.
Bender, H. & K. Wallenfels (1961) *Biochem. Z. 334*, 79.
Benoit, H. (1947) *J. Chim. Phys. 44*, 18.
Bersohn, R. & I. Isenberg (1961) *J. chem. Phys. 35*, 1640.
Bice, C. W., M. M. MacMasters & G. E. Hilbert (1945) *Cereal Chem. 22*, 463.
Bird, R. & R. H. Hopkins (1954) *Biochem. J. 56*, 86.
Bittiger, H. & E. Husemann (1968) in *Solution Properties of Natural High Polymers*, Special Publication 23, 159. London: The Chemical Society.
Bittiger, H., E. Husemann & B. Pfannemüller (1971) *Stärke 23*, 113.
Black, R. C., J. D. Loerch, F. J. McArdle & R. G. Creech (1966) *Genetics 53*, 661.
Blackwell, J., A. Sarko & R. H. Marchessault (1969) *J. mol. Biol. 42*, 379.
Blandamer, M. J. (1970) *Q. Rev. 24*, 169.
Borch, J., A. Sarko & R. H. Marchessault (1969) *Stärke 21*, 279.
Bourne, E. J. & W. J. Whelan (1950) *Nature 166*, 258.
Brammer, C. L., M. A. Rougvie & D. French (1972) *Carbohydrate Res. 24*, 343.
Brant, D. A. & W. L. Dimpfl (1970) *Macromolecules 3*, 655.
Brant, D. A. & K. D. Goebel (1971) *Macromolecules 5*, 536.
Brant, D. A. & B. K. Min (1969) *Macromolecules 2*, 1.
Bridgman, W. (1942) *J. Amer. chem. Soc. 64*, 2349.
Brinkman, H. C. (1947) *Physica 13*, 447.
Briones, V. P., L. G. Magbanua & B. O. Juliano (1968) *Cereal Chem. 45*, 351.
Brown, D. H. & B. Illingworth (1962) *Proc. Natl. Acad. Sci. U.S. 48*, 1783.
Brown, D. H., B. Illingworth & C. F. Cori (1961) *Proc. Natl. Acad. Sci. U.S. 47*, 479.
——— (1963) *Nature 197*, 980.
Bryce, W. A. J., J. M. G. Cowie, C. T. Greenwood & I. G. Jones (1958) *J. chem. Soc.*, 3558.
Bryce, W. A. J., C. T. Greenwood & I. G. Jones (1958) *J. chem. Soc.*, 3845.
Bryce, W. A. J., C. T. Greenwood, I. G. Jones & D. J. Manners (1958) *J. chem. Soc.*, 711.
Bueding, E. & S. A. Orrell (1961) *J. biol. Chem. 236*, 2854.
——— (1964) *J. biol. Chem. 239*, 4018.
Bünning, E. & C. Hess (1954) *Naturwiss. 41*, 339.

Burchard, W. (1960) *Makromol. Chem. 42*, 151.
——— (1961) *Makromol. Chem. 50*, 20.
——— (1963) *Makromol. Chem. 64*, 110.
——— (1963a) *Makromol. Chem. 67*, 182.
——— (1968) in *Solution Behaviour of Natural High Polymers*, Special Publication 23, 135. London: The Chemical Society.
Burr, B. & O. E. Nelson (1973) *Ann. N.Y. Acad. Sci. 210*, 129.
Buttrose, M. S. (1960) *J. Ultrastruct. Res. 4*, 231.
——— (1962) *J. cell. Biol. 14*, 159.
——— (1963) *Aust. J. Biol. Sci. 16*, 305.
——— (1966) *Stärke 18*, 122.

## C

Cabantchik, Z. & M. Schramm (1965) *Israel J. Chem. 3*, 105 P.
Casu, B., N. Reggioni, G. G. Gallo & A. Vigevani (1966) *Tetrahedron 22*, 3061.
——— (1968) in *Solution Behaviour of Natural High Polymers*, Special Publication 23, 217. London: The Chemical Society.
Cerf, R. (1948) *Compt. Rend. 227*, 1221.
Chandorkar, K. R. & N. P. Badenhuizen (1966) *Stärke 18*, 91.
Charbonnière, R., C. Mercier, M-T. Tollier & A. Guilbot (1968) *Stärke 20*, 75.
Chipman, D. M. & N. Sharon (1969) *Science 165*, 454.
Chu, S. S. C. & G. A. Jeffrey (1967) *Acta Crystallogr. 23*, 1038.
Cori, C. F. (1958) in *Polysaccharides in Biology* (ed. G. F. Springer), 65. Proc. 3rd Conf., Madison, NJ: Madison Printing Co.
——— (1973) *Ann. N.Y. Acad. Sci. 210*, 7.
Cori, G. T. & J. Larner (1951) *J. biol. Chem. 188*, 17.
Cowie, J. M. G. (1961a) *Makromol. Chem. 42*, 230.
——— (1961b) *J. polymer Sci. 49*, 455.
——— (1962) *Makromol. Chem. 53*, 13.
Cowie, J. M. G. & S. Bywater (1965) *Polymer 6*, 197.
Cowie, J. M. G., I. D. Fleming, C. T. Greenwood & D. J. Manners (1957) *J. chem. Soc.*, 4430.
——— (1958) *J. chem. Soc.*, 697.
Cowie, J. M. G. & C. T. Greenwood (1957a) *J. chem. Soc.*, 2862.
——— (1957b) *J. chem. Soc.*, 4640.
——— (1957c) *J. chem. Soc.*, 2658.
Cox, M. J., M. M. MacMasters & G. E. Hilbert (1944) *Cereal Chem. 21*, 447.
Cozzone, P., L. Pasero & G. Marchis-Mouren (1970) *Biochim. Biophys. Acta 200*, 590.
Cramer, F. (1951) *Ber. 84*, 855.
——— (1952a) *Naturwiss. 39*, 256.
——— (1952b) *Angew. Chem. 64*, 437.
——— (1956) *Rec. Trav. Chim. 75*, 891.
Creech, R. G. (1965) *Genetics 52*, 1175.
Cronan, C. L. & F. W. Schneider (1969) *J. phys. Chem. 73*, 3990.
Cunningham, W. L., D. J. Manners, A. Wright & I. D. Fleming (1960) *J. chem. Soc.*, 2602.
Czaja, A. T. (1954) *Planta 43*, 379.

## D

Davies, M. & E. Gwynne (1952) *J. Amer. chem. Soc. 74*, 2748.
Deatherage, W. L., M. M. MacMasters, M. L. Vineyard & R. P. Bear (1954) *Cereal Chem. 31*, 50.

Debye, P. (1946) *J. chem. Phys. 14*, 636.
Debye, P. & A. M. Bueche (1948) *J. chem. Phys. 16*, 573.
De Fekete, M. A. R. & C. E. Cardini (1964) *Arch. Biochem. Biophys. 104*, 173.
De Fekete, M. A. R., L. F. Leloir & C. E. Cardini (1960) *Nature 187*, 918.
Del Re, G. (1958) *J. chem. Soc.*, 4031.
Del Re, G., B. Pullman & T. Yonezawa (1963) *Biochim. Biophys. Acta* 75, 153.
Delsarte, J. & G. Weill (1974) *Macromolecules 7*, 450.
De Wulf, H. & H. G. Hers (1972) in *Biochemistry of the Glycosidic Linkage, an Integrated View* (eds R. Piras & H. G. Pontis), PAABS Symposium *2*, 399. Academic Press.
Dickinson, D. B. & J. Preiss (1969) *Plant Physiol. 44*, 1058.
Dintzis, F. R. & R. Tobin (1969) *Biopolymers 7*, 581.
Doi, K. & A. Doi (1969) *J. Japanese Soc. Starch Sci. 17*, 89.
Doi, A., K. Doi & Z. Nikuni (1964) *Biochem. Biophys. Acta 92*, 628.
Doppert, H. L. & A. J. Staverman (1966a) *J. polymer Sci. (A–1) 4*, 2367.
——— (1966b) *J. polymer Sci. (A–1) 4*, 2373.
Drochmans, P. (1963) *J. Ultrastruct. Res. 6*, 141.
——— (1973) *Ann. N.Y. Acad. Sci. 210*, 46.
Drummond, G. S., E. E. Smith & W. J. Whelan (1970) *FEBS Letters 9*, 136.
——— (1972) *European J. Biochem. 26*, 168.
Drummond, G. S., E. E. Smith, W. J. Whelan & H. Tai (1969) *FEBS Letters 5*, 85.
Duffus, C. M. (1969) *Phytochemistry 8*, 1205.
Dvonch, W., H. J. Yearian & R. L. Whistler (1950) *J. Amer. chem. Soc. 72*, 1748.
Dyar, M. T. (1950) *Amer. J. Bot. 37*, 786.
Dygert, S. (1971) PhD diss., Indiana University. Quoted by Thoma, Spradlin & Dygert (1971).

## E

Eliassaf, J. & M. Lewin (1963) *Israel J. Chem. 1*, 261.
Erlander, S. R. (1958) *Enzymologia 19*, 273.
——— (1960) *Cereal Chem. 37*, 81.
Erlander, S. R. & D. French (1958a) *J. Amer. chem. Soc. 80*, 4413.
——— (1958b) *J. polymer Sci. 20*, 7.
Erlander, S. R. & R. M. Purvinas (1968) *Stärke 20*, 37.
Erlander, S. R., R. M. Purvinas & H. L. Griffin (1968) *Cereal Chem. 45*, 140.
Espada, J. (1962) *J. biol. Chem. 237*, 3577.
Everett, W. W. & J. F. Foster (1959a) *J. Amer. chem. Soc. 81*, 3459.
——— (1959b) *J. Amer. chem. Soc. 81*, 3464.
Evers, A. D. (1971) *Stärke 23*, 157.
——— (1973) *Stärke 25*, 303.
Evers, A. D., B. M. Gough & J. N. Pybus (1971) *Stärke 23*, 16.
Evers, A. D., C. T. Greenwood, D. D. Muir & C. Venables (1974) *Stärke 26*, 331.
Evers, A. D. & E. E. McDermott (1970) *Stärke 22*, 23.

## F

Fergason, V. L., J. I. Helm & M. S. Zuber (1966) *Crop Sci. 6*, 273.
Finkelstein, R. S. & A. Sarko (1972) *Biopolymers 11*, 881.
Fischer, E. H. & W. Settele (1953) *Helv. chim. Acta 36*, 811.
Flory, P. J. (1949) *J. chem. Phys. 17*, 303.
——— (1953) *Principles of Polymer Chemistry*. New York: Cornell University Press.
——— (1966) *Makromol. Chem. 98*, 128.
——— (1969) *Statistical Mechanics of Chain Molecules*. New York: John Wiley.

Flory, P. J. & T. G. Fox (1951) *J. Amer. chem. Soc. 73*, 1904.
Flory, P. J. & L. Mandelkern (1952) *J. chem. Phys. 20*, 212.
Foster, J. F. (1965) in *Starch Chemistry and Technology* vol. 1 (eds R. L. Whistler & E. F. Paschall), 371. New York: Academic Press.
Fox, J., L. D. Kennedy, J. S. Hawker, J. L. Ozbun, E. Greenberg, C. Lammel & J. Preiss (1973) *Ann. N.Y. Acad. Sci. 210*, 90.
Frank, H. S. & W. Y. Wen (1957) *Discuss. Faraday Soc. 24*, 133.
Frederick, J. F. (1964) *Phyton* (Buenos Aires) *21*, 85.
——— (1971) *Physiol. Plantarum 24*, 32, 55.
——— (1973) *Ann. N.Y. Acad. Sci. 210*, 254.
French, D. (1957) *Baker's Digest 31* (August), 24.
——— (1961) *Nature 190*, 445.
——— (1973) *Ann. N.Y. Acad. Sci. 210*, 38.
French, D. & M. Abdullah (1966a) *Cereal Chem. 43*, 555.
——— (1966b) *Biochem. J. 100*, 6P.
French, D. & S. Kikumoto (1973) *Arch. Biochem. Biophys. 156*, 794.
French, D., D. W. Knapp & J. H. Pazur (1950) *J. Amer. chem. Soc. 72*, 1866.
French, D., M. L. Levine, J. H. Pazur & E. Norberg (1950) *J. Amer. chem. Soc. 72*, 1746.
French, D., A. O. Pulley & W. J. Whelan (1963) *Stärke 15*, 349.
French, D., E. E. Smith & W. J. Whelan (1972) *Carbohydrate Res. 22*, 123.
French, D. & G. M. Wild (1953) *J. Amer. chem. Soc. 75*, 4490.
French, D. & R. W. Youngquist (1963) *Stärke 15*, 425.
Freudenberg, K., E. Schaaf, G. Dumpert & T. Ploetz (1939) *Naturwiss. 27*, 850.
Frey-Wyssling, A. & M. S. Buttrose (1961) *Makromol. Chem. 44*, 173.
Frydman, R. B. (1963) *Arch. Biochem. Biophys. 102*, 242.
Frydman, R. B. & C. E. Cardini (1964a) *Biochem. Biophys. Res. Commun. 14*, 353.
——— (1964b) *Biochem. Biophys. Res. Commun. 17*, 407.
——— (1967) *J. biol. Chem. 242*, 312.
Frydman, R. B., B. C. de Souza & C. E. Cardini (1966) *Biochim. Biophys. Acta 113*, 620.
Frydman, R. B. & E. Slabnik (1973) *Ann. N.Y. Acad. Sci. 210*, 153.
Fukui, T. (1958) *Mem. Inst. Sci. & Ind. Res.* (Osaka University) *15*, 201.
Fukui, T. & A. Kamogawa (1969) *J. Japanese Soc. Starch Sci. 17*, 117.
Fuwa, H. (1957) *Arch. Biochem. Biophys. 70*, 157.

## G

Gahan, L. C. & H. E. Conrad (1968) *Biochemistry 7*, 3979.
Gallant, D., M. Degrois, C. Sterling & A. Guilbot (1972) *Stärke 24*, 116.
Gallant, D., A. Derrien, A. Aumaitre & A. Guilbot (1973) *Stärke 25*, 56.
Gallant, D. & A. Guilbot (1969) *Stärke 21*, 156.
Gallant, D., C. Mercier & A. Guilbot (1972) *Cereal Chem. 49*, 354.
Geddes, R. (1965) PhD thesis, Edinburgh University.
——— (1968) *Carbohydrate Res. 7*, 493.
——— (1969) *Q. Rev. 23*, 57.
Geddes, R. & C. T. Greenwood (1969) *Stärke 21*, 148.
Geddes, R., C. T. Greenwood, A. W. MacGregor, A. R. Procter & J. Thomson (1964) *Makromol. Chem. 79*, 189.
Geddes, R., C. T. Greenwood & S. L. Mackenzie (1965) *Carbohydrate Res. 1*, 71.
Ghosh, H. P. & J. Preiss (1965a) *J. biol. Chem. 240*, PC960.
——— (1965b) *Biochemistry 4*, 1354.
——— (1966) *J. biol. Chem. 241*, 4491.

Gilbert, G. A. & J. V. R. Marriott (1948) *Trans. Faraday Soc. 44*, 84.
Glusker, D. L. & H. W. Thompson (1955) *J. chem. Soc.*, 471.
Goebel, K. D. & D. A. Brant (1970) *Macromolecules 3*, 634.
Goebel, C. V., W. L. Dimpfl & D. A. Brant (1970) *Macromolecules 3*, 644.
Goldemberg, S. H. (1962) *Biochim. Biophys. Acta 56*, 357.
Gordon, M. (1957) *Trans. Faraday Soc. 53*, 1662.
——— (1960) *J. phys. Chem. 64*, 19.
Gordon, M. & L. R. Shenton (1959) *J. polymer Sci. 38*, 157, 179.
Gordon, R. B., D. H. Brown & B. I. Brown (1972) *Biochim. Biophys. Acta 289*, 97.
Gorin, E., W. Kauzmann & J. Walter (1939) *J. chem. Phys. 7*, 327.
Goring, D. A. I. (1966) *Pulp Paper Mag. Canada 67*, T519.
Gough, B. M. & J. N. Pybus (1971) *Stärke 23*, 210.
——— (1973) *Stärke 25*, 123.
Green, D. E. & P. K. Stumpf (1942) *J. biol. Chem. 142*, 355.
Greenberg, E. & J. Preiss (1964) *J. biol. Chem. 239*, 4314.
——— (1965) *J. biol. Chem. 240*, 2341.
Greenwood, C. T. (1956) *Adv. Carbohydrate Chem. 11*, 335.
——— (1960) *Stärke 12*, 169.
Greenwood, C. T. & P. C. Das Gupta (1958) *J. Chem. Soc.*, 703.
Greenwood, C. T., D. J. Hourston & A. R. Procter (1970) *European J. Polymer Sci. 6*, 293.
Greenwood, C. T. & A. W. MacGregor (1965) *J. Inst. Brewing 71*, 405.
——— (1969) *Stärke 21*, 199.
Greenwood, C. T., A. W. MacGregor & E. A. Milne (1965a) *Carbohydrate Res. 1*, 229.
——— (1965b) *Carbohydrate Res. 1*, 303.
——— (1965c) *Stärke 17*, 219.
——— (1965d) *Arch. Biochem. Biophys. 112*, 466.
Greenwood, C. T. & S. L. MacKenzie (1963) *Stärke 15*, 251.
——— (1966) *Carbohydrate Res. 3*, 7.
Greenwood, C. T. & E. A. Milne (1968a) *Adv. Carbohydrate Chem. 23*, 281.
——— (1968b) *Stärke 20*, 101.
——— (1968c) *Stärke 20*, 139.
Greenwood, C. T., E. A. Milne & G. R. Ross (1968) *Arch. Biochem. Biophys. 126*, 244.
Greenwood, C. T. & J. S. M. Robertson (1954) *J. chem Soc.*, 3769.
Greenwood, C. T. & H. Rossotti (1958) *J. Polymer Sci. 27*, 481.
Greenwood, C. T. & J. Thomson (1959) *J. Inst. Brewing 65*, 346.
——— (1960) *Chem. & Ind.*, 1110.
——— (1961a) *J. Inst. Brewing 67*, 64.
——— (1961b) *J. Chem. Soc.*, 1534.
——— (1962a) *J. Chem. Soc.*, 222.
——— (1962b) *Biochem. J. 82*, 156.
Griffin, H. L., S. R. Erlander & F. R. Senti (1967) *Stärke 19*, 8.
Griffin, H. L. & Y. V. Wu (1968) *Biochemistry 7*, 3063.
——— (1971) *Biochemistry 10*, 4330.
Guilbot, A., R. Charbonnière & R. Drapron (1961) *Stärke 13*, 204.
Gunja, Z. H. & D. J. Manners (1959) *Chem. & Ind.*, 1017.
Gunja, Z. H., D. J. Manners & K. Maung (1960) *Biochem. J. 75*, 441.
——— (1961) *Biochem. J. 81*, 392.
Gunja-Smith, Z., J. J. Marshall, C. Mercier, E. E. Smith & W. J. Whelan (1970) *FEBS Letters 12*, 101.
Gunja-Smith, Z., J. J. Marshall & E. E. Smith (1971) *FEBS Letters 13*, 309.

Gunja-Smith, Z., J. J. Marshall, E. E. Smith & W. J. Whelan (1970) *FEBS Letters 12*, 96.

## H

Hall, D. M. & J. G. Sayre (1970) *Textile Res. J. 40*, 147.
Hamori, E. & M. B. Senior (1973) *Ann. N.Y. Acad. Sci. 210*, 34.
Hanes, C. S. (1937) *New Phytologist 36*, 189.
Hanssen, E., E. Dodt & E. G. Niemann (1953) *Kolloid-Z 130*, 19.
Harada, T., A. Misaki, H. Akai, K. Yokobayashi & K. Sugimoto (1972) *Biochim. Biophys. Acta 268*, 497.
Harada, T., K. Yokobayashi & A. Misaki (1968) *Appl. Microbiol. 16*, 1439.
Harris, G. & I. C. MacWilliam (1958) *Cereal Chem. 35*, 82.
Hassid, W. Z. & R. M. McCready (1943) *J. Amer. chem. Soc. 65*, 1157.
Haszeldine, R. N. (1954) *J. chem. Soc.*, 4145.
Hawker, J. S., J. L. Ozbun, H. Ozaki, E. Greenberg & J. Preiss (1974) *Arch. Biochem. Biophys. 160*, 530.
Hawker, J. S., J. L. Ozbun & J. Preiss (1972) *Phytochemistry 11*, 1287.
Haworth, W. N. (1947) *Nature 160*, 901.
Haworth, W. N., E. L. Hirst & F. A. Isherwood (1937) *J. chem. Soc.*, 577.
Haworth, W. N., S. Peat & E. J. Bourne (1944) *Nature 154*, 236.
Hearst, J. E. & W. H. Stockmayer (1962) *J. chem. Phys. 37*, 1425.
Hehre, E. J., A. C. Carlson & D. M. Hamilton (1949) *J. biol. Chem. 177*, 289.
Hehre, E. J., G. Okada & D. S. Genghof (1969) *Arch. Biochem. Biophys. 135*, 75.
Heller, J. & M. Schramm (1964) *Biochim. Biophys. Acta 81*, 96.
Herzberg, G. (1950) *Spectra of Diatomic Molecules*. Princetown, NJ: D. Van Nostrand Inc.
Hess, K., H. Mahl & E. Gütter (1955) *Mikroskopie 10*, 329.
Heyn, A. N. (1959) *Textile Res. J. 29*, 366.
Higginbotham, R. S. (1949a) *Shirley Inst. Memoirs 23*, 159.
——— (1949*b*) *Shirley Inst. Memoirs 23*, 171.
Higginbotham, R. S. & G. A. Morrison (1949) *J. Textile Inst. 40*, T201.
Hilbert, G. E. & M. M. MacMasters (1946) *J. biol. Chem. 162*, 229.
Hirst, E. L. & D. J. Manners (1954) *Chem. & Ind.*, 224.
Hizukuri, S. (1969) *J. Japanese Soc. Starch Sci. 17*, 73.
Hizukuri, S., M. Fujii & Z. Nikuni (1961) *Nature 192*, 239.
——— (1965) *Stärke 17*, 40.
Hobson, P. N., W. J. Whelan & S. Peat (1950) *J. chem. Soc.*, 3566.
——— (1951) *J. chem. Soc.*, 1451.
Hodge, J. E., S. A. Karjala & G. E. Hilbert (1951) *J. Amer. chem. Soc. 73*, 3312.
Hodge, J. E., E. M. Montgomery & G. E. Hilbert (1948) *Cereal Chem. 25*, 19.
Hodges, H. F., R. G. Creech & J. D. Loerch (1969) *Biochim. Biophys. Acta 185*, 70.
Holló, J. & J. Szejtli (1957) *Stärke 9*, 109.
——— (1958a) *Kolloid-Z. 20*, 229.
——— (1958b) *Periodica Polytechnica 2*, 25.
——— (1958c) *Stärke 10*, 49.
——— (1962) *Stärke 14*, 75.
——— (1968) in *Starch and its Derivatives* (ed. J. A. Radley), 203. London: Chapman and Hall Ltd.
Holló, J., J. Szejtli & M. Toth (1961) *Stärke 13*, 222.
Hopkins, R. H. & R. Bird (1953) *Nature 172*, 492.
Hopkins, R. H. & B. Jelinek (1948) *Nature 164*, 955.

Huggins, M. L. (1942) *J. Amer. chem. Soc. 64*, 2716.
Hughes, R. C., E. E. Smith & W. J. Whelan (1963) *Biochem. J. 88*, 63 P.
Hull, G. A. & T. J. Schoch (1959) *Tappi 42*, 438.
Husemann, E., W. Burchard, B. Pfannemüller & R. Werner (1961) *Stärke 13*, 196.
Husemann, E., B. Fritz, R. Lippert & B. Pfannemüller (1958) *Makromol. Chem. 26*, 199.
Husemann, E. & B. Pfannemüller (1961) *Makromol. Chem. 43*, 156.
——— (1963) *Makromol. Chem. 69*, 74.
——— (1965) *Makromol. Chem. 87*, 139.
Hybl, A., R. E. Rundle & D. E. Williams (1965) *J. Amer. chem. Soc. 87*, 2779.

## I

Illingworth, B., D. H. Brown & C. F. Cori (1961) *Proc. Natl. Acad. Sci. U.S. 47*, 469.
Inagaki, H., H. Suzuki & M. Kurata (1966) *J. polymer Sci. C15*, 409.

## J

Jackobs, J. J., R. R. Bumb & B. Zaslow (1968) *Biopolymers 6*, 1659.
Johnson, W. S., J. A. Margrave, V. J. Bauer, M. Frisch, L. H. Dreger & W. N. Hubbard (1961) *J. Amer. chem. Soc. 83*, 606.
Jones, G. & B. B. Kaplan (1928) *J. Amer. chem. Soc. 50*, 1845.
Jones, I. G. (1959) PhD thesis, Edinburgh University.

## K

Kainuma, K. & D. French (1969) *FEBS Letters 5*, 257.
——— (1970) *FEBS Letters 6*, 182.
——— (1971) *Biopolymers 10*, 1673.
Kainuma, K., S. Kobayashi, T. Ito & S. Suzuki (1972) *FEBS Letters 26*, 281.
Kamide, K. & Y. Inamoto (1965) *Chem. High Polymers Tokyo 22*, 691.
Kamogawa, A., T. Fukui & Z. Nikuni (1968) *J. Biochem. 63*, 361.
Katz, J. R. (1937) *Rec. trav. chim. Pays-Bas 56*, 785.
Katz, J. R. & J. C. Derksen (1933) *Z. Physic. Chem. A 165*, 228.
Katz, J. R. & Th. B. van Itallie (1930) *Z. Physic Chem. A 150*, 90.
Katzbeck, W. J. & R. W. Kerr (1950) *J. Amer. chem. Soc. 72*, 3208.
Keller, A. (1959) *J. polymer Sci. 39*, 151.
Kerr, R. W. (1946) *Chemistry and Industry of Starch*, 146. New York: Academic Press.
——— (1952) *Stärke 4*, 39.
Killean, R. C. G., W. G. Ferrier & D. W. Young (1962) *Acta Crystallogr. 15*, 911.
Killion, P. J. & J. F. Foster (1960) *J. polymer Sci. 46*, 65.
Kiribuchi, S. & M. Nakamura (1973) *J. agric. chem. Soc. Japan 47*, 333.
Kirkwood, J. G. & J. Riseman (1948) *J. chem. Phys. 16*, 565.
——— (1956) in *Rheology, Theory and Applications* (ed. F. R. Eirich), 495. New York: Academic Press.
Kitaigorodsky, A. I. (1961) *Tetrahedron 14*, 230.
Kjølberg, O. & D. J. Manners (1963) *Biochem. J. 86*, 258.
Klein, B., J. A. Foreman & R. L. Searcy (1969) *Anal. Biochem. 31*, 412.
Kobayashi, T. (1969) *J. Japanese Soc. Starch Sci. 17*, 61.
Koenig, J. L. & P. D. Vasko (1970) *J. macromol. Sci.-Phys. B4*, 347.
Kooiman, P. (1963) *Antonie van Leuvwenhock J. Microbiol. Serol 29*, 169.
Kramer, H. H., R. P. Bear & M. S. Zuber (1958) *Agron. J. 50*, 229.
Kramer, H. H., P. L. Pfahler & R. L. Whistler (1965) *Agron. J. 50*, 207.

Kratky, O. & G. Porod (1949) *Rec. Trav. Chim. 68*, 1106.
Kreger, D. R. (1946) *Nature 158*, 199.
——— (1947) *Nature 160*, 369.
——— (1951) *Biochim. Biophys. Acta 6*, 406.
Krisman, C. R. (1972) *Biochem. Biophys. Res. Commun. 46*, 1206.
——— (1973) *Ann. N. Y. Acad. Sci. 210*, 81.
Kuge, T. & S. Ono (1960a) *Bull. chem. Soc. Japan 33*, 1269.
——— (1960b) *Bull. chem. Soc. Japan 33*, 1273.
——— (1961) *Bull. chem. Soc. Japan 34*, 1264.
Kuhn, H. (1949) *J. chem. Phys. 17*, 1198.
——— (1930) *Ber. 63*, 1503.
——— (1936) *Kolloid-Z. 76*, 258.
——— (1939) *Kolloid-Z. 87*, 3.
Kuhn, W. & H. Kuhn (1943) *Helv. Chim. Acta 26*, 1394.
Kuhn, W., H. Kuhn & A. Silberberg (1954) *J. polymer Sci. 14*, 193.
Kung, J. T., V. M. Hanrahan & M. L. Caldwell (1953) *J. Amer. chem. Soc. 75*, 5548.
Kurata, M., W. H. Stockmayer & A. Roig (1960) *J. chem. Phys. 33*, 151.
Kurata, M. & W. H. Stockmayer (1963) *Fortsch. Hochpolym.-Forsch. 3*, 196.

## L

La Berge, D. E., A. W. MacGregor & W. O. S. Meredith (1971) *Canad. J. Plant Sci. 51*, 469.
Lang, N. J. (1963) *J. Protozool. 10*, 333.
Lansky, S., M. Kooi & T. J. Schoch (1949) *J. Amer. chem. Soc. 71*, 4066.
Larner, J. (1953) *J. biol. Chem. 202*, 491.
Larner, J., B. Illingworth, G. T. Cori & C. F. Cori (1952) *J. biol. Chem. 199*, 641.
Larner, J., B. R. Ray & H. F. Crandall (1956) *J. Amer. chem. Soc. 78*, 5890.
Larner, J. & L. H. Schliselfeld (1956) *Biochim. Biophys. Acta 20*, 53.
Laskov, R. & E. Margoliash (1963) *Bull. Res. Council Israel A 11*, 351.
Lavintman, N. (1966) *Arch. Biochem. Biophys. 116*, 1.
Leach, H. W., L. D. McCowen & T. J. Schoch (1959) *Cereal Chem. 36*, 534.
Leach, H. W. & T. J. Schoch (1961) *Cereal Chem. 38*, 34.
——— (1962) *Cereal Chem. 39*, 318.
Lee, E. Y. C. (1971) *Arch. Biochem. Biophys. 146*, 488.
Lee, E. Y. C. & J. J. Braun (1973) *Ann. N. Y. Acad. Sci. 210*, 115.
Lee, E. Y. C., J. H. Carter, L. D. Nielsen & E. H. Fischer (1970) *Biochemistry 9*, 2347.
Lee, E. Y. C., J. J. Marshall & W. J. Whelan (1971) *Arch. Biochem. Biophys. 143*, 365.
Lee, E. Y. C., C. Mercier & W. J. Whelan (1968) *Arch. Biochem. Biophys. 125*, 1028.
Lee, E. Y. C., L. D. Nielsen & E. H. Fischer (1967) *Arch. Biochem. Biophys. 121*, 245.
Lee, E. Y. C. & W. J. Whelan (1966) *Arch. Biochem. Biophys. 116*, 162.
——— (1971) in *Enzymes*, vol. 5, 3rd edition (ed. P. D. Boyer), 191. New York: Academic Press.
Lee, Y. (1960) *Biochim. Biophys. Acta 43*, 18.
Leloir, L. F. (1964) in *Control of Glycogen Metabolism* (eds W. J. Whelan & M. P. Cameron). London: J. & A. Churchill.
Leloir, L. F. & C. E. Cardini (1957) *J. Amer. chem. Soc. 79*, 6340.
Leloir, L. F., M. A. R. de Fekete & C. E. Cardini (1961) *J. biol. Chem. 236*, 636.
Leloir, L. F., J. M. Olavarria, S. H. Goldemberg & H. Carminatti (1959) *Arch. Biochem. Biophys. 81*, 508.
Loewus, F. A. & D. R. Briggs (1957) *J. Amer. chem. Soc. 79*, 1494.

# M

McCracken, D. A. & N. P. Badenhuizen (1970) *Stärke 22*, 289.
McCracken, D. A. & J. L. Dodd (1971) *Science 174*, 419.
McCready, R. M. & W. Z. Hassid (1943) *J. Amer. chem. Soc. 65*, 1154.
McDonald, T. R. & C. A. Beevers (1952) *Acta Crystallogr. 5*, 654.
MacGregor, A. W. (1964) PhD thesis, Edinburgh University.
MacGregor, A. W., A. G. Gordon, W. O. S. Meredith & L. Lacroix (1972) *J. Inst. Brewing 78*, 174.
MacGregor, A. W., D. E. La Berge & W. O. S. Meredith (1971) *Cereal Chem. 48*, 255.
McLaren, A. D. (1963) *Enzymologia 26*, 237.
McLaren, A. D. & L. Packer (1970) *Adv. Enzymol. 33*, 245.
MacMasters, M. M. & G. E. Hilbert (1944) *Ind. Engng. Chem. 36*, 958.
MacWilliam, I. C. & G. Harris (1959) *Arch. Biochem. Biophys. 84*, 442.
Machell, G. & G. N. Richards (1958) *J. chem. Soc.*, 1199.
Madsen, N. B. & C. F. Cori (1958) *J. biol. Chem. 233*, 1251.
Mangat, B. S. & N. P. Badenhuizen (1970) *Stärke 22*, 329.
——— (1971) *Canad. J. Bot. 49*, 1787.
Manley, R. St. J. (1964) *J. polymer Sci. A 2*, 4503.
Manners, D. J. (1968) in *Starch and its Derivatives*, 4th edition (ed. J. A. Radley), 66. London: Chapman & Hall Ltd.
——— (1971) *Nature New Biol. 234*, 150.
Manners, D. J., J. J. Marshall & D. Yellowlees (1970) *Biochem. J. 116*, 539.
Manners, D. J. & K. Maung (1955) *Chem. & Ind.*, 950.
Manners, D. J., G. A. Mercer, J. R. Stark & J. F. Ryley (1965) *Biochem. J. 96*, 530.
Manners, D. J. & K. L. Rowe (1969) *Carbohydrate Res. 9*, 107.
Manners, D. J., J. J. M. Rowe & K. L. Rowe (1968) *Carbohydrate Res. 8*, 72.
Manners, D. J. & K. L. Sparra (1966) *J. Inst. Brewing 72*, 366.
Manners, D. J. & A. Wright (1960) *Biochem. J.*, 10P.
Maquenne, L. & E. Roux (1905) *C.R. Acad. Sci., Paris, 140*, 1303.
Marchessault, R. H., K. Okamura & C. J. Su (1970) *Macromolecules 3*, 735.
Marshall, J. J. (1970) *Anal. Biochem. 37*, 466.
——— (1974) *Adv. in Carbohydrate Chem. & Biochem. 30*, 257.
Marshall, J. J. & W. J. Whelan (1970) *FEBS Letters 9*, 85.
——— (1974) *Arch. Biochem. Biophys. 161*, 234.
Maruo, B. & T. Kobayashi (1951) *Nature 167*, 606.
Matheson, N. K. (1971) *Phytochemistry 10*, 3213.
Matheson, N. K. & J. M. Wheatley (1962) *Aust. J. biol. Sci. 15*, 445.
May, L. H. & M. S. Buttrose (1959) *Aust. J. biol. Sci. 12*, 146.
Maywald, E. C., H. W. Leach & T. J. Schoch (1968) *Stärke 20*, 189.
Meeuse, B. J. D. & D. M. Hall (1973) *Ann. N.Y. Acad. Sci. 210*, 39.
Mencik, Z., R. H. Marchessault & A. Sarko (1971) *J. molec. Biol. 55*, 193.
Mercier, C. (1973) *Stärke 25*, 78.
Mercier, C., R. Charbonnière, D. Gallant & A. Guilbot (1970) *Stärke 22*, 9.
Mercier, C. & W. J. Whelan (1970) *European J. Biochem. 16*, 579.
Merritt, N. R. & J. T. Walker (1969a) *Nature 221*, 482.
——— (1969b) *J. Inst. Brewing 75*, 156.
Mes, M. J. & I. Menge (1954) *Physiol. Plantarum 7*, 637.
Meyer, K. H. (1952) *Experientia* 8, 405.
Meyer, K. H. & P. Bernfeld (1940) *Helv. chim. Acta 23*, 875.
——— (1941) *Helv. chim. Acta 24*, 359E.
Meyer, K. H., P. Bernfeld, R. A. Boissonnas, P. Guërtler & G. Noelting (1949) *J. phys. Chem. 53*, 319.

Meyer, K. H., P. Bernfeld & W. Hohenemser (1940) *Helv. chim. Acta 23*, 885.
Meyer, K. H., P. Bernfeld & E. Wolff (1940) *Helv. chim. Acta 23*, 854.
Meyer, K. H. & W. F. Gonon (1951) *Helv. chim. Acta 34*, 294.
Meyer, K. H. & R. Menzi (1953) *Helv. chim. Acta 36*, 702.
Meyer, K. H., M. Wertheim & P. Bernfeld (1940) *Helv. chim. Acta 23*, 865.
Mikus, F. F., R. M. Hixon & R. E. Rundle (1946) *J. Amer. chem. Soc. 68*, 1115.
Milne, E. A. (1966) PhD thesis, Edinburgh University.
Montgomery, E. M. & F. R. Senti (1958) *J. polymer Sci. 28*, 1.
Montgomery, E. M., K. R. Sexson, R. J. Dimler & F. R. Senti (1964) *Stärke 16*, 314
Montgomery, E. M., K. R. Sexson & F. R. Senti (1961) *Stärke 13*, 215.
Montgomery, E. M., F. B. Weakley & G. E. Hilbert (1949) *J. Amer. chem. Soc. 71*, 1682.
Mordoh, J., C. R. Krisman & L. F. Leloir (1966) *Arch. Biochem. Biophys. 113*, 265.
Mould, D. L. (1954) *Biochem. J. 58*, 593.
Mould, D. L. & R. L. M. Synge (1954) *Biochem. J. 58*, 585.
Muetgeert, J. (1961) *Adv. Carbohydrate Chem. 16*, 299.
Muir, D. D. (1973) PhD thesis, Edinburgh University.
Mukherjee, S. & S. Bhattacharyya (1946) *J. Indian chem. Soc. 23*, 459.
Mulliken, R. S. (1950) *J. Amer. chem. Soc. 72*, 600.
Murakami, H. (1954) *J. chem. Phys. 22*, 367.
——— (1955) *J. chem. Phys. 23*, 1979.
Murata, T. & T. Akazawa (1964) *Biochem. Biophys. Res. Commun. 16*, 6.
——— (1966) *Arch. Biochem. Biophys. 114*, 76.
Murata, T., T. Sugiyama & T. Akazawa (1964) *Arch. Biochem. Biophys. 107*, 92.
——— (1965) *Biochem. Biophys. Res. Commun. 18*, 371.
Murata, T., T. Sugiyama, T. Minamikawa & T. Akazawa (1966) *Arch. Biochem. Biophys. 113*, 34.
Myrbäck, K. (1950) *Arkiv Kemi 2*, 417.
Myrbäck, K. & N. O. Johansson (1945) *Arkiv Kemi Mineral. Geol. 20 A*, 1.
Myrbäck, K. & L. G. Sillén (1949) *Acta chem. Scand. 3*, 190.

## N

Nagai, K. (1961) *J. chem. Phys. 34*, 887.
Nagashima, H., S. Nakamura, K. Nisizawa & T. Hori (1971) *Plant & Cell Physiol. 12*, 243.
Neely, W. B. (1961) *J. org. Chem. 26*, 3015.
Nelson, D. E. & H. W. Rines (1962) *Biochem. Biophys. Res. Commun. 9*, 297.
Nemethy, G. & H. A. Scheraga (1967) *J. chem. Phys. 36*, 3382.
Neufeld, E. F. & W. Z. Hassid (1955) *Arch. Biochem. Biophys. 59*, 405.
Nielsen, J. P. & P. C. Gleason (1945) *Ind. Engng. Chem. Anal. Ed. 17*, 131.
Nishimura, S. (1930) *Biochem. Z. 223*, 161; *225*, 264.
——— (1931) *Biochem. Z. 232*, 156; *237*, 133.
Nordin, P. & D. French (1958) *J. Amer. chem. Soc. 80*, 1445.
Nordin, P. & Y. S. Kim (1960) *J. Amer. chem. Soc. 82*, 4604.
Nordin, P., H. Moser, G. Rao, N. Giri & T. Liang (1970) *Stärke 22*, 256.
Nussenbaum, S. & W. Z. Hassid (1952) *J. biol. Chem. 196*, 785.

## O

Ohashi, K. (1959) *Nippon Nogei Kagaku Kaishi 33*, 576.
——— (1961) *J. agric. chem. Soc. Japan 35*, 1065.
Oka, S. (1942) *Proc. phys. math. Soc. Japan 24*, 657.

Okada, G. & E. J. Hehre (1974) *J. biol. Chem.* *249*, 126.
Ono, S. (1969) *J. Japanese Soc. Starch Sci.* *17*, 51.
Ono, S., S. Tsuchihashi & T. Kuge (1953) *J. Amer. chem. Soc.* *75*, 3601.
Orrell, S. A., E. Bueding & M. Reissig (1964) in *Control of Glycogen Metabolism* (eds W. J. Whelan & M. P. Cameron). London: J. & A. Churchill.
Overbeck, J. Th. G. & H. G. B. de Jong (1949) *Colloid Science*, vol. 11, 209, Elsevier.
Ozbun, J. L., J. S. Hawker & J. Preiss (1971a) *Biochem. Biophys. Res. Commun.* *43*, 631.
——— *Plant Physiol.* *48*, 765.
——— (1972) *Biochem. J.* *126*, 953.

## P

Painter, T. J. (1963) *J. chem. Soc.*, 779.
Palmer, R. P. & A. J. Cobbold (1964) *Makromol. Chem.* *74*, 174.
Patel, J. R., C. K. Patel & R. D. Patel (1967) *Stärke* *19*, 330.
Pauling, L., R. B. Corey & H. R. Branson (1951) *Proc. Natl. Acad. Sci. U.S.* *37*, 241, 2105.
Paulson, C. M. (1965) PhD thesis, University of California at Berkeley; quoted by Cronan and Schneider (1969).
Pazur, J. H. (1953) *J. biol. Chem.* *205*, 75.
Pazur, J. H. & T. Budovich (1955) *Science* *121*, 702.
Pazur, J. H., D. French & D. W. Knapp (1950) *Proc. Iowa Acad. Sci.* *57*, 203.
Pazur, J. H. & S. Okada (1966) *J. biol. Chem.* *241*, 4146.
Peat, S., S. J. Pirt & W. J. Whelan (1952a) *J. chem. Soc.*, 705.
——— (1952b) *J. chem. Soc.*, 714.
Peat, S. & W. J. Whelan (1953) *Nature* *172*, 494.
Peat, S., W. J. Whelan & J. M. Bailey (1953) *J. chem. Soc.*, 1422.
Peat, S., W. J. Whelan & S. J. Pirt (1949) *Nature* *164*, 499.
Peat, S., W. J. Whelan & W. R. Rees (1953) *Nature* *172*, 158.
Peat, S., W. J. Whelan & G. J. Thomas (1952a) *J. chem. Soc.*, 722.
——— (1952b) *J. chem. Soc.*, 4546.
——— (1956) *J. chem. Soc.*, 3025.
Peat, S., W. J. Whelan & J. R. Turvey (1956) *J. chem. Soc.*, 2317.
Perlin, A. S. (1958) *Canad. J. Chem.* *36*, 810.
Peterlin, A. & M. Čopič (1956) *J. appl. Phys.* *27*, 434.
Peticolas, W. L. (1963) *Nature* *197*, 898.
Pfahler, P. L., H. H. Kramer & R. L. Whistler (1957) *Science* *125*, 442.
Pfannemüller, B. (1968) *Stärke* *20*, 1.
Pfannemüller, B., H. Mayerhöffer & R. C. Schulz (1969) *Makromol. Chem.* *121*, 147.
Pflüger, E. F. W. (1904) *Pflügers Arch. ges. Physiol.* *102*, 305.
Porter, H. K. (1950) *Biochem. J.* *47*, 476.
Posternak, T. (1950) *J. biol. Chem.* *188*, 317.
Potter, A. L. & W. Z. Hassid (1951) *J. Amer. chem. Soc.* *73*, 593.
Potter, A. L., V. Silveira, R. M. McCready & H. S. Owens (1953) *J. Amer. chem. Soc.* *75*, 1335.
Preiss, J. & E. Greenberg (1965) *Biochemistry* *4*, 2328.
Preiss, J., J. L. Ozbun, J. S. Hawker, E. Greenberg & C. Lammel (1973) *Ann. N.Y. Acad. Sci.* *210*, 265.
Preiss, J., L. Shen, E. Greenberg & N. Gentner (1966) *Biochemistry* *5*, 1833.
Price, F. P. (1959) *J. polymer Sci.* *37*, 71.
Ptitsyn, D. B. & Y. Y. Eisner (1959) *Vysokomol. Soedineriya* *1*, 1200.

## R

Rao, V. S. R. & J. F. Foster (1963a) *J. phys. Chem. 67*, 951.
——— (1963b) *Biopolymers 1*, 527.
——— (1965) *J. phys. Chem. 69*, 636.
Rao, V. S. R. & P. R. Sundararajan (1968) in *Solution Properties of Natural High Polymers*, Special Publication 23, 173. London: The Chemical Society.
Rao, V. S. R., P. R. Sundararajan, C. Ramakrishnan & G. N. Ramachandran (1967) in *Conformation of Biopolymers* (ed. G. N. Ramachandran), 721. London: Academic Press.
Rao, V. S. R., N. Yathindra & P. R. Sundararajan (1969) *Biopolymers 8*, 325.
Recondo, E. & L. F. Leloir (1961) *Biochem. Biophys. Res. Commun. 6*, 85.
Reddy, J. M., K. Knox & M. B. Robin (1964) *J. chem. Phys. 40*, 1082.
Rees, D. A. (1970) *J. chem. Soc. B*, 877.
——— (1973) in *MTP International Review of Science: Carbohydrates*. Organic Chemistry (ed. D. H. Hey), series 1 vol. 7 (ed. G. O. Aspinall), 255. Butterworths University Park Press.
Reeves, R. E. (1949) *J. Amer. chem. Soc. 71*, 215.
——— (1950) *J. Amer. chem. Soc. 72*, 1499.
——— (1954) *J. Amer. chem. Soc. 76*, 4595.
Reichert, E. T. (1913) in *The Differentiation and Specificity of Starches in Relation to Genera, Species, etc.* Washington: Carnegie Institute.
Richter, M. & J. Szejtli (1966) *Stärke 18*, 95.
Rickson, F. R. (1973) *Ann. N.Y. Acad. Sci. 210*, 104.
Riseman, J. & J. G. Kirkwood (1950) *J. chem. Phys. 18*, 512.
Roberts, E. A. & B. E. Proctor (1954) *Science 119*, 509.
Roberts, P. J. P. & W. J. Whelan (1960) *Biochem. J. 76*, 246.
Robin, J. P., C. Mercier, R. Charbonnière & A. Guilbot (1974) *Cereal Chem. 51*, 389.
Robin, M. B. (1964) *J. chem. Phys. 40*, 3369.
Robyt, J. F. & R. J. Ackerman (1971) *Arch. Biochim. Biophys. 145*, 105.
Robyt, J. F. & D. French (1963) *Arch. Biochem. Biophys. 100*, 451.
——— (1964) *Arch. Biochem. Biophys. 104*, 338.
——— (1967) *Arch. Biochem. Biophys. 122*, 8.
——— (1970) *J. biol. Chem. 245*, 3917.
——— (1971) *Arch. Biochem. Biophys. 138*, 662.
Rossotti, H. (1959) *J. polymer Sci. 36*, 557.
Rouse, P. E. (1955) *J. chem. Phys. 21*, 1272.
Rowe, J. J. M., J. Wakim & J. A. Thoma (1968) *Anal. Biochem. 25*, 206.
Rundle, R. E. (1947) *J. Amer. chem. Soc. 69*, 1769.
Rundle, R. E. & R. R. Baldwin (1943) *J. Amer. chem. Soc. 65*, 554.
Rundle, R. E., L. Daasch & D. French (1944) *J. Amer. chem. Soc. 66*, 130.
Rundle, R. E. & F. C. Edwards (1943) *J. Amer. chem. Soc. 65*, 2200.
Rundle, R. E., J. F. Foster & R. R. Baldwin (1944) *J. Amer. chem. Soc. 66*, 2116.
Rundle, R. E. & D. French (1943a) *J. Amer. chem. Soc. 65*, 558.
——— (1943b) *J. Amer. chem. Soc. 65*, 1707.

## S

Sair, L. & W. Fetzer (1944) *Ind. Engn. Chem. 36*, 205.
Salema, R. & N. P. Badenhuizen (1967) *J. Ultrastruct. Res. 20*, 383.
Samec, M., C. Nučič & V. Pirkmaier (1941) *Kolloid-Z. 94*, 350.
Sandstedt, R. M. (1946) *Cereal Chem. 23*, 337.
——— (1954) *Cereal Chem. 31*, 17 (suppl.).
——— (1955) *Cereal Chem. 32*, 17 (suppl.).

Sandstedt, R. M., M. J. Blish, D. K. Meecham & C. E. Bode (1937) *Cereal Chem. 14*, 17.
Saric, S. P. & R. K. Schofield (1946) *Proc. Roy. Soc. A 185*, 431.
Sarko, A. & R. H. Marchessault (1967) *J. Amer. chem. Soc. 89*, 6454.
Savage, R. R., W. L. Deatherage, M. M. MacMasters & F. R. Senti (1958) *Cereal Chem. 35*, 392.
Schiefer, S. (1973) *Fed. Proc. Fed. Amer. Soc. Exp. Biol. 32*, 603.
Schiefer, S., E. Y. C. Lee & W. J. Whelan (1973) *FEBS Letters 30*, 129.
Schneider, F. W., C. L. Cronan & S. K. Podder (1968) *J. phys. Chem. 72*, 4563.
Schoch, T. J. (1942a) *J. Amer. chem. Soc. 64*, 2957.
——— (1942b) *J. Amer. chem. Soc. 64*, 2954.
——— (1945) *Adv. Carbohydrate Chem. 1*, 247.
——— (1961) *Proc. Amer. Soc. Brewing Chem.*, 83.
Schoch, T. J. & E. C. Maywald (1956) *Anal. Chem. 28*, 382.
——— (1968) *Stärke 20*, 362.
Schulz, G. V. (1939) *Z. Phys. Chem. Abt. B 43*, 25.
Schwimmer, S. (1945) *J. biol. Chem. 161*, 219.
Scott, R. A. & H. A. Scheraga (1966a) *Biopolymers 4*, 237.
——— (1966b) *J. chem. Phys. 44*, 3054.
Seidemann, J. (1966) in *Stärke-Atlas*, 175. Berlin: Paul Parey.
Senior, M. B. & E. Hamori (1973) *Biopolymers 12*, 65.
Senti, F. R. & L. P. Witnauer (1946) *J. Amer. chem. Soc. 68*, 2407.
——— (1948) *J. Amer. chem. Soc. 70*, 1438.
Shannon, J. C. & R. G. Creech (1973) *Ann. N.Y. Acad. Sci. 210*, 279.
Shannon, J. C., R. G. Creech & J. D. Loerch (1970) *Plant Physiol. 45*, 163.
Shibaoka, T., N. Suetsugu, K. Hiromi & S. Ono (1971) *FEBS Letters 16*, 33.
Slabnik, E. & R. B. Frydman (1970) *Biochem. Biophys. Res. Commun. 30*, 709.
Smith, E. E., G. S. Drummond, J. J. Marshall & W. J. Whelan (1970) *Federation Proc. 29*, 930.
Spark, L. C. (1952) *Biochim. Biophys. Acta 8*, 101.
Stacy, C. J. & J. F. Foster (1956) *J. polymer Sci. 20*, 57.
Stamberg, O. E. & C. H. Bailey (1939) *Cereal Chem. 16*, 319.
Staudinger, H. & E. Husemann (1937) *Ann. 527*, 195.
Stein, R. S. (1964) in *Newer Methods of Polymer Characterisation* (ed. B. Ke), 185. New York: Wiley Interscience.
Stein, R. S. & J. J. Keane (1956) *J. polymer Sci. 20*, 327.
Stein, R. S., D. A. Keedy & M. B. Rhodes (1962) *J. polymer Sci. 62*, 573.
Stein, R. S. & M. B. Rhodes (1960) *J. appl. Phys. 31*, 1873.
——— (1962) *J. polymer Sci. 62*, 584.
——— (1963) *Polymer Letters 1*, 663.
Stein, R. S. & R. E. Rundle (1948) *J. chem. Phys. 16*, 195.
Sterling, C. (1968) in *Starch and its Derivatives*, 4th edition (ed. J. A. Radley), 139. London: Chapman and Hall.
——— (1971) *Stärke 23*, 193.
Sterling, C. & J. Pangborn (1960) *Amer. J. Bot. 47*, 577.
Stetten, M. R. & H. M. Katzen (1961) *J. Amer. chem. Soc. 83*, 2912.
Stetten, M. R., H. M. Katzen & D. Stetten (1956) *J. biol. Chem. 222*, 587.
Stevens, D. J. & G. A. H. Elton (1971) *Stärke 23*, 8.
Stockmayer, W. H. & A. C. Albrecht (1958) *J. polymer Sci. 32*, 215.
Stockmayer, W. H. & M. Fixman (1963) *J. polymer Sci. C 1*, 137.
Summer, R. & D. French (1956) *J. biol. Chem. 222*, 469.
Svanborg, K. & K. Myrbäck (1953) *Arkiv Kemi 6*, 113.

Swanson, M. A. (1948) *J. biol. Chem. 172*, 805, 825.
Swanson, M. A. & C. Cori (1948) *J. biol. Chem. 172*, 815.
Szejtli, J. (1966) *Stärke 18*, 274.
Szejtli, J. & S. Augustat (1966) *Stärke 18*, 38.
Szejtli, J., S. Augustat & M. Richter (1967) *Biopolymers 5*, 17.
Szejtli, J., M. Richter & S. Augustat (1967a) *Biopolymers 5*, 5.
——— (1967b) *Biopolymers 5*, 27.

T

Takagi, T., H. Toda & T. Isemura (1971) in *Enzymes*, vol. 5 (ed. P. D. Boyer), 115. New York: Academic Press.
Tanaka, Y., S. Minagawa & T. Akazawa (1967) *Stärke 19*, 206.
Tanford, C. (1961) *Physical Chemistry of Macromolecules*. New York: Wiley.
Thoma, J. A. (1968) *J. theor. Biol. 19*, 297.
Thoma, J. A., C. Brothers & J. E. Spradlin (1970) *Biochemistry 9*, 1768.
Thoma, J. A. & D. French (1958) *J. Amer. chem. Soc. 80*, 6142.
——— (1960) *J. Amer. chem. Soc. 82*, 4144.
——— (1961) *J. phys. Chem. 65*, 1825.
Thoma, J. A. & D. E. Koshland (1960) *J. Amer. chem. Soc. 82*, 3329.
Thoma, J. A., G. V. K. Rao, C. Brothers, J. E. Spradlin & L. H. Li (1971) *J. biol. Chem. 246*, 5621.
Thoma, J. A. & J. E. Spradlin (1970) *Brewers Digest 2*, 58, 66.
Thoma, J. A., J. E. Spradlin & S. Dygert (1971) in *Enzymes*, vol. 5, 3rd edition (ed P. D. Boyer), 115. New York: Academic Press.
Thompson, A. & M. L. Wolfrom (1951) *J. Amer. chem. Soc. 73*, 5849.
Thompson, J. C. & E. Hamori (1969) *Biopolymers 8*, 689.
——— (1971) *J. phys. Chem. 75*, 272.
Tonelli, A. E. (1974) *Macromolecules 7*, 628.
Tsai, C. Y. & O. E. Nelson (1966) *Science 151*, 341.
——— (1969a) *Genetics 61*, 813.
——— (1969b) *Plant Physiol. 44*, 159.

U

Ueda, M. & K. Kajitani (1967) *Makromol. Chem. 109*, 22.
Ulmann, B. & B. Wendt (1954) *Makromol. Chem. 12*, 155.

V

Valletta, R. M., F. J. Germino, R. E. Lang & R. J. Moshy (1964) *J. polymer Sci. A 2*, 1085.
Vasko, P. D. & J. L. Koenig (1972) *J. Macromol. Sci.-Phys. B 6*, 117.

W

Wakim, J., M. Robinson & J. A. Thoma (1969) *Carbohydrate Res. 10*, 487.
Walker, G. J. (1968) *Biochem. J. 108*, 33.
Walker, G. J. & P. M. Hope (1963) *Biochem. J. 86*, 452.
Walker, G. J. & W. J. Whelan (1959) *Nature 183*, 46.
——— (1960) *Biochem. J. 76*, 264.
Watanabe, T., K. Ogawa & S. Ono (1970) *Bull. chem. Soc. Japan 43*, 950.
Watanabe, T., S. Kawamura, M. Tanno & K. Matsuda (1968) *Chem. Abstr. 68*, 36169.
Watson, J. D. & F. H. C. Crick (1953) *Nature 171*, 738.
Weintraub, M. S. & D. French (1970) *Carbohydrate Res. 15*, 241.

Welker, N. E. & L. L. Campbell (1967) *J. Bacteriol. 94*, 1131.
West, C. D. (1947) *J. chem. Phys. 15*, 689.
Whelan, W. J. (1958) in *Encyclopaedia of Plant Physiology*, vol. 6 (ed. W. Ruhland), 154. Berlin: Springer-Verlag.
——— (1960a) *Stärke 12*, 358.
——— (1960*b*) *Ann. Rev. Biochem. 29*, 105.
——— (1963) *Stärke 15*, 247.
——— (1971) *Biochem. J. 122*, 609.
Whelan, W. J. & J. M. Bailey (1954) *Biochem. J. 58*, 560.
Whistler, R. L. & J. N. BeMiller (1962) *Arch. Biochem. Biophys. 98*, 120.
Whistler, R. L. & G. E. Hilbert (1945) *J. Amer. chem. Soc. 67*, 1161.
Whittington, S. G. (1971) *Macromolecules 4*, 569.
Whittington, S. G. & R. M. Glover (1972) *Macromolecules 5*, 55.
Witnauer, L. P., F. R. Senti & M. D. Stern (1955) *J. polymer Sci. 16*, 1.
Wolf, M. J., M. M. MacMasters, J. E. Hubbard & C. E. Rist (1948) *Cereal Chem. 25*, 312.
Wolf, R. & R. C. Schulz (1967) *Tetrahedron Letters*, 1799.
——— (1968) *J. Macromol. Sci.-Chem. A 2*, 821.
Wolff, I. A., B. T. Hofreiter, P. R. Watson, W. L. Deatherage & M. M. MacMasters (1955) *J. Amer. chem. Soc. 77*, 1654.
Wood, H. L. (1960) *Aust. J. agric. Res. 11*, 673.

## Y

Yamakawa, H. & M. Kurata (1958) *J. phys. Soc. Japan 13*, 78.
Yamashita, Y. & N. Hirai (1965) *J. polymer Sci. A3*, 3251.
——— (1966) *J. polymer Sci. A2, 4*, 161.
Yamashita, Y. & K. Monobe (1971) *J. polymer Sci. A2, 9*, 1471.
Yin, H. C. & C. N. Sun (1947) *Science 105*, 650.
Yokobayashi, K., A. Misaki & T. Harada (1969) *Agr. biol. Chem. 33*, 625.
——— (1970) *Biochem. Biophys. Acta 212*, 458.

## Z

Zaslow, B. (1963) *Biopolymers 1*, 165.
Zaslow, B. & R. L. Miller (1961) *J. Amer. chem. Soc. 83*, 4378.
Ziese, W. (1934) *Z. Physiol. Chem. 229*, 213.
——— (1935) *Z. Physiol. Chem. 230*, 235.
Zimm, B. H. (1948) *J. chem. Phys. 16*, 1099.
——— (1956) *J. chem. Phys. 24*, 269.
Zimm, B. H. & R. W. Kilb (1959) *J. polymer Sci. 37*, 19.
Zobel, H. F., T. M. Cotton & F. R. Senti (1964) *Cereal Sci. Today 9*, 128.
Zobel, H. F., A. D. French & M. E. Hinkle (1967) *Biopolymers 5*, 837.

## Author Index

G



H

## Subject Index

## Q



## R



## S



## T



## U